basic
arithmetic

basic arithmetic

Larry Lewis

Thomas Nelson Community College
Hampton, VA

Vera Goetz Smyth

Berkeley County School District
Moncks Corner, SC

Prentice-Hall, Inc., Englewood Cliffs, New Jersey 07632

Library of Congress Cataloging in Publication Data

Lewis, Larry (date)
    Basic arithmetic.

    Includes index.
    1. Arithmetic—1961–   . I. Smyth, Vera Goetz,
(date)   . II. Title.
QA107.L5 1984   513   83-17744
ISBN 0-13-056804-X

Editorial/production supervision and interior design: Kathleen M. Lafferty
Editorial Assistant: Susan Pintner
Cover and chapter opening designs: Christine Gehring-Wolf
Frontmatter design: A Good Thing, Inc.
Manufacturing buyer: Anthony Caruso

Printed in the United States of America

10 9 8 7 6 5 4 3 2 1

ISBN 0-13-056804-X

Prentice-Hall International, Inc., *London*
Prentice-Hall of Australia Pty. Limited, *Sydney*
Editora Prentice-Hall do Brasil, Ltda., *Rio de Janeiro*
Prentice-Hall Canada Inc., *Toronto*
Prentice-Hall of India Private Limited, *New Delhi*
Prentice-Hall of Japan, Inc., *Tokyo*
Prentice-Hall of Southeast Asia Pte. Ltd., *Singapore*
Whitehall Books Limited, *Wellington, New Zealand*

contents

UNIT 2 *fractions and mixed numbers*   135

UNIT 3 *decimal numbers* 283

UNIT 4 *ratio, proportion, and percent* 367

UNIT 5 *measurement* 435

UNIT 6 *introduction to algebra*   481

index   511

preface

Basic Arithmetic is dedicated to students who do not have a firm foundation in or an understanding of the skills and applications of basic arithmetic. It was written for those who need to study arithmetic in order to qualify for a curriculum that requires proficiency in computation or for those who need a more complete understanding of arithmetic before they begin the study of algebra.

Calculators have eliminated the need for people who can perform long and difficult calculations without error time after time. Even so, students still need an understanding of the basic operations, a sense of when to use them, and the habit of judging the reasonableness of an answer. In addition to presenting and practicing these skills, this text introduces the use of calculators and provides optional practice with applications.

Writing Style: The approach involves a readable, nonthreatening, conversational writing style with emphasis on examples. This text contains more problems in the form of examples with complete solutions than comparable texts.

Skills are broken into very small components with smooth and easy progressions between them, which, along with exercises and tests, promote an opportunity for early detection of student difficulties.

Organization   The organization of the text aims to simulate the activity between students and teacher in a classroom.

Unit Pretests: Students are given an opportunity to find out how much they know of the material in each chapter of the unit.

Examples: First the topics are introduced and clarified in the text material. The explanations and examples are clear and detailed, yet never condescending.

Examples–Solutions Section: Students are asked to work problems similar to those introduced in the text. The material supports the student in that the steps and the solutions are given in full. This section does not begin with the last or most advanced problem but reworks the development of the section, similar to a review.

Now-Try-These Section: Next, students are asked to perform without the aid of the complete solution and to decide if they are ready for the exercise section or if they need to talk with an instructor and study the section again.

Exercises: Finally, students practice the new skills in the exercise section. In the Part I exercises, all answers are given at the end of the unit. Answers are not given to Part II exercises.

Review Skills: Each review skill consists of a brief explanation of a specific aspect of a skill introduced in the chapter, a worked out example, an exercise set with all answers given at the end of the unit, and a shorter exercise set with no answers. These individual skill drills can be used by the student before the chapter review exercise or, diagnostically, after the chapter review and before the chapter post-test.

Review Exercises: Each chapter has a review section in which the exercises are grouped by section so that students can easily refer back to review specific skills. All answers are provided.

Chapter Post-Test: Students are given the opportunity to test their mastery of the skills contained in the chapter. Again, all answers are provided.

Unit Review: Each unit has a review section in which the skills presented throughout the unit are summarized.

Unit Post-Test: Students are given the opportunity to see how well they have mastered the cumulative skills representing the topics presented in each of the chapters contained in the unit.

Instructor's Manual

The instructor's manual is primarily intended for use with lab or self-paced courses. It includes five forms of each unit test, five forms of a final test, and three forms of each chapter test. In addition, objectives are provided as well as answers to any exercise for which answers are not given in the text.

Acknowledgments

We wish to thank the following reviewers for their valuable suggestions: Sharon W. Bird, Richland College, Dallas, Texas; Joan M. Haig, Anchorage Community College, Anchorage, Alaska; Robert L. Maynard, Rockingham Community College, Wentworth, North Carolina; Gerald Skidmore, Alvin Community College, Alvin, Texas; and Edward B. Wright, Western Oregon State College, Monmouth, Oregon.

Larry Lewis
Vera Goetz Smyth

basic
arithmetic

UNIT 1
whole numbers

**UNIT ONE
PRETEST**

This test is given *before* you read the material to see what you already know. If you cannot work each sample problem, you need to read the indicated chapters carefully. Answers to this test can be found on page 130. You may wish to read each chapter even when you are familiar with the information, but you may omit reading those chapters for which you answered *each* sample problem correctly.

CHAPTER 1

1. In the number 503,124 name the digit with the indicated position.
   (a) ten-thousands   **(b)** tens   **(c)** ones

2. Write 1,523,507 as you would read it.

3. Write the numeral for one hundred sixty-three thousand, five hundred twenty-three.

4. Round each number to the nearest hundred.
   (a) 5,753   **(b)** 6,248   **(c)** 9,579

CHAPTER 2

5. Find the indicated sums.
   (a) 120 + 63 + 98   **(b)**   1,593
                                           2,589
                                    +5,320

6. Find the total bill for the following items: carbon paper for $2, typing paper for $6, typewriter rental for $69, and typewriter ribbon for $8.

CHAPTER 3

7. Find the indicated differences.
   (a) 1,895 – 957   **(b)**   7,602
                                     – 2,387

8. A book is 543 pages long. If you read 289 pages, how many pages remain to be read?

CHAPTER 4

9. Find the indicated products.
   (a) 27 × 15   **(b)**   207
                               × 89

10. Evaluate 4^3.

11. A box contains 8 pens. How many pens are in 15 boxes?

CHAPTER 5

12. Find the indicated quotients.
    (a) 3,987 ÷ 15   **(b)** 125 $\overline{)5,937}$

13. Harry types 732 words in 12 minutes. How many words does he type per minute?

14. Average the following numbers: 15, 45, and 72.

1

whole numbers

WHOLE NUMBERS AND PLACE VALUE

$$0, 1, 2, 3, 4, 5, 6, 7, 8, 9, 10, 11, 12, \ldots$$

This section explains **place value** of whole numbers. After studying this section, you will understand all of the terms in this example.

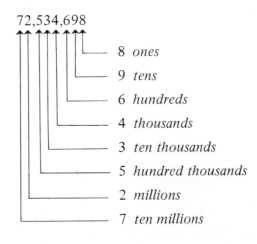

72,534,698

- 8 *ones*
- 9 *tens*
- 6 *hundreds*
- 4 *thousands*
- 3 *ten thousands*
- 5 *hundred thousands*
- 2 *millions*
- 7 *ten millions*

3

If you think you know this information, turn to the exercises and work *each* problem. If you have difficulty with the exercise section, turn back and read this section completely.

Counting   What do these two groups have in common?

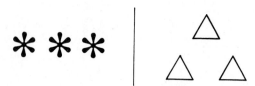

Although the objects in each group are different, each group has *three* objects. The numeral 3 is used to represent this common number idea of threeness.

A **number** is an abstract idea, whereas a **numeral** is the symbol used to name the number idea. Do not worry too much about this distinction, since most people (including instructors) use the words interchangeably (and so shall I).

Whole Numbers   You are already familiar with the **whole numbers** represented by the numerals

$$0, 1, 2, 3, 4, 5, 6, 7, 8, 9, 10, 11, 12, \ldots$$

You know these numbers because they are the number ideas that are associated with counting.

$$0, 1, 2, 3, 4, 5, 6, 7, 8, 9, 10, 11, 12, \ldots$$

The three dots following the list indicate that there is no last number in the list; the counting continues without end.

Place Value   Although you are familiar with whole numbers, you may have forgotten why you read and write them the way you do now. The reason is based on the concept of **place value**. Our numeration system uses ten **digits**:

$$0, 1, 2, 3, 4, 5, 6, 7, 8, 9$$

 The *position* (or place) in which a digit appears in a numeral determines the *value* of that digit.

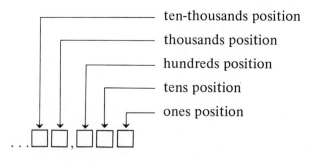

I shall use place value in the next chapters to explain how arithmetic works with whole numbers. For now, you must learn each of the different names used and the positions they occupy.

EXAMPLE 1:   Let us look at the values for the digits in the seven-digit number 1,964,352.

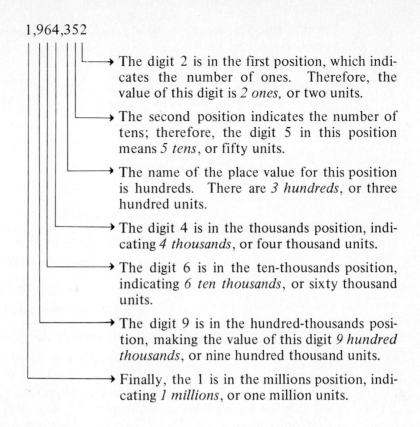

1,964,352

→ The digit 2 is in the first position, which indicates the number of ones. Therefore, the value of this digit is *2 ones,* or two units.

→ The second position indicates the number of tens; therefore, the digit 5 in this position means *5 tens*, or fifty units.

→ The name of the place value for this position is hundreds. There are *3 hundreds,* or three hundred units.

→ The digit 4 is in the thousands position, indicating *4 thousands*, or four thousand units.

→ The digit 6 is in the ten-thousands position, indicating *6 ten thousands*, or sixty thousand units.

→ The digit 9 is in the hundred-thousands position, making the value of this digit *9 hundred thousands*, or nine hundred thousand units.

→ Finally, the 1 is in the millions position, indicating *1 millions*, or one million units.

EXAMPLE 2:   Give the value of the 2 in the numbers 205, 2,394, and 592.

The same digit can have different values, depending on its position. Because of the position of the digit 2 in these numbers, the value of the 2 changes.

205	2,394	592
↓	↓	↓
2 hundreds	2 thousands	2 ones
(two hundred)	(two thousand)	(two)

Up to now I have used words to indicate place value. However, place value can be described in two ways:

1. By using the name to indicate the position a digit occupies, or
2. By using a number to indicate the position.

The place values of the first 10 positions are shown in the chart, using both methods.

PLACE VALUE	NAME OF POSITION
1	ones
10	tens
100	hundreds
1,000	thousands
10,000	ten thousands
100,000	hundred thousands
1,000,000	millions
10,000,000	ten millions
100,000,000	hundred millions
1,000,000,000	billions

Notice that the value of each position is *10 times the value of the previous position.*

EXAMPLE 3: Similarly, the values of digits can be expressed in two ways. I can write the value of the digits in 72,534,698 using either words or numerals.

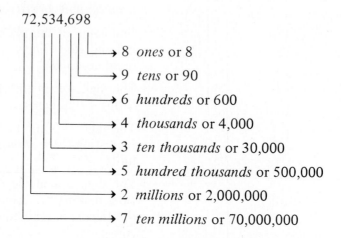

72,534,698

→ 8 *ones* or 8
→ 9 *tens* or 90
→ 6 *hundreds* or 600
→ 4 *thousands* or 4,000
→ 3 *ten thousands* or 30,000
→ 5 *hundred thousands* or 500,000
→ 2 *millions* or 2,000,000
→ 7 *ten millions* or 70,000,000

Because you have worked with this numeration system over a period of years, you may no longer think of numbers in this way. But it is important that you do so in order to understand what follows. Therefore, practice by trying to work the following problems, covering up the right side until you think you know the answer; then uncover the correct solution and compare it with your solution.

Examples	Solutions
Name the position of the digit 3 in each of the following numbers. (a) 7,830 (b) 3,975 (c) 793 (d) 35,207	Do not read this until you have tried to find the answer yourself; then let my solution help! (a) 7,830 ones tens The digit 3 is in the *tens* place. (b) 3,975 ones tens hundreds thousands The digit 3 is in the *thousands* place. (c) 793 ones The digit 3 is in the *ones* place.

Examples	Solutions

	(d) 35,207
	
	     ones
	     tens
	     hundreds
	     thousands
	     ten thousands
	
	The digit 3 is in the *ten-thousands* place.

Give the value of the digit 7 in each of the following numbers.
 	
(a) 7,830
(b) 3,975
(c) 793
(d) 35,207
	(a) 7,830
	
	     → 0 ones
	     → 3 tens
	     → 8 hundreds
	     → 7 thousands
	
	The value of the digit 7 is 7 thousands, written 7,000.
	
	(b) 3,975
	
	     → 5 ones
	     → 7 tens
	
	The value of the digit 7 is 7 tens, or 70.
	
	(c) 793
	
	     → 3 ones
	     → 9 tens
	     → 7 hundreds
	
	The value of the digit 7 is 7 hundreds, or 700.
	
	(d) 35,207
	
	     → 7 ones
	
	The value of the digit 7 is 7 ones, or 7.

In the number 2,415,973 name the digit with the indicated position.
 	
(a) tens
(b) ten-thousands
(c) hundreds
(d) ones
	2,415,973
	
	     ones
	     tens
	     hundreds
	     thousands

Examples	Solutions
	2,415,973 ↑ ↑↑ | | ten thousands | hundred thousands millions (a) The digit 7 is in the tens position. (b) The digit 1 is in the ten-thousands position. (c) The digit 9 is in the hundreds position. (d) The digit 3 is in the ones position.
In the number 8,592 name the position of the indicated digit. (a) 9 (b) 8 (c) 5 (d) 2	8,592 || ↓ ones | ↓ tens ↓ hundreds thousands (a) 9 is in the tens position. (b) 8 is in the thousands position. (c) 5 is in the hundreds position. (d) 2 is in the ones position.

Now try these—this time only the answer is given!

Name the position of the digit 4 in each of the numbers. (a) 34,597 (b) 3,754 (c) 5,479	(a) thousands (b) ones (c) hundreds
In the number 3,571,204, name the digit with the indicated position. (a) hundreds (b) ones (c) thousands (d) hundred-thousands	(a) 2 (b) 4 (c) 1 (d) 5

If you had difficulty with these two examples, please talk with an instructor or read this section again *before* working the exercises that follow.

Exercises Part I

Answers to these exercises can be found on page 130.

1. In the number 6,792 name the position of the following digits.
   (a) 7   (b) 9   (c) 2

2. In the following numbers, name the position of the digit 5.
   (a) 52,437   (b) 1,507   (c) 753

3. In the number 5,607,291 name the digit with the indicated position.
   (a) thousands   (b) millions   (c) ten-thousands

(d) ones   **(e)** tens   **(f)** hundred-thousands
(g) hundreds

4. In the number 15,372 name the digit with the indicated position.
   (a) ten-thousands   **(b)** tens   **(c)** thousands

Exercises
Part II

1. In the number 5,973 name the position of the indicated digits.
   (a) 5   **(b)** 7   **(c)** 3

2. In the number 6,897,532 name the digit with the indicated position.
   (a) hundreds   **(b)** ten-thousands   **(c)** millions

SECTION 1.2
READING AND WRITING WHOLE NUMBERS

341

Sept. 5, 19 86

PAY TO THE ORDER OF __AB Automotive__ $ | 716 |

__Seven hundred sixteen__ ———————————— DOLLARS

BANK OF TNCC

MEMO _____   _John Jones_

⑆021 244 21 056 2⑈ 402 100538 2⑉ 200096⑆

This section explains reading and writing whole numbers. After studying this section, you will understand all the terms in the following chart.

	Period Names										
	Billion			Million			Thousand			Ones	
Place Value Names	Ten billions	Billions	Hundred millions	Ten millions	Millions	Hundred thousands	Ten thousands	Thousands	Hundreds	Tens	Ones
	2	7 ,	5	4	2 ,	3	9	4 ,	5	0	2

Read: twenty-seven billion, five hundred forty-two million, three hundred ninety-four thousand, five hundred two.

If you think you know this information turn to the exercises and work *each* problem. If you have difficulty with the exercise section, turn back and read this section completely.

Periods Because you have just finished the first section, you should be familiar with writing and reading whole numbers. But, as in the first section, you are so familiar with these techniques that you may have forgotten why!

We read and write whole numbers in groups, or **periods**, of three-digit numbers, counting from the right. In written form, we usually separate these periods with commas. Although the comma is not necessary, it makes reading the number easier.

Consider the number 1,072,534,698. There are four periods separated by commas.

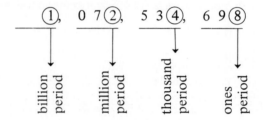

The name of each period uses the position name of the rightmost digit in the group, which is circled in the following number.

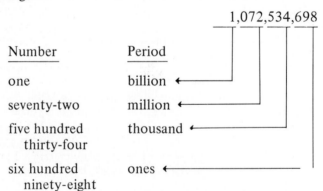

Reading a Whole Number   EXAMPLE 1:   To read the number 1,072,534,698 read the three-digit number within a given period and say the name of the period. Begin reading with the period farthest left and move to the right. Of course, the leftmost period may not have three digits.

1,072,534,698

Number	Period
one	billion ←
seventy-two	million ←
five hundred thirty-four	thousand ←
six hundred ninety-eight	ones ←

Read: one billion, seventy-two million, five hundred thirty-four thousand, six hundred ninety-eight.

Things to remember when reading a whole number.

1. Read each three-digit number within the period and use the period name. The leftmost period may or may not have three digits.
2. Do not say *and* between any of the periods or in reading the three-digit numbers.
3. Do not say *ones* for the last period.

EXAMPLE 2: Write the number 1,964,352 the way you would read it. There are three periods, counting from the right.

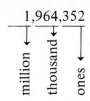

1,964,352

Use these period names along with the numbers in each period.

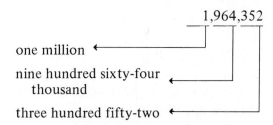

1,964,352

one million

nine hundred sixty-four thousand

three hundred fifty-two

Read: one million, nine hundred sixty-four thousand, three hundred fifty-two.

EXAMPLE 3: The period and place value names for the 11-digit number 27,542,394,502 are shown in the following chart.

	Period Names										
	Billion		Million			Thousand			Ones		
Place Value Names	Ten billions	Billions	Hundred millions	Ten millions	Millions	Hundred thousands	Ten thousands	Thousands	Hundreds	Tens	Ones
	2	7	5	4	2	3	9	4	5	0	2

Read: twenty-seven billion, five hundred forty-two million, three hundred ninety-four thousand, five hundred two.

Writing a Whole Number

You should also be able to write a number that someone reads aloud. For example, I shall write the following as a numeral: two million, two hundred seventy thousand, five hundred twenty-two.

Remember that the number is read in periods of three digits, with the possible exception of the leftmost period.

2,   two million

2,270,   two hundred seventy thousand

2,270,522   five hundred twenty-two

Therefore, the numeral is 2,270,522.

Writing a whole number is the reverse procedure of reading a whole number. To write a whole number use a separate three-digit numeral for each period. Again, the left-most period may or may not have three digits. Commas are generally used to separate the periods.

EXAMPLE 4: I can write the number three million, sixty-seven thousand, six hundred thirty-two in three steps, one for each period.

3,	three million
3,067,	sixty-seven thousand
3,067,632	six hundred thirty-two

Therefore, the numeral is 3,067,632.

Notice that in this last example 67 was written as 067, in order to occupy three spaces in the thousand period.

Now you try some problems. Remember to try to answer *before* looking at the solution.

Examples	Solutions
Fill in the chart with the period names.	

				Million	Thousand	Ones
5 7	2 5 0	7 1 3		5 7	2 5 0	7 1 3

Examples	Solutions
Write the number 2,502,713 as you would read it.	2,502,713 million · thousand · ones two million, five hundred two thousand, seven hundred thirteen
Write the number 50,002,015 as you would read it.	50,002,015 million · thousand · ones fifty million, two thousand, fifteen
Write the numeral for twenty million, fifty-two thousand, forty-seven.	20,___,___ Twenty million 20,052,___ Fifty-two thousand *Notice the use of zero to show a three-digit number for 52.*

Examples	Solutions
	20,052,047 Forty-seven
	Again, notice the use of zero.
	Numeral: 20,052,047
Write the numeral for five million, four hundred seventy-three thousand, seven.	5,_ _ _,_ _ _ Five million
	5,473,_ _ _ Four hundred seventy-three thousand
	5,473,007 Seven
	Numeral: 5,473,007
Write the numeral for one million, fifty-two.	1,_ _ _,_ _ _ One million
	There are no thousands. This is shown by using zeros.
	1,000,_ _ _
	1,000,052 Fifty-two
	Numeral: 1,000,052

Now try these—No hints given!

Write the number 6,079,150 as you would read it.	six million, seventy-nine thousand, one hundred fifty
Write the numeral for two million, five hundred three thousand, seventy-six.	2,503,076

If you had difficulty with these two examples, please talk with an instructor or read this section again *before* working the exercises.

Exercises Part I

Answers to these exercises can be found on page 130.

Write the following numbers as you would read them.

1. 523
2. 502
3. 57
4. 1,521
5. 12,501
6. 123,123
7. 150,027
8. 15,469,523
9. 257,305,240
10. 105,005,004

Write the numeral for each of the following.

11. fifty-seven thousand, four hundred twenty-nine
12. one hundred seventy-two thousand, fifty-two
13. twenty million, ninety-five thousand, three
14. seven million, thirty-five thousand, sixty-three
15. ten million, four thousand three

Write in words the amounts on the following checks.

16.

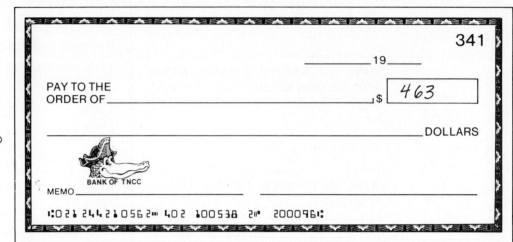

17.

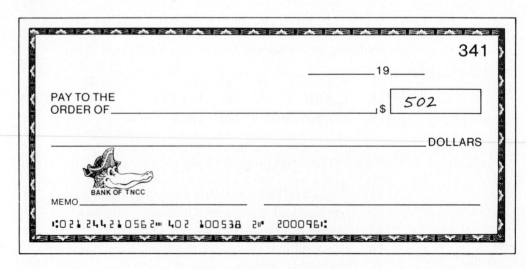

18.

Write, in numerals, the amounts on the following checks.

19.

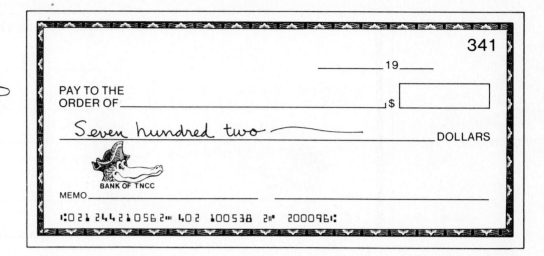

20.

**Exercises
Part II**   Write these numbers as you would read them.

   1. 151   **2.** 25   **3.** 3,705   **4.** 29,023   **5.** 16,109,572

Write the numeral for each of the following.

   6. fifty-three thousand, forty-three
   7. five million, two hundred thirteen thousand, five hundred twenty-one
   8. seventy million, three hundred five thousand, fifteen

Complete the entries on each check.

9.

10.

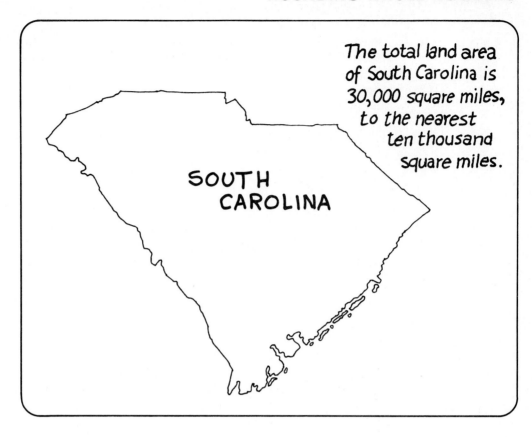

The total land area of South Carolina is 30,000 square miles, to the nearest ten thousand square miles.

SOUTH CAROLINA

This section explains rounding whole numbers. After studying this section you will understand the following rule for rounding.

> **Rounding**
>
> Drop all digits to the right of the rounding position and replace them with zeros.
>
> 1. If the first digit dropped to the immediate right of the rounding position is 5 or greater (5, 6, 7, 8, 9), round up by adding 1 to the digit in the rounding position.
> 2. If the first digit dropped is less than 5 (0, 1, 2, 3, 4), leave the digit in the rounding position unchanged.

If you think you know this information, turn to the exercises and work *each* problem. If you have difficulty, turn back and read this section completely.

Introduction If you were to ask me how many people I invited to an election day party, I might remember the number to be 212 or 213. But if I am sure you are not interested in the *exact* amount, I would probably reply that *about* 200 guests were asked. In a department store, I always consider an item tagged $29.95 to cost $30.00. Both of these situations are examples of rounding.

Rounding   In **rounding**, you replace a number with a new number close to it in value.

EXAMPLE 1:   Round 2,759 to the nearest thousand.

First, when rounding to the nearest thousand, the thousands place is called the **rounding position**. To round, I must determine to which thousand 2,759 is closer.

Suppose I count to 2,759. Counting to 2,759, I pass 2,000 and continue more than halfway to 3,000. Notice that 2,500 is halfway and 2,759 is greater than that.

... 2,000 .... 2,500 ... 2,759 ... 3,000 ...
↓
halfway
point

That is, 2,759 is between 2,000 and 3,000, but it is closer to 3,000.

To round 2,759 to the nearest thousand, I drop all digits after the thousands position and replace them with zeros. Then I add 1 to the digit 2 in the thousands position to round up to the closer thousand.

000
2,~~759~~   Drop and replace with zeros.
3,000   Add 1 to the rounding position.

The number 2,759 rounded to the nearest thousand is 3,000.

EXAMPLE 2:   Round 6,250 to the nearest thousand.

Again, the rounding position is thousands. Although in counting to 6,250 I pass 6,000, I do not count more than halfway to 7,000, because 250 is less than 500.

... 6,000 ... 6,250 ... 6,500 .... 7,000 ...
↓
halfway
point

That is, 6,250 is between 6,000 and 7,000, but it is closer to 6,000.

Again, I replace all digits after the thousands position with zeros, but this time I leave the digit 6 in the thousands position unchanged.

000
6,~~250~~   Drop and replace with zeros.
6,000   Leave 6 unchanged.

The number 6,250 rounded to the nearest thousand is 6,000.

EXAMPLE 3:   Round 3,500 to the nearest thousand.

... 3,000 ... 3,500 ... 4,000 ...
↓
halfway
point

In this example, 500 is the halfway point. The original number is closer to neither 3,000 nor 4,000; it is exactly halfway between them.

There are *different rules* for rounding in this situation. Do not be concerned if another text or instructor uses a different rule. The simplest (and most common) rule is to round up. That is, replace all digits after the thousands position with zeros and round up by changing the digit 3 to the next thousand, 4.

000
3,~~500~~   Drop and replace with zero.
4,000   Add 1 to the rounding position.

With this rule for rounding, 3,500 rounded to the nearest thousand is 4,000.

Approximating a Number

The symbol ≅ is used to indicate that although the numbers are not equal, they are close (to some place value position) or *approximately equal*. After it is rounded, the number is roughly equal to but not as precise as the given number.

From the first three examples,

$$2{,}759 \text{ rounds to } 3{,}000 \quad \text{or} \quad 2{,}759 \cong 3{,}000$$

$$6{,}250 \text{ rounds to } 6{,}000 \quad \text{or} \quad 6{,}250 \cong 6{,}000$$

$$3{,}500 \text{ rounds to } 4{,}000 \quad \text{or} \quad 3{,}500 \cong 4{,}000$$

Notice that in two examples the digit was rounded up in the thousands position, whereas in the second example, the digit 6 was unchanged.

The decision to round up or to leave the digit unchanged can be based on the first digit to the immediate right of the rounding position.

1. From the first problem,

$$\overset{000}{2{,}\cancel{759}} \cong 3{,}000$$

In this example, the digit to the immediate right of the thousands position is 7, the first digit dropped. Because 7 is larger than 5, it indicates that I am more than halfway to the next thousand, which is 3,000. To round in this example, I add 1 to the thousands position.

$$\overset{000}{2{,}\cancel{759}} \cong 3{,}000$$
$$\underset{\text{add } 1}{\llcorner\qquad\lrcorner}$$

2. In the second example, the first digit dropped is 2.

$$\overset{000}{6{,}\cancel{250}} \cong 6{,}000$$

Because 2 is less than 5, it indicates that I am not halfway to the next thousand. Therefore 6,250 is closer to 6,000 than to 7,000. To round in this example, I leave the thousands position unchanged.

$$\overset{000}{6{,}\cancel{250}} \quad \cong \quad 6{,}000$$
$$\underset{\text{unchanged}}{\llcorner\qquad\qquad\lrcorner}$$

3. In the third example, I use the rule to round up when a 5 is the first digit dropped, by adding 1 to the thousands position.

$$\overset{000}{3{,}\cancel{500}} \cong 4{,}000$$
$$\underset{\text{add } 1}{\llcorner\qquad\lrcorner}$$

Rounding

Drop all digits to the right of the rounding position and replace them with zeros.

1. If the first digit dropped to the immediate right of the rounding position is 5 or greater (5, 6, 7, 8, 9), round up by adding 1 to the digit in the rounding position.
2. If the first digit dropped is less than 5 (0, 1, 2, 3, 4), leave the digit in the rounding position unchanged.

EXAMPLE 4:   Round each number to the nearest hundred.   In each part, the rounding position is hundreds.

(a) Round 625 to the nearest hundred.

First, I drop all digits after the hundreds position and replace them with zeros.

$$\overset{00}{6\,2\cancel{5}}$$

Because the first digit dropped (the 2) is smaller than 5, I leave the digit 6 unchanged.

$$\overset{00}{6\,2\cancel{5}} \cong 600$$
$$\underset{\text{unchanged}}{\llcorner}\!\!\longrightarrow\!\!\urcorner$$

The number 625 rounded to the nearest hundred is 600. That is, 625 is closer to 600 than it is to 700.

(b) Round 685 to the nearest hundred.

Again, I drop digits after the hundreds position.

$$\overset{00}{6\,8\cancel{5}}$$

Because 8 is larger than 5, I round up by adding 1 to the hundreds position.

$$\overset{00}{6\,8\cancel{5}} \cong 700$$
$$\underset{\text{add 1}}{\llcorner}\!\!\longrightarrow\!\!\urcorner$$

The number 685 rounded to the nearest hundred is 700. That is, 685 is closer to 700 than it is to 600.

(c) Round 653 to the nearest hundred.

Because I am rounding to the nearest hundred, I drop all digits after the rounding position.

$$\overset{00}{6\,5\cancel{3}}$$

The first digit dropped is 5; the rule is to round up.

$$\overset{00}{6\,5\cancel{3}} \cong 700$$
$$\underset{\text{add 1}}{\llcorner}\!\!\longrightarrow\!\!\urcorner$$

The number 653 rounded to the nearest hundred is 700.

(d) Round 1,763 to the nearest hundred.

Drop digits after the hundreds position.

$$\overset{00}{1,7\cancel{63}}$$

Because 6 is larger than 5, round up.

$$\overset{00}{1,7\cancel{63}} \cong 1,800$$
$$\underset{\text{add 1}}{\llcorner}\!\!\longrightarrow\!\!\urcorner$$

The number 1,763 rounded to the nearest hundred is 1,800.

(e) Round 15,324 to the nearest hundred.

Drop digits and replace them with zeros.

$$\begin{matrix}00\\15,3\cancel{24}\end{matrix}$$

Because 2 is smaller than 5, leave the 3 unchanged.

15,3$\cancel{24}$ ≅ 15,300

└unchanged┘

The number 15,324 rounded to the nearest hundred is 15,300.

Now it is your turn. Try to work the following problems before looking at the solutions.

Examples	Solutions
Round 293 to the nearest hundred.	$$\begin{matrix}00\\\cancel{293}\cong 300\end{matrix}$$ └add 1┘ Replace all digits after the hundreds position with zeros, and round up the hundreds position because 9 is larger than 5.
Round 2,359 to the nearest thousand.	Drop all digits after the thousands position. $$\begin{matrix}000\\2,\cancel{359}\cong 2,000\end{matrix}$$ └unchanged┘ Because the first digit dropped is less than 5, leave the thousands position unchanged.
Round 5,325 to the nearest ten.	The digit 2 is in the rounding position. $$\begin{matrix}0\\5,32\cancel{5}\cong 5,330\end{matrix}$$ └add 1┘ When the first digit dropped is 5, round up.

Now you try these—no hints given.

Round 6,385 to the (a) nearest thousand; (b) nearest hundred; (c) nearest ten.	(a) 6,000 (b) 6,400 (c) 6,390

Ready to continue? If not, please talk with an instructor or read this section again.

Two more examples will illustrate the last situation in rounding.

EXAMPLE 5: Round 3,197 to the nearest ten.
First drop the digit after the tens position.

$$\begin{array}{c} 0 \\ 3,19\cancel{7} \end{array}$$

Because 7 is larger than 5, I must add 1 to round up. But this makes 3,190 change to 3,200.

$$\begin{array}{c} 0 \\ 3,19\cancel{7} \cong 3,200 \\ \underline{\qquad}\text{add 1}\underline{\qquad} \end{array}$$

With a 9 in the rounding position, it is possible that digits before the rounding position can change.

$$\begin{array}{c} 0 \\ 3,19\cancel{7} \quad \cong \quad 3,200 \\ \underline{\quad}1 \text{ changed}\underline{\quad} \\ \text{to 2} \end{array}$$

EXAMPLE 6: When it is stated that the population of Asia was 2,229,700,000 in 1973, it is understood that the number has been rounded to the nearest hundred thousand. What was the population of Asia in 1973 to the nearest million?

$$\begin{array}{c} 000,000 \\ 2,229,\cancel{700,000} \cong 2,230,000,000 \\ \underline{\qquad}\text{add 1}\underline{\qquad} \end{array}$$

Notice that by adding one, 2,229,000,000 became 2,230,000,000.

Your turn again!

Examples	Solutions
The state of Alaska has a land area of 566,432 square miles. What is the area to the nearest thousand square miles?	$\begin{array}{c} 000 \\ 566,\cancel{432} \cong 566,000 \\ \text{unchanged} \end{array}$ 566,000 square miles
The state of Hawaii has an area of 6,425 square miles. Round this to the nearest ten square miles.	$\begin{array}{c} 0 \\ 6,42\cancel{5} \cong 6,430 \\ \text{add 1} \end{array}$ 6,430 square miles
Maine has 30,933 square miles of land. Round this to the nearest thousand square miles.	$\begin{array}{c} 000 \\ 30,\cancel{933} \cong 31,000 \\ \text{add 1} \end{array}$ 31,000 square miles
Nevada has 109,889 square miles of land. What is the area to the nearest thousand square miles?	$\begin{array}{c} 000 \\ 109,\cancel{889} \cong 110,000 \\ \text{add 1} \end{array}$ (109,000 became 110,000) 110,000 square miles

Examples	Solutions
Now try these—no hints given.	

Round 35,298 to the nearest
  (a) ten thousand;
  (b) thousand;
  (c) hundred;
  (d) ten.
	(a) 40,000
	(b) 35,000
	(c) 35,300
	(d) 35,300

If you had difficulty with these problems, please talk with an instructor or read this entire section again.

Exercises Part I

Answers to these exercises can be found on page 130.

Round each number to the nearest hundred.

1. 235   2. 754   3. 893   4. 1,552
5. 2,621   6. 3,481   7. 3,901   8. 4,962

Round each number in these statements to the nearest thousand.

9. The Nile River is 4,132 miles long.
10. The Amazon River is 3,900 miles long.
11. The Ganges River is 1,550 miles long.
12. Iceland has 39,800 square miles of land area.

Exercises Part II

Round each number to the nearest ten.

1. 322   2. 456   3. 785   4. 301   5. 798   6. 509

REVIEW MATERIAL

Review Skill Reading and Writing Whole Numbers

Whole numbers are read and written in periods of three digits. For example,

(a)   8,309,271   Read: eight million, three hundred nine thousand, two hundred seventy-one.

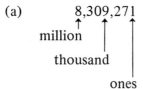

million
thousand
ones

(b)   One million, two hundred thirty-five thousand, twenty-three is written

1,235,023

Exercise A

Answers to this exercise can be found on page 130.

Write each of the following numbers as you would read them.

1. 640   2. 56
3. 4,688   4. 42,273
5. 7,951,008   6. 31,858,521

7. 46	**8.** 65
9. 8,864	**10.** 37,224
11. 8,001,597	**12.** 125,858,310
13. 748	**14.** 77,000
15. 59,001	

Write the numerals for the following numbers.

16. thirty-four	**17.** nineteen
18. eight thousand, two hundred seventy-four	**19.** six hundred twenty-six thousand, six hundred three
20. four hundred thousand, three hundred four	

Exercise B No answers are given.

Write each of the following numbers as you would read them.

1. 530	**2.** 75
3. 4,582	**4.** 53,734
5. 7,856,002	**6.** 64
7. 31,563,623	**8.** 57
9. 4,689	**10.** 224,437
11. 1,597,008	**12.** 85,831,125
13. 86,000	**14.** 64,070

Write the numerals for the following numbers.

15. four hundred thirty-four	**16.** four hundred forty
17. eleven thousand, three hundred thirty-five	**18.** thirty-six thousand, seventy-seven

Review Skill Rounding Whole Numbers

Drop all digits to the right of the rounding position and replace them with zeros.

1. If the first digit dropped to the immediate right of the rounding position is 5 or greater, round up by adding 1 to the digit in the rounding position.

$$8,\underline{8}64 \cong 8,\underline{9}00 \qquad \text{To the nearest } hundred.$$

2. If the first digit dropped is smaller than 5, leave the digit in the rounding position unchanged.

$$8,8\underline{6}4 \cong 8,8\underline{6}0 \qquad \text{To the nearest } ten.$$

Exercise C Answers to this exercise can be found on page 130.

Round to the indicated position.

1. 384, nearest ten	**2.** 348, nearest ten
3. 4,688, nearest thousand	**4.** 4,285, nearest thousand
5. 528, nearest hundred	**6.** 582, nearest hundred
7. 629,771, nearest ten thousand	**8.** 172,823, nearest ten thousand

9. 5,752, nearest hundred
11. 29,637, nearest ten
13. 4,582, nearest thousand
15. 16,927, nearest thousand

10. 5,953, nearest hundred
12. 2,395, nearest ten
14. 4,324, nearest thousand

Exercise D No answers are given.

Round to the indicated position.

1. 528, nearest ten
3. 4,582, nearest thousand
5. 384, nearest hundred
7. 102,635, nearest ten thousand
9. 6,853, nearest hundred
11. 12,649, nearest ten
13. 5,492, nearest ten
15. 15,672, nearest thousand

2. 582, nearest ten
4. 4,388, nearest thousand
6. 348, nearest hundred
8. 169,573, nearest ten thousand
10. 7,952, nearest hundred
12. 1,596, nearest ten
14. 5,294, nearest thousand

REVIEW EXERCISES

Refer to the section listed if you have difficulty. Answers to these exercises can be found on pages 130–131.

Section 1.1

1. In the number 15,793, name the position of the following digits.
   (a) 5   (b) 3   (c) 7   (d) 1
2. In the following numbers, name the position of the digit 5.
   (a) 15,493   (b) 1,452   (c) 5,493,721   (d) 52,179
3. In the number 163,524, name the digit in the following position.
   (a) hundred-thousands   (b) hundreds   (c) ones

Section 1.2 Write the following numbers as you would read them.

4. 27,542   5. 1,052,002   6. 13,524   7. 10,503   8. 15,402,523

Write the numeral for the following.

9. one hundred sixty-two thousand, five hundred forty-nine
10. seven million, forty-eight thousand, two

Section 1.3 Round each number to the nearest ten.

11. 239   12. 1,252   13. 2,575   14. 2,598

Round each number in these statements to the nearest hundred.

15. Texas has 262,970 square miles of land.
16. Utah has 82,381 square miles of land.
17. Oregon has 96,209 square miles of land.

CHAPTER POST-TEST

This test should determine if you have mastered the topics in Chapter 1. Answers to this test can be found on page 131.

1. In the number 27,563,418, name the position of the following digits.
   (a) 7   (b) 6   (c) 1   (d) 4   (e) 3   (f) 8   (g) 2
   (h) 5

2. In the number 365,814, name the digit in the following position.
   (a) ten-thousands   (b) tens   (c) hundreds

3. Write each number as you would read it.
   (a) 5,062,103   (b) 3,102,359

4. Write the numeral for the following.
   (a) fifty million, two hundred twenty-seven thousand, forty-three
   (b) one hundred five thousand, seven

5. Round each number to the nearest thousand.
   (a) 43,503   (b) 47,814   (c) 149,502   (d) 42,328   (e) 149,405
   (f) 55,555

addition of whole numbers

$$5 + 7 = \square$$

$$8 + 4 = \square$$

$$\square + 6 = 9$$

This section explains and allows you to practice basic addition facts, such as

$$\begin{array}{r} 2 \\ + 3 \\ \hline 5 \end{array} \qquad \begin{array}{r} 5 \\ + 7 \\ \hline 12 \end{array} \qquad \begin{array}{r} 9 \\ + 5 \\ \hline 14 \end{array}$$

If you think you know these facts, turn to the exercises for practice, working each problem. If you have difficulty, turn back and read this section completely.

Basic Facts Suppose I have 6 points in a card game, and on the next play I receive 3 more points. What is my new total?

Starting at 6 and counting three more—

$$7, 8, 9$$

—I get a total of 9 points.

Instead of counting I could **add** 6 and 3; that is,

$$6 + 3 = 9$$

The numbers 6 and 3 are called **addends**; 9 is called the **sum.**

$$6 + 3 = 9$$ Read: 6 plus 3 equals 9.

addends   sum

The statement $6 + 3 = 9$ is one of the basic addition facts. The basic addition facts are all the problems in which the addends are pairs of single-digit whole numbers. It is essential that you know these facts thoroughly before continuing in this text. For example,

$$5 + 4 = 9$$

This sum, like any other basic addition fact, can be found by counting.

$$5 + 4 = ?$$

Count: $1, 2, 3, 4, 5, 6, 7, 8, 9$

5   plus   4 more   equals   9
5   +   4   =   9

Counting also explains why *you can add in either order*; that is,

$$5 + 4 = 9$$

$$4 + 5 = 9$$ Reverse order of addends.

However, I am sure you do not want to count every addition problem. You should know the basic addition facts so well that you recall the answer without counting. This comes only from practice.

This chart contains the basic addition facts.

$$\begin{array}{ccccccccc}
1 & 1 & 1 & 1 & 1 & 1 & 1 & 1 & 1 \\
+1 & +2 & +3 & +4 & +5 & +6 & +7 & +8 & +\,9 \\
\hline
2 & 3 & 4 & 5 & 6 & 7 & 8 & 9 & 10 \\
\end{array}$$

$$\begin{array}{ccccccccc}
2 & 2 & 2 & 2 & 2 & 2 & 2 & 2 & 2 \\
+1 & +2 & +3 & +4 & +5 & +6 & +7 & +\,8 & +\,9 \\
\hline
3 & 4 & 5 & 6 & 7 & 8 & 9 & 10 & 11 \\
\end{array}$$

$$\begin{array}{ccccccccc}
3 & 3 & 3 & 3 & 3 & 3 & 3 & 3 & 3 \\
+1 & +2 & +3 & +4 & +5 & +6 & +\,7 & +\,8 & +\,9 \\
\hline
4 & 5 & 6 & 7 & 8 & 9 & 10 & 11 & 12 \\
\end{array}$$

4	4	4	4	4	4	4	4	4
+1	+2	+3	+4	+5	+6	+7	+8	+9
5	6	7	8	9	10	11	12	13

5	5	5	5	5	5	5	5	5
+1	+2	+3	+4	+5	+6	+7	+8	+9
6	7	8	9	10	11	12	13	14

6	6	6	6	6	6	6	6	6
+1	+2	+3	+4	+5	+6	+7	+8	+9
7	8	9	10	11	12	13	14	15

7	7	7	7	7	7	7	7	7
+1	+2	+3	+4	+5	+6	+7	+8	+9
8	9	10	11	12	13	14	15	16

8	8	8	8	8	8	8	8	8
+1	+2	+3	+4	+5	+6	+7	+8	+9
9	10	11	12	13	14	15	16	17

9	9	9	9	9	9	9	9	9
+1	+2	+3	+4	+5	+6	+7	+8	+9
10	11	12	13	14	15	16	17	18

Addition with Zero

The following rule is an important additional fact:

> The sum of a number and zero is the number itself.
>
> $$9 + 0 = 0 + 9 = 9$$

Now try to add the following pairs of single-digit whole numbers. If you need to count or think too long to get the answer, practice this exercise over a period of days or weeks until you increase your speed. Remember to cover each solution until you think you know the answer.

Examples	Solutions
1. $2 + 7 = ?$	9
2. $5 + 3 = ?$	8
3. $8 + 4 = ?$	12
4. $7 + 1 = ?$	8
5. $0 + 9 = ?$	9
6. $2 + 5 = ?$	7
7. $5 + 7 = ?$	12
8. $8 + 8 = ?$	16
9. $7 + 2 = ?$	9
10. $8 + 0 = ?$	8

Examples	Solutions
How are you doing? Remember, there is no way to get these facts in your head other than by practicing them.	
11. $\begin{array}{r} 6 \\ +7 \\ \hline \end{array}$	13
12. $\begin{array}{r} 9 \\ +8 \\ \hline \end{array}$	17
13. $\begin{array}{r} 9 \\ +4 \\ \hline \end{array}$	13
14. $\begin{array}{r} 8 \\ +6 \\ \hline \end{array}$	14
15. $\begin{array}{r} 7 \\ +9 \\ \hline \end{array}$	16
16. $\begin{array}{r} 8 \\ +7 \\ \hline \end{array}$	15
17. $8 + 1 = ?$	9
18. $8 + 3 = ?$	11
19. $9 + 5 = ?$	14
20. $9 + 2 = ?$	11
21. $6 + 5 = ?$	11
22. $6 + 3 = ?$	9
23. $2 + 6 = ?$	8
24. $2 + 4 = ?$	6
25. $2 + 8 = ?$	10
26. $3 + 9 = ?$	12
27. $3 + 7 = ?$	10
28. $3 + 9 = ?$	12
29. $4 + 5 = ?$	9
30. $4 + 7 = ?$	11
31. $5 + 8 = ?$	13
32. $5 + 5 = ?$	10
33. $6 + 8 = ?$	14
34. $6 + 9 = ?$	15
35. $4 + 2 = ?$	6

Examples	Solutions
36. 3 + 3 = ?	6
37. 4 + 3 = ?	7
38. 4 + 6 = ?	10
39. 4 + 4 = ?	8
40. 6 + 6 = ?	12
41. 7 + 7 = ?	14
42. 9 + 1 = ?	10
43. 5 + 0 = ?	5
44. 9 + 9 = ?	18

Please practice these over and over to improve your speed and accuracy!

Exercises Part I
1. On a sheet of paper number from 1 to 44. Use this sheet to cover the right side of the Examples-Solutions section above and rework the section as a drill. Time yourself! You should be able to complete the problems correctly in less than a minute.

Practice them over and over until you can!

Exercises Part II
1. Purchase a set of flash cards. Discover the facts that give you trouble by removing them from the deck. Practice these facts each day.

SECTION 2.2
ADDITION OF WHOLE NUMBERS

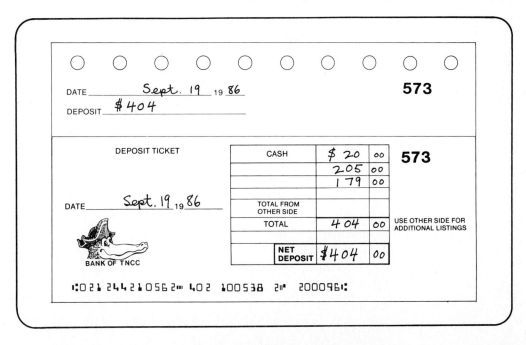

This section explains addition of whole numbers using a procedure called *carrying*, as illustrated in the following example.

$$\begin{array}{r} {\scriptstyle 2\ 1} \\ 1,392 \\ 4,157 \\ +\ \ 9,194 \\ \hline 14,743 \end{array}$$

If you think you know this procedure, turn to the exercises for practice. If you have difficulty, turn back and read this section completely.

Addition of Whole Numbers

When adding whole numbers, first arrange the numbers in columns. Line up digits having the same place value.

EXAMPLE 1:   Add: 253 + 514.
First, write the problem in vertical form.

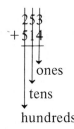

Digits having the same place value are lined up.

ones

tens

hundreds

Now think of these numbers in terms of place value.

$$\begin{array}{l} 253 = 2 \text{ hundreds and } 5 \text{ tens and } 3 \text{ ones} \\ +514 = 5 \text{ hundreds and } 1 \text{ ten } \text{ and } 4 \text{ ones} \end{array}$$

Next, use the basic addition facts to find the sum for each place.

$$\begin{array}{l} 253 = 2 \text{ hundreds and } 5 \text{ tens and } 3 \text{ ones} \\ +514 = 5 \text{ hundreds and } 1 \text{ ten } \text{ and } 4 \text{ ones} \\ \hline 7 \text{ hundreds and } 6 \text{ tens and } 7 \text{ ones} \end{array} \quad \Big\downarrow \text{ add}$$

As you can see (or know), this means that you can add the digits in column form.

$$\begin{array}{r} {\scriptstyle |\ |\ |} \\ 253 \\ +514 \\ \hline 767 \\ {\scriptstyle \downarrow\downarrow\downarrow} \end{array}$$

A good habit to develop with arithmetic is to estimate (make a good guess) what the answer will be before you calculate. In this example, round each number to the leading position.

253 is approximately equal to 300 $253 \cong 300$

514 is approximately equal to 500 $514 \cong 500$

Therefore, the sum should be approximately equal to 300 plus 500,

Actual	Estimation
253	300
+514	+500
	800

or approximately equal to 800.

Do not expect the actual answer to be 800, but do expect it to be close to it in value. In this example, the actual answer 767 is close to the guess of 800. If my guess and the actual answer were very different, I would repeat my calculations to find my error.

Now try these!

Examples	Solutions
Add 27 and 51.	First, arrange in columns. Line up digits having same place value. 27 +51 Next, estimate the answer by rounding each number to the leading position and quickly calculating the sum. Actual Estimation 27 30 +51 +50 80 Finally, add the actual numbers. 27 +51 78 This answer agrees with the original guess of 80 because my guess and the actual answer are not very different.
Add 23 and 55.	Arrange in columns. Line up digits having the same place value and add. Actual Estimation 23 20 +55 +60 78 80 The actual sum is 78.
Add 123 and 54.	Be careful to line up digits having the same place value. A frequent error is to carelessly arrange the digits *incorrectly*. For example, 123 +54 663

Examples	Solutions
	I would see that this result is incorrect if I estimated before finding the actual sum. The sum of 123 + 54 is approximately equal to 100 + 50, or 150. This estimate is significantly different from the erroneous sum of 663 I first obtained. I need to recheck my work to find the error, which—without estimation—I might not realize had occurred.

Arrange the digits carefully.

$$
\begin{array}{r}
|\,|\,| \\
123 \\
+\ 54 \\
\hline
\end{array}
$$

ones

tens

hundreds

Now add.

$$
\begin{array}{r}
123 \\
+\ 54 \\
\hline
177
\end{array}
$$

Add 108 and 51.	Actual	Estimation
	$\begin{array}{r} 108 \\ +\ 51 \\ \hline 159 \end{array}$	$\begin{array}{r} 100 \\ +\ 50 \\ \hline 150 \end{array}$
Add 54 and 123.	Actual	Estimation
	$\begin{array}{r} 54 \\ +123 \\ \hline 177 \end{array}$	$\begin{array}{r} 50 \\ +100 \\ \hline 150 \end{array}$

Ready to continue?

EXAMPLE 2:   Add: 259 + 514.

First, write in vertical form, then add digits in each place.

$$
\begin{array}{l}
259 = 2 \text{ hundreds and } 5 \text{ tens and } 9 \text{ ones} \\
+514 = 5 \text{ hundreds and } 1 \text{ ten and } 4 \text{ ones} \\
\hline
\quad\ 7 \text{ hundreds and } 6 \text{ tens and } 13 \text{ ones}
\end{array}
$$

But 13 ones is the same as 1 ten and 3 ones. Therefore, add the additional ten to the tens position.

$$
\begin{array}{l}
259 = 2 \text{ hundreds and } 5 \text{ tens and } 9 \text{ ones} \\
+514 = 5 \text{ hundreds and } 1 \text{ ten and } 4 \text{ ones} \\
\hline
\quad\ 7 \text{ hundreds and } 6 \text{ tens and } \cancel{13 \text{ ones}} \\
\qquad\qquad\qquad\qquad 1 \text{ ten and } 3 \text{ ones} \\
\hline
\quad\ 7 \text{ hundreds and } 7 \text{ tens and } 3 \text{ ones} = 773
\end{array}
$$

The renaming of 13 ones and placing the 1 in the next position is called **carrying**.

Here is the same example:

$$\overset{1}{259}$$
$$+514$$
$$\overline{3}$$

Think: (Starting in the ones position) 9 plus 4 equals 13 ones. 13 ones means 1 ten and 3 ones.

Write: Record the 3 in the ones position, and carry the 1 into the tens position.

Continue the addition with the tens position, followed by the hundreds position.

$$\overset{1}{259}$$
$$+514$$
$$\overline{773}$$

Do not forget to add the carried value in the tens position.

$$1 + 5 + 1 = 7$$

 Because of the possibility of a carried value, it is important that you start adding (in column form) in the ones position and work to the left.

A carried value from *any* position is treated in the same way.

EXAMPLE 3: Add: 1,529 + 3,718.
Start the addition in the ones position and work to the left.

Step 1: Think: 9 plus 8 equals 17.
 Write: Record the 7 and carry the 1 into the tens position.

$$\overset{1}{1,529}$$
$$+3,718$$
$$\overline{7}$$

Now add the digits in the tens position, including the carried value.

Step 2: Think: 1 plus 2 equals 3, plus 1 more equals 4.
 Write: Record the 4 in the tens position.

$$\overset{1}{1,529}$$
$$+3,718$$
$$\overline{47}$$

Next, add the digits in the hundreds position.

Step 3: Think: 5 plus 7 equals 12. 12 hundreds = 1200 = 1 thousand and 2 hundreds.
 Write: Record the 2 in the hundreds position and carry the 1 into the thousands position.

$$\overset{11}{1,529}$$
$$+3,718$$
$$\overline{247}$$

Finally, add the digits in the thousands position (including the carried value).

Step 4: Think: 1 plus 1 equals 2, plus 3 more equals 5.
 Write: Record the 5 in the thousands position.

$$\overset{11}{1,529}$$
$$+3,718$$
$$\overline{5,247}$$

Done! 1,529 + 3,718 = 5,247.

When you add more than two whole numbers, the same procedure applies, but the carried value may be other than 1.

EXAMPLE 4: Add: 1,392 + 4,157 + 9,194.
Do not forget to estimate.

$$1,394 \cong 1,000$$

$$4,157 \cong 4,000$$

$$9,194 \cong 9,000$$

Therefore, the sum should be approximately equal to 1,000 + 4,000 + 9,000, or 14,000.

Step 1: Think: Starting in the ones position, 2 plus 7 plus 4 equals 13.

Write: Record the 3 and carry the 1.

$$\begin{array}{r} 1 \\ 1{,}392 \\ 4{,}157 \\ +9{,}194 \\ \hline 3 \end{array}$$

Step 2: Think: Now add the digits in the tens position. 1 plus 9 plus 5 plus 9 equals 24.

Write: Record the 4 and carry the 2.

$$\begin{array}{r} 2\,1 \\ 1{,}392 \\ 4{,}157 \\ +9{,}194 \\ \hline 43 \end{array}$$

Step 3: Think: Next, add the digits in the hundreds position. 2 plus 3 plus 1 plus 1 equals 7.

Write: Record the 7; there is no carried value.

$$\begin{array}{r} 2\,1 \\ 1{,}392 \\ 4{,}157 \\ +9{,}194 \\ \hline 743 \end{array}$$

Step 4: Think: Finally, add the digits in the thousands position. 1 plus 4 plus 9 equals 14.

Write: Record the 4 and carry the 1 to the next position (in the answer).

$$\begin{array}{r} 2\,1 \\ 1{,}392 \\ 4{,}157 \\ +\ 9{,}194 \\ \hline 14{,}743 \end{array}$$

That is it! 1,392 + 4,157 + 9,194 = 14,743, which agrees with the estimation.

If you understood this example, you can now add whole numbers! Practice by working the following problems.

Examples	Solutions	
Add: 123 + 456.	$$\begin{array}{r} |\ |\ | \\ 123 \\ +456 \\ \hline 579 \\ \downarrow\downarrow\downarrow \end{array}$$	Arrange in columns. Line up digits having the same place value. Add the digits in each place.
Add: 654 + 123.	$$\begin{array}{r} 654 \\ +123 \\ \hline 777 \end{array}$$	There is no carried value.
Add: 564 + 129.	$$\begin{array}{r} 1 \\ 564 \\ +129 \\ \hline 3 \end{array}$$	Think: 9 plus 4 equals 13. Write: Record 3 and carry 1.
	$$\begin{array}{r} 1 \\ 564 \\ +129 \\ \hline 693 \end{array}$$	Continue adding the digits in each column. Work from right to left.

Examples	Solutions

Add: 987
   +715
	$\overset{1\ 1}{987}$
	+ 715
	1,702
	
	   7 + 5 = 12;
	   record 2, carry 1.
	   1 + 8 + 1 = 10;
	   record 0, carry 1.
	   1 + 9 + 7 = 17.

Add: 789
   +809
	Actual Estimation
	    $\overset{1}{789}$ 800
	+ 809 + 800
	 1,598 1,600

Add: 506
   111
  +394
	Actual Estimation
	    $\overset{1\ 1}{506}$ 500
	   111 100
	+ 394 + 400
	 1,011 1,000

Add: 299 + 37 + 451.	$\overset{1\ 1}{299}$ 37 ←—— Arrange carefully! +451 787

Add: 706 + 75 + 8,878.	Actual Estimation
	    $\overset{1\ 1}{706}$ 700
	  $\overset{1}{\ }$75 80
	+8,878 +9,000
	 9,659 9,780

Now try these—no hints given!	
Add: 17 + 34.	51
Add: 821 + 777.	1,598
Add: 75 + 315 + 218.	608

If you had difficulty with these last examples, please talk with an instructor or read this section again.

Answers to these exercises can be found on page 131.

Find the indicated sums.

1. $34 + 558$ 2. $8 + 15$

3. $90 + 18$ 4. $19 + 80$

5. $89 + 67$ 6. $987 + 213$

7. $817 + 746$ 8. $555 + 83$

9. $987 + 54$ 10. $678 + 715$

11.
$$\begin{array}{r} 308 \\ 666 \\ +691 \\ \hline \end{array}$$
12.
$$\begin{array}{r} 755 \\ 111 \\ +307 \\ \hline \end{array}$$

13.
$$\begin{array}{r} 299 \\ 9 \\ +451 \\ \hline \end{array}$$
14. $61,304 + 218 + 51,816$

15. $8 + 8,481 + 1,816$ 16. $4,667 + 4,311 + 7,105$

17. $319 + 21 + 31,565$ 18.
$$\begin{array}{r} 315 \\ 75 \\ +1,209 \\ \hline \end{array}$$

19.
$$\begin{array}{r} 75 \\ 123 \\ +\ \ 8 \\ \hline \end{array}$$
20.
$$\begin{array}{r} 761 \\ 57 \\ 48,513 \\ +\ \ 8,787 \\ \hline \end{array}$$

Find the indicated sums.

1. $22 + 35$ 2. $197 + 19$ 3. $693 + 572$

4.
$$\begin{array}{r} 211 \\ 597 \\ +810 \\ \hline \end{array}$$
5.
$$\begin{array}{r} 45 \\ 162 \\ +999 \\ \hline \end{array}$$
6.
$$\begin{array}{r} 207 \\ 95 \\ +563 \\ \hline \end{array}$$

7. $566 + 27 + 1,563$ 8. $19 + 15,524 + 1,750$ 9.
$$\begin{array}{r} 120 \\ 19 \\ 653 \\ +299 \\ \hline \end{array}$$

10.
$$\begin{array}{r} 1,763 \\ 2,595 \\ 3,799 \\ +9,500 \\ \hline \end{array}$$

AB Automotive 124 First Street			Name ___John Jones___ Address ___174 Marrow___ City ___Charleston, S.C.___		
PARTS			DESCRIPTION OF WORK		
1	Fuel Pump	23 | 00	Replace Fuel Pump		
1	gasket	2 | 00	and check for leaks	44	00
			Labor Only	44	00
			Parts	25	00
			Gas, Oil, Grease		
			Tax	1	00
Total Parts		25 | 00	TOTAL	70	00

What is the total?

How many in all?

How many all together?

What do they add up to?

How much for everything?

All these questions require an addition. My examples will come from situations in a department store. Estimations will not be shown, but do not forget them in your work!

EXAMPLE 1:   What is the total purchase price of the following items: shirt for $15, pants for $27 and belt for $7?

$$\begin{array}{r} \overset{1}{15} \\ 27 \\ +\ 7 \\ \hline 49 \end{array} \text{ dollars}$$

For everything:

EXAMPLE 2:   It is inventory time. On the shelf there are 39 rolls of black-and-white film, 157 rolls of color-print film, and 215 rolls of color-slide film. How many rolls of film are there all together?

$$\begin{array}{r} \overset{2}{_{1}39} \\ 157 \\ +215 \\ \hline 411 \end{array} \text{ rolls of film}$$

All together:

EXAMPLE 3:   There are three entrances to the department store. Each has an electronic counter. How many people in all came to the store if the first counter reads 653, the second reads 275, and the third reads 1,119?

$$
\begin{array}{r}
\overset{1\ 1}{653} \\
\overset{1}{275} \\
+\,1,119 \\
\hline
2,047 \quad \text{people}
\end{array}
$$

In all:

EXAMPLE 4:   If Sylvia Dwyer receives $200 salary for a week and $60 commission for sales, what is her total salary for the week?

$$
\begin{array}{r}
200 \\
+\ 60 \\
\hline
260 \quad \text{dollars}
\end{array}
$$

Total:

EXAMPLE 5:   Complete this deposit ticket.

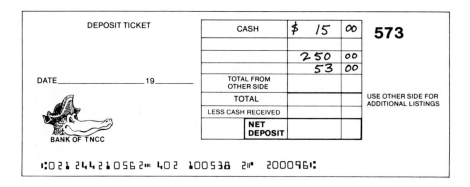

Totaling the cash and checks, I get $318, which I list next to the word *total*. Because no cash was received, the net deposit is also $318.

Now you try some problems!

Examples	Solutions
What is the total bill for the following items: circular saw for $62, saw blades for $10, and saw table for $59?	Total: $$\begin{array}{r}\overset{1}{62}\\10\\+\ 59\\\hline 131 \quad \text{dollars}\end{array}$$
There are three parking lots for a department store. The first lot has 1,524 spaces; the second has 568 spaces; the third has 1,408. How many spaces are there all together?	All together: $$\begin{array}{r}\overset{1\ 12}{1,524}\\568\\+\,1,408\\\hline 3,500 \quad \text{spaces}\end{array}$$

Examples	Solutions

José has $27, Maria has $39, and the bill is for $1 more than both José and Maria have together. (a) How much money do José and Maria have together? (b) How much is the bill?

(a)
$$
\begin{array}{r}
\text{José} \quad \overset{1}{2}7 \\
+\text{Maria} \quad 39 \\
\hline
66 \quad \text{dollars}
\end{array}
$$

José and Maria together have $66.

(b) The bill is $1 more than José and Maria have together ($66).

$$
\begin{array}{r}
66 \\
+\ 1 \\
\hline
67 \quad \text{dollars}
\end{array}
$$

The bill is $67.

Larry has saved $5; Mary has saved $12. If they need $3 more to buy a basketball, how much does it cost?

Cost:

$$
\begin{array}{r}
\text{Larry} \quad 5 \\
+\ \text{Mary} \quad 12 \\
\hline
17 \quad \text{dollars saved} \\
+\text{Amount needed} \quad 3 \\
\hline
20 \quad \text{dollars}
\end{array}
$$

The basketball costs $20.

Find the total distance around the building shown in the figure.

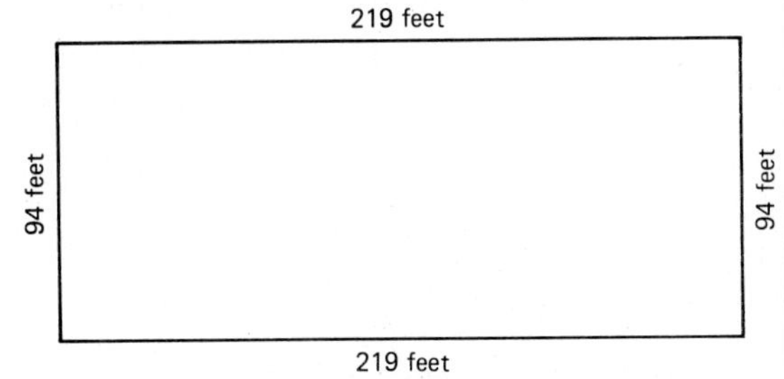

219 feet

94 feet · 94 feet

219 feet

Total distance:

$$
\begin{array}{r}
219 \\
94 \\
219 \\
+\ 94 \\
\hline
626 \quad \text{feet}
\end{array}
\left.\right\} \text{four sides}
$$

Now try these—no hints given.

Find the total bill for these items: math book for $19, English book for $8, and history book for $12.

$39

What is the distance around the figure?

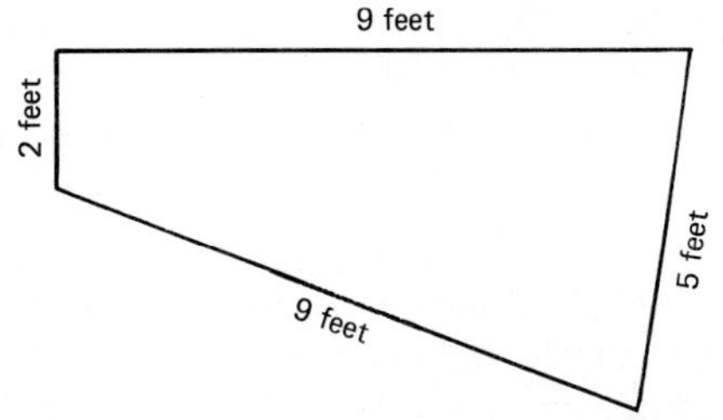

9 feet

2 feet

9 feet

5 feet

28 feet

Anne has paid $204 to Ruth Willis for typing. If she still owes $150, what was the total bill?

$354

How did you do? Please read this section again or talk with an instructor if you had any difficulty.

Answers to these exercises can be found on page 131.

Exercises
Part I

1. Find the total bill for these items: hammer for $18, screwdriver for $6, and wrench for $9.

2. Ruth Willis charged $8 for proofreading, $12 for materials, and $200 for typing a manuscript. What was the total bill?

3. Find the distance around this figure.

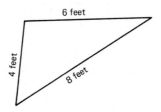

4. Find the distance around this figure.

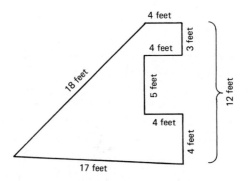

5. Muriel paid the dentist $50 in January and $65 in February. If she still owes $125, how much was the total bill?

6. Sue put 5 gallons of gas in the car. If it takes 9 more gallons to fill the tank, how much gas does the car hold?

7. Tom saved $69, Fred saved $52, but Susan saved $10 more than Tom and Fred combined.
   (a) How much did Tom and Fred save?
   (b) How much did Susan save?
   (c) How much did all three save together?

8. It takes Larry 15 minutes to drive from home to the babysitter. From the babysitter to work takes 18 minutes. How long is his total trip from home to work?

9. Jane froze a roast that she had baked for 2 hours covered and 1 hour uncovered. If she baked it in the oven for another hour before serving, how long did the roast bake?

10. North Carolina has 48,880 square miles of land. South Carolina has 30,280 square miles. What is the total land area of the two states?

Complete the following deposit tickets.

11.

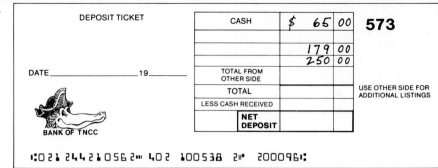

12.

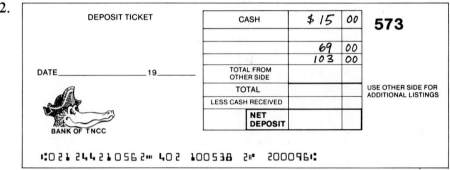

Exercises Part II

1. What is the total bill for these items: lawn mower for $500, grass catcher for $37, and a maintenance agreement for $58?

2. Sadie paid $12 and $20 on her layaway items. If she still owes $39, how much was the total bill?

3. Find the distance around this figure.

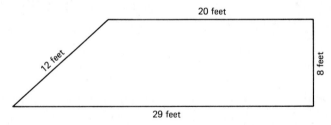

4. Raoul has $8 and Kim has $7. If they need $9 more for the game they want, what is the price of the game?

REVIEW MATERIAL

**Review Skill
Addition of
Whole Numbers
(Part I)**

When adding whole numbers first arrange the numbers in columns. Line up digits having the same place value.

$59 + 73 =$

$$\begin{array}{r} \overset{1}{59} \\ + 73 \\ \hline 2 \end{array}$$

Think: 9 plus 3 equals 12.
Write: Record 2 and carry 1.

$$\begin{array}{r} \overset{1}{59} \\ + \ 73 \\ \hline 132 \end{array}$$

Think: 1 plus 5 plus 7 equals 13.
Write: Record 13.

Exercise A Answers to this exercise can be found on page 131.

Add.

1.   24
      +32

2.   38
      +29

3. 25 + 68

4. 37 + 97

5. 32 + 88   6. 30 + 79   7. 23 + 78   8. 21 + 89
9. 26 + 58   10. 29 + 92   11. 27 + 9   12. 7 + 93
13. 15 + 15   14. 17 + 86   15. 2 + 19   16. 8 + 25
17. 90 + 15   18. 65 + 30   19. 10 + 50   20. 80 + 90

Exercise B No answers are given.

Add.

1.   55
      +82

2.   53
      +50

3. 51 + 31

4. 59 + 63

5. 56 + 24   6. 55 + 95   7. 42 + 76   8. 48 + 47
9. 43 + 68   10. 47 + 19

Review Skill Addition of Whole Numbers (Part II)

$3,789 + 5,696 =$

$$\begin{array}{r} \overset{1\ 1\ 1}{3789} \\ +5696 \\ \hline 9485 \end{array}$$

Think: 9 plus 6 equals 15; record 5 and carry 1.
Think: 1 plus 8 plus 9 equals 18; record 8 and carry 1.
Think: 1 plus 7 plus 6 equals 14; record 4 and carry 1.
Think: 1 plus 3 plus 5 equals 9; record 9.

Exercise C Answers to this exercise can be found on page 131.

Add.

1.   737
      +763

2.   577
      +382

3. 666 + 546

4. 967 + 242   5. 3,618 + 1,036   6. 3,163 + 8,365
7. 1,874 + 5,427   8. 5,789 + 4,583   9. 5,936 + 572
10. 457 + 6,582   11. 23,456 + 8,679   12. 345 + 34,999
13. 34,004 + 3,098   14. 54,601 + 4,509   15. 40,005 + 12,405

Exercise D No answers are given.

Add.

1.   787
      +849

2.   471
      +575

3. 626 + 213

4. $639 + 910$ **5.** $9,674 + 8,428$ **6.** $8,625 + 8,172$

7. $5,033 + 2,144$

**Review Skill
Addition of
Whole Numbers
(Part III)**

The same procedure applies when you add more than two whole numbers, but the carried value may be other than 1.

$$48 + 9 + 27 = \qquad \overset{2}{4}8$$
$$9$$
$$+27$$
$$\overline{84}$$

— Think: 8 plus 9 plus 7 equals 24; record 4 and carry 2.

— Think: 2 plus 4 plus 2 equals 8; record 8.

Do not forget to estimate to see if your actual answer is reasonable.

$$48 + 9 + 27 \cong 50 + 9 + 30$$

$$\cong 89$$

The actual sum is approximately equal to the estimate.

Exercise E   Answers to this exercise can be found on page 131.

Add.

1.   15   **2.**   39
         95   71
      $+97$ $+65$

3. $885 + 569 + 42 + 231$ **4.** $526 + 586 + 467 + 766$

5. $8,110 + 6,759 + 3,841 + 6,352$ **6.** $9,672 + 5,739 + 952 + 4,518$

7. $5,689 + 456 + 1,456$ **8.** $45,790 + 304 + 2,300$

9. $7,890 + 40,009 + 399$ **10.** $1,234 + 1,234 + 1,234$

Exercise F   No answers are given.

Add.

1.   48   **2.**   36
         84   72
      $+26$ $+83$

3. $643 + 734 + 56 + 688$ **4.** $928 + 197 + 785 + 527$

5. $5,659 + 7,810 + 3,498 + 3,643$ **6.** $2,116 + 7,849 + 248 + 9,819$

REVIEW EXERCISES

Refer to the section listed if you have difficulty. Answers to these exercises can be found on page 131.

Section 2.2 Find the indicated sums.

1. $29 + 52$ 2. $20 + 59 + 63$ 3. $128 + 235$

4. $569 + 809$ 5. $1,256 + 3,445$ 6. $\begin{array}{r} 103 \\ + \ 29 \\ \hline \end{array}$

7. $\begin{array}{r} 1,593 \\ 270 \\ + \ \ 53 \\ \hline \end{array}$ 8. $59 + 653 + 8$ 9. $10,563 + 275 + 5,955$

10. $2,885 + 3,772 + 553$

Section 2.3

11. Find the total bill for these items: carbon paper for $3, typing paper for $5, typewriter rental for $58, and typewriter ribbon for $7.

12. Find the distance around this figure.

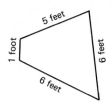

13. Tom has paid $16,048 on his mortgage. If he still owes $52,563, how much was the total mortgage?

CHAPTER POST-TEST

This test should determine if you have mastered the topics in Chapter 2. Answers to this test can be found on page 131.

Find the indicated sums.

1. $27 + 36 + 8$
2. $127 + 405 + 28$
3. $1,684 + 2,593 + 5,763$
4. $\begin{array}{r} 6,481 \\ 5,932 \\ +7,635 \\ \hline \end{array}$ 5. $\begin{array}{r} 1,563,425 \\ 3,642,563 \\ +2,524,365 \\ \hline \end{array}$

6. Find the total bill for items costing $25, $36, and $29.
7. Find the distance around this figure.

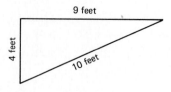

8. John paid $6 on a layaway item. If it will take $15 more to get it out, what was the total cost of the item?

3

subtraction of whole numbers

5 − 3 = □

7 − □ = 5

8 − 5 = □

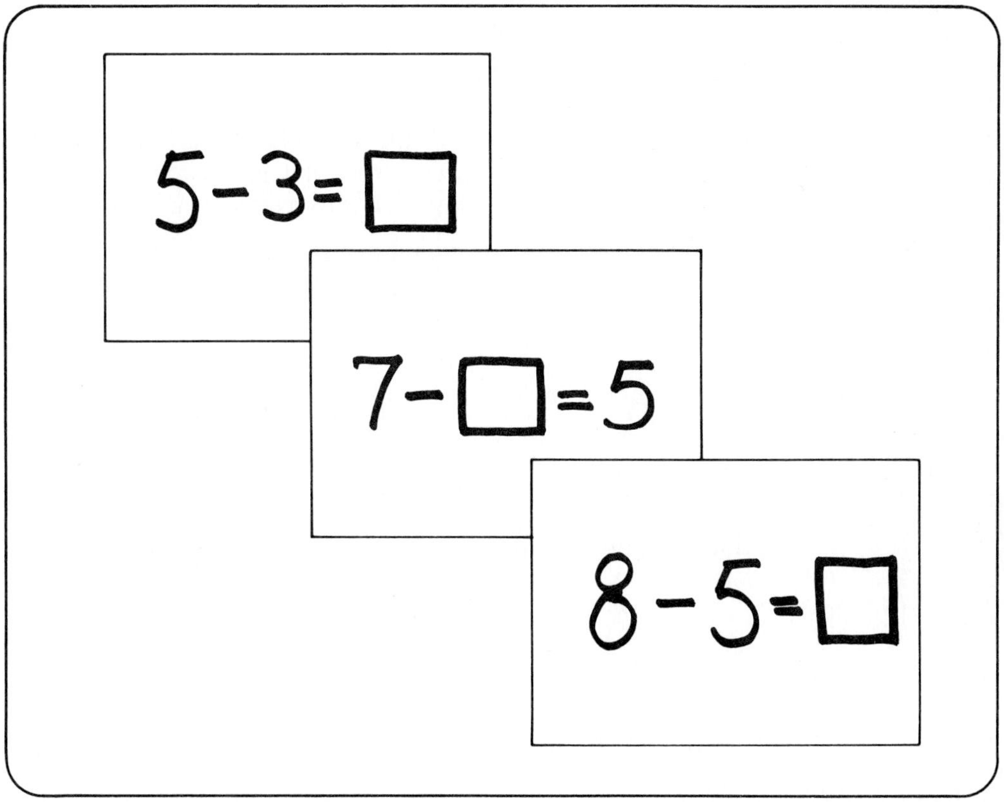

This section explains and allows you to practice basic substraction facts, such as

$$
\begin{array}{r} 5 \\ -3 \\ \hline 2 \end{array} \qquad \begin{array}{r} 9 \\ -6 \\ \hline 3 \end{array} \qquad \begin{array}{r} 8 \\ -4 \\ \hline 4 \end{array}
$$

If you think you know these facts, turn to the exercises for practice. If you have difficulty, turn back and read this section completely.

Basic Facts Suppose I have \$15 and spend \$7 on a record album on sale. How much money do I have left?

Starting at 15 and counting backward 7 whole numbers—

$$14, 13, 12, 11, 10, 9, 8$$

—I am left with \$8.

Instead of counting I could **subtract** 7 from 15; that is,

$$15 - 7 = 8$$

The number 15 is called the **minuend**; 7 is called the **subtrahend**; 8 is called the **difference**.

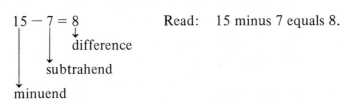

Read: 15 minus 7 equals 8.

The statement $15 - 7 = 8$ is one of the basic subtraction facts.

As in addition, it is essential that you know the basic subtraction facts so well that you know the answers without counting backward.

To add you count forward, and to subtract you count backward. Therefore, addition and subtraction are considered *opposite* operations. Knowing this will help you learn the basic subtraction facts.

For example, to answer the question,

$$15 - 9 = ? \quad \text{or} \quad \begin{array}{r} 15 \\ -\ 9 \\ \hline ? \end{array}$$

think an addition question: What (?) added to 9 equals 15?

$$
\left.\begin{array}{r} 15 \\ -\ 9 \\ \hline ? \end{array}\right\} \quad \begin{array}{l} \text{equals 15} \\ \text{? plus 9} \end{array}
$$

Because $6 + 9 = 15$, $15 - 9 = 6$.

$$
\left.\begin{array}{r} 15 \\ -\ 9 \\ \hline 6 \end{array}\right\} \quad \begin{array}{l} \text{equals 15} \\ \text{6 plus 9} \end{array}
$$

When you forget a basic fact, you can count backward or use addition to find the answer, but with practice you will remember them all.

Subtraction with Zero

The following example illustrates an important subtraction fact.

$$5 - 0 = 5 \quad \text{because} \quad 5 + 0 = 5$$

$$\left.\begin{array}{r} 5 \\ -0 \\ \hline 5 \end{array}\right\} \quad \begin{array}{l} \text{equals 5} \\ \text{5 plus 0} \end{array}$$

> Subtraction of zero from a number gives the number itself.
>
> $$7 - 0 = 7 \quad \text{or} \quad \begin{array}{r} 7 \\ -0 \\ \hline 7 \end{array}$$

However, subtraction from zero will for now be considered impossible in arithmetic. Later, in algebra, such situations will be discussed. Thus, questions such as

$$0 - 5 = ? \quad \text{or} \quad 0 - 7 = ?$$

will be considered to have no answer.

Now try to subtract the following pairs of whole numbers. If you need to count or think too long to get the answers, practice this exercise over a period of days or weeks. Remember to cover the answers as you work.

Examples	Solutions
1. $2 - 1 = ?$	1
2. $4 - 3 = ?$	1 (Think: What added to 3 is 4?)
3. $6 - 5 = ?$	1
4. $7 - 6 = ?$	1
5. $\begin{array}{r} 9 \\ -5 \\ \hline \end{array}$	4
6. $\begin{array}{r} 2 \\ -2 \\ \hline \end{array}$	0
7. $\begin{array}{r} 4 \\ -4 \\ \hline \end{array}$	0
8. $\begin{array}{r} 8 \\ -4 \\ \hline \end{array}$	4
9. $\begin{array}{r} 10 \\ -\ 6 \\ \hline \end{array}$	4
10. $7 - 2 = ?$	5
11. $7 - 5 = ?$	2 (Think: What added to 5 is 7?)
12. $10 - 2 = ?$	8

Examples	Solutions
13. $5 - 5 = ?$	0
14. $9 - 3 = ?$	6
15. $8 - 6 = ?$	2
16. $3 - 2 = ?$	1
17. $5 - 2 = ?$	3
18. $5 - 4 = ?$	1
19. $8 - 5 = ?$	3
20. $10 - 3 = ?$	7
21. $9 - 6 = ?$	3
22. $7 - 7 = ?$	0
23. $6 - 0 = ?$	6
24. $6 - 3 = ?$	3
25. $4 - 1 = ?$	3
26. $1 - 1 = ?$	0
27. $\begin{array}{r} 5 \\ -3 \\ \hline \end{array}$	2
28. $\begin{array}{r} 7 \\ -4 \\ \hline \end{array}$	3
29. $\begin{array}{r} 9 \\ -2 \\ \hline \end{array}$	7
30. $\begin{array}{r} 9 \\ -8 \\ \hline \end{array}$	1
31. $\begin{array}{r} 8 \\ -7 \\ \hline \end{array}$	1
32. $10 - 4 = ?$	6
33. $8 - 3 = ?$	5
34. $6 - 4 = ?$	2
35. $3 - 1 = ?$	2
36. $4 - 2 = ?$	2
37. $7 - 1 = ?$	6
38. $6 - 1 = ?$	5
39. $9 - 4 = ?$	5
40. $6 - 6 = ?$	0

Examples	Solutions
41. $8 - 2 = ?$	6
42. $3 - 0 = ?$	3
43. $5 - 1 = ?$	4
44. $7 - 3 = ?$	4
45. $6 - 2 = ?$	4
46. $12 - 8 = ?$	4
47. $16 - 8 = ?$	8
48. $12 - 5 = ?$	7
49. $13 - 6 = ?$	7
50. $17 - 8 = ?$	9
51. $13 - 4 = ?$	9
52. $14 - 6 = ?$	8
53. $16 - 7 = ?$	9
54. $15 - 8 = ?$	7
55. $11 - 3 = ?$	8
56. $14 - 9 = ?$	5
57. $11 - 2 = ?$	9
58. $11 - 6 = ?$	5
59. $12 - 9 = ?$	3
60. $12 - 3 = ?$	9
61. $11 - 7 = ?$	4
62. $13 - 5 = ?$	8
63. $14 - 8 = ?$	6
64. $15 - 6 = ?$	9
65. $12 - 6 = ?$	6
66. $14 - 7 = ?$	7
67. $18 - 9 = ?$	9

Practice will improve speed and accuracy!

Exercises Part I

1. On a sheet of paper number from 1 to 67. Use this sheet to cover the right side of the Examples-Solutions section above. Time yourself! You should be able to complete the problems correctly in less than a minute and a half.

1. Purchase a set of flash cards. Discover the facts that give you trouble by removing them from the deck. Practice these facts each day.

SECTION 3.2
SUBTRACTION OF WHOLE NUMBERS

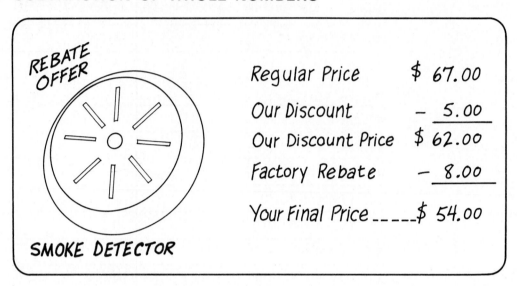

This section explains subtraction of whole numbers using a procedure called *borrowing*, as illustrated in the following example.

$$
\begin{array}{r}
\overset{6}{\cancel{7}}\overset{9}{\cancel{0}}\overset{1}{4} \\
-\ \ 86 \\
\hline
618
\end{array}
$$

If you think you know this procedure, turn to the exercises for practice. Turn back and read this section completely if you have difficulty with the exercises.

**Subtraction of
Whole Numbers**

> When subtracting whole numbers first arrange the numbers in columns. Line up digits having the same place value.

EXAMPLE 1: Subtract: 875 − 232.
First write the problem in vertical form.

$$
\begin{array}{r}
875 \\
-232 \\
\hline
\end{array}
$$
Line up digits having the same place value.

Now think of these numbers in terms of place value, and subtract for each place by using the basic subtraction facts.

$$
\begin{array}{rl}
875 =& 8 \text{ hundreds and } 7 \text{ tens and } 5 \text{ ones} \\
-232 =& 2 \text{ hundreds and } 3 \text{ tens and } 2 \text{ ones} \\
\hline
=& 6 \text{ hundreds and } 4 \text{ tens and } 3 \text{ ones}
\end{array}
$$

As you can see, this means you can subtract the digits in column form.

$$\overset{|\ |\ |}{875}$$
$$-232$$
$$\overline{643}$$
$$\downarrow\downarrow\downarrow$$

Check the subtraction by addition to see if you are right. As before,

$$\left.\begin{array}{r} 875 \\ -232 \\ \hline 643 \end{array}\right\} \quad \begin{array}{l} \text{equals } 875 \\ 643 \text{ plus } 232 \end{array}$$

 Most people do this check in the vertical form by drawing another line under the answer and adding.

$$\begin{array}{r} 875 \\ -232 \\ \hline \end{array} \Bigg]\ \text{add}$$

Answer: $\overline{643}\ \downarrow$

Check: $\overline{875}$ If the answer is correct, this addition gives you the minuend.

You can also make a rough check using estimation.

Actual	Estimation
875	900
-232	-200
	700

This means the actual answer should be close to 700. Answers close to 7, 70, or 7000 should be checked for mistakes.

You try these!

Examples	Solutions

Subtract: $698 - 325$.	First, arrange the numbers in columns. Line up digits having the same place value.
	
	    Actual Estimation
	     698 700
	   -325 -300
	                400
	
	Next, subtract in column form.
	
	          698
	      -325
	         373

Subtract 105 from 397.	Arrange in columns. Line up digits having the same place value.
	
	    Actual Estimation
	     397 400
	   -105 -100
	     292 300

Examples	Solutions
Subtract: $584 - 23$.	Arrange carefully!

Solutions continued:

	Actual	Estimation
	584	600
	$-\ 23$	$-\ 20$
	561	580

Ready to continue?

EXAMPLE 2:   Subtract: $554 - 219$.

$$554 = 5 \text{ hundreds and } 5 \text{ tens and } 4 \text{ ones}$$
$$-219 = 2 \text{ hundreds and } 1 \text{ ten and } 9 \text{ ones}$$

Subtraction is not possible in the ones position because 9 is larger than 4. I must rename the tens and ones positions in 554.

5 tens and 4 ones

I take 1 ten from the 5 tens, leaving 4 tens,

4 tens
~~5 tens~~ and 4 ones

and add the 1 ten (or 10 ones) to the ones position.

4 tens and 10 ones }
~~5 tens~~ and 4 ones }   4 tens and 14 ones

Now I can subtract.

$$
\begin{array}{lll}
 & 4 \text{ tens} + 10 \text{ ones} \rbrace & \\
554 = 5 \text{ hundreds} + \text{~~5 tens~~} + \ 4 \text{ ones} \rbrace & = 5 \text{ hundreds} + 4 \text{ tens} + 14 \text{ ones} \\
-219 = 2 \text{ hundreds} + 1 \text{ ten} + \ 9 \text{ ones} & = 2 \text{ hundreds} + 1 \text{ ten} + \ \ 9 \text{ ones} \\
\hline
 & \ \ 3 \text{ hundreds} + 3 \text{ tens} + \ \ 5 \text{ ones}
\end{array}
$$

This renaming in subtraction is called **borrowing**.
  Here is the same example.

$$
\begin{array}{l}
^{4} \\
5\!\!\not{5}4 \\
-219 \\
\hline
\end{array}
$$
Think: Borrow 1 ten from the 5 tens, leaving 4 tens.
Write: Indicate this by crossing out the 5 and replacing it with 4.

$$
\begin{array}{l}
^{4} \\
5\!\!\not{5}^{1}4 \\
-219 \\
\hline
\end{array}
$$
Think: Add the ten to the ones, giving 14 ones.
Write: Indicate this by writing the 1 to the upper left of the 4.

$$
\begin{array}{l}
^{4}{}^{1} \\
5\!\!\not{5}4 \\
-219 \\
\hline
335
\end{array}
$$
Finally, subtract in column form.

$14 - 9 = 5$
$4 - 1 = 3$
$5 - 2 = 3$

Because of the possibility of borrowing, it is important that you start subtraction in the ones position and work to the left.

A borrowed value from any position is treated in the same way.

EXAMPLE 3: Subtract: 6,392 − 2,735.

Step 1: Think: The subtraction is not possible in the ones position. Borrow 1 ten from the 9 tens, leaving 8 tens.

Write: Cross out the 9 and replace it with an 8.

$$\begin{array}{r} \overset{8}{6,3\!\!\!/92} \\ -2,735 \\ \hline \end{array}$$

Step 2: Think: 1 ten equals 10 ones; add this to the ones position, giving 12 ones.

Write: Indicate this by writing 1 to the upper left of the digit 2.

$$\begin{array}{r} \overset{8}{6,3\!\!\!/9\,^{1}2} \\ -2,735 \\ \hline \end{array}$$

Step 3: Think: Next, subtract the numbers in the ones position. (12 − 5 = 7)

Write: Record the 7.

$$\begin{array}{r} \overset{8}{6,3\!\!\!/9\,^{1}2} \\ -2,735 \\ \hline 7 \end{array}$$

Step 4: Think: Next, subtract in the tens position. (8 − 3 = 5)
Write: Record the 5.

$$\begin{array}{r} \overset{8}{6,3\!\!\!/9\,^{1}2} \\ -2,735 \\ \hline 57 \end{array}$$

The digits in the hundreds position cannot be subtracted because 7 is larger than 3.

Step 5: Think: Borrow 1 thousand from the 6 thousand, leaving 5 thousand.

Write: Cross out to show the borrowing.

$$\begin{array}{r} \overset{5}{6\!\!\!/},\overset{8}{3\!\!\!/9\,^{1}2} \\ -2,735 \\ \hline 57 \end{array}$$

Step 6: Think: But 1 thousand equals 10 hundreds. Add the 1 thousand that was borrowed to the hundreds position as 10 hundreds, giving a total of 13 hundreds.

Write: Place 1 next to the digit 3.

$$\begin{array}{r} \overset{5}{6\!\!\!/}\,^{1}\overset{8}{3\!\!\!/9\,^{1}2} \\ -2,735 \\ \hline 57 \end{array}$$

Step 7: Think: Now subtract in the hundreds position. (13 − 7 = 6)

Write: Record the 6.

$$\begin{array}{r} \overset{5}{6\!\!\!/}\,^{1}\overset{8}{3\!\!\!/9\,^{1}2} \\ -2,735 \\ \hline 657 \end{array}$$

Step 8: Think: Finally, subtract in the thousands position. (5 − 2 = 3)

Write: Record the 3.

$$\begin{array}{r} \overset{5}{6\!\!\!/}\,^{1}\overset{8}{3\!\!\!/9\,^{1}2} \\ -2,735 \\ \hline 3,657 \end{array}$$

The preceding examples illustrate the borrowing procedure.

Borrowing Procedure

Borrowing 1 from a digit to the left reduces the borrowing position by 1 and adds 10 to the original position.

EXAMPLE 4: Subtract: 9,562 − 5,936.

The subtraction is not possible in the ones position.

$$\begin{array}{r} 9,562 \\ -5,936 \\ \hline \end{array}$$

Borrow 1 from the tens place and add 10 to the digit 2.

$$
\begin{array}{r}
9,5\overset{\overset{5}{1}}{\cancel{6}}2 \\
-5,936 \\
\end{array}
$$

Subtract in the ones, then in the tens position. $12 - 6 = 6$; record the 6. $5 - 3 = 2$; record the 2.

$$
\begin{array}{r}
9,5\overset{\overset{5}{1}}{\cancel{6}}2 \\
-5,936 \\
\hline
26 \\
\end{array}
$$

Notice that the subtraction in the hundreds position is not possible. Borrow 1 from the thousands place and add 10 to the digit 5.

$$
\begin{array}{r}
\overset{8}{\cancel{9}},\overset{\overset{1}{5}}{}\overset{\overset{5}{1}}{\cancel{6}}2 \\
-5,936 \\
\hline
26 \\
\end{array}
$$

Continue the subtraction in the hundreds and thousands positions.

$$
\begin{array}{r}
\overset{8}{\cancel{9}},\overset{\overset{1}{5}}{}\overset{\overset{5}{1}}{\cancel{6}}2 \\
-5,936 \\
\hline
3,626 \\
\end{array}
$$

To check the answer, add.

checks!
$$
\begin{array}{r}
9,562 \\
-5,936 \\
\hline
3,626 \\
9,562 \\
\end{array}
\quad \text{add}
$$

Correct. I got the minuend.

Again, you might want to estimate to obtain a rough check.

$$9,562 - 5,936 \cong 10,000 - 6,000$$
$$\cong 4,000$$

This estimate is close to the actual difference, 3,636, so it tells you that your answer is reasonable.

Try these.

Examples	Solutions
Subtract: $95 - 39$.	$\begin{array}{r} 95 \\ -39 \\ \hline \end{array}$
	Borrow 1 from 9, leaving 8. Add 10 to the 5, giving 15.
	$\begin{array}{r} \overset{8}{\cancel{9}}\overset{\overset{1}{}}{5} \\ -39 \\ \hline \end{array}$

Examples	Solutions
	Now subtract.
	$$\begin{array}{r} \overset{8}{\cancel{9}}\overset{1}{5} \\ -39 \\ \hline 56 \end{array}$$
Subtract and check: $5{,}162 - 91$.	$$\begin{array}{r} 5{,}162 \\ -91 \\ \hline \end{array}$$ ⟵ —— Arrange carefully!
	I do not need to borrow in the ones position.
	$$\begin{array}{r} 5{,}162 \\ -91 \\ \hline 1 \end{array}$$
	Borrow 1 from 1, leaving 0. Add 10 to 6, giving 16.
	$$\begin{array}{r} 5{,}\overset{0}{\cancel{1}}\overset{1}{}62 \\ -91 \\ \hline 5{,}071 \end{array}$$
	$$\begin{array}{r} 5{,}162 \\ -91 \\ \hline 5{,}071 \\ \hline 5{,}162 \end{array}\Bigg\downarrow \text{ add}$$ Checks!
Subtract 1,659 from 7,292.	Actual　　　Estimation $$\begin{array}{rr} 7{,}292 & 7{,}000 \\ -1{,}659 & -2{,}000 \\ \hline & 5{,}000 \end{array}$$
	$$\begin{array}{r} 7{,}2\overset{8}{\cancel{9}}\overset{1}{2} \\ -1{,}659 \\ \hline \end{array}$$　Borrow 1 from 9, leaving 8. Add 10 to 2.
	$$\begin{array}{r} 7{,}2\overset{8}{\cancel{9}}\overset{1}{2} \\ -1{,}659 \\ \hline 33 \end{array}$$　Now subtract in the ones and tens positions.
	$$\begin{array}{r} \overset{6}{\cancel{7}}{,}2\overset{8}{\cancel{9}}\overset{1}{2} \\ -1{,}659 \\ \hline 33 \end{array}$$　Borrow 1 from 7, leaving 6. Add 10 to 2.
	$$\begin{array}{r} \overset{6}{\cancel{7}}{,}2\overset{8}{\cancel{9}}\overset{1}{2} \\ -1{,}659 \\ \hline 5{,}633 \end{array}$$　Now subtract in the hundreds and thousands positions.
	This is a reasonable answer with an estimate of 5,000.

Ready to continue?

EXAMPLE 5: Subtract: 704 − 86.

I need to borrow for the ones position.

$$
\begin{array}{r}
704 \\
-\ 86 \\
\hline
\end{array}
$$

In this example, there are no tens from which to borrow. I shall borrow from the hundreds first and then borrow from the tens.

Borrow 1 from the 7 hundreds and add 10 to the next position. Now there are 10 tens.

$$
\begin{array}{r}
{}^{6}\!\!\!\!{}_{1}\!\!7\,04 \\
-\ 86 \\
\hline
\end{array}
$$

Borrow 1 from the 10 tens, leaving 9 tens, and add 10 to the next position. Now there are 14 ones.

$$
\begin{array}{r}
{}^{6}\!\!\!\!{}^{9}\!\!\!\!{}_{1}\!\!7\,04 \\
-\ 86 \\
\hline
\end{array}
$$

Now subtract

$$
\begin{array}{r}
{}^{6}\!\!\!\!{}^{9}\!\!\!\!{}_{1}\!\!7\,04 \\
-\ 86 \\
\hline
618 \\
\end{array}
$$

      14 − 6 = 8
      9 − 8 = 1
      6 − 0 = 6

EXAMPLE 6: Subtract: 7,800 − 179.

Because of the zeros, the 8 is the first digit available for borrowing.

$$
\begin{array}{r}
7{,}800 \\
-\ 179 \\
\hline
\end{array}
$$

Borrow 1 from the 8 hundreds and add 10 to the next position.

$$
\begin{array}{r}
{}^{7}\!\!\!\!{}_{1}\!\!7{,}8\,00 \\
-\ 179 \\
\hline
\end{array}
$$

Borrow 1 from the 10 tens (leaving 9) and add 10 to the last position.

$$
\begin{array}{r}
{}^{7}\!\!\!\!{}^{9}\!\!\!\!{}_{1}\!\!7{,}8\,0\,0 \\
-\ 179 \\
\hline
\end{array}
$$

Now subtract!

$$
\begin{array}{r}
{}^{7}\!\!\!\!{}^{9}\!\!\!\!{}_{1}\!\!7{,}8\,0\,0 \\
-\ 179 \\
\hline
7{,}621 \\
\end{array}
$$

Check by adding.

$$\begin{array}{r} 7{,}800 \\ -179 \\ \hline 7{,}621 \\ 7{,}800 \end{array} \Bigg\downarrow \text{ add}$$

checks!

Try these.

Examples	Solutions
Subtract: $805 - 119$.	$\begin{array}{r} 805 \\ -119 \\ \hline \end{array}$
	$\begin{array}{r} \overset{7}{\underset{1}{8}}05 \\ -119 \\ \hline \end{array}$ Borrow 1 from 8, leaving 7. Add 10 to 0, giving 10.
	$\begin{array}{r} \overset{7}{\underset{1}{8}}\overset{9}{\underset{1}{0}}5 \\ -119 \\ \hline \end{array}$ Borrow 1 from 10, leaving 9. Add 10 to 5, giving 15.
	$\begin{array}{r} \overset{7}{\underset{1}{8}}\overset{9}{\underset{1}{0}}5 \\ -119 \\ \hline 686 \end{array}$ Now subtract.
Subtract: $8{,}800 - 158$.	$\begin{array}{r} 8{,}800 \\ -158 \\ \hline \end{array}$
	$\begin{array}{r} 8{,}\overset{7}{\underset{1}{8}}00 \\ -158 \\ \hline \end{array}$ Borrow 1 from 8, leaving 7. Add 10 to 0.
	$\begin{array}{r} 8{,}\overset{7}{\underset{1}{8}}\overset{9}{\underset{1}{0}}0 \\ -158 \\ \hline 8{,}642 \end{array}$ Borrow 1 from 10, leaving 9. Add 10 to 0, then subtract.
Subtract and check: $9{,}005 - 2{,}759$.	$\begin{array}{r} 9{,}005 \\ -2{,}759 \\ \hline \end{array}$
	$\begin{array}{r} \overset{8}{9}{,}\overset{9}{\underset{1}{0}}\overset{9}{\underset{1}{0}}5 \\ -2{,}759 \\ \hline 6{,}246 \end{array}$ Borrow from successive positions.
	$\begin{array}{r} 9{,}005 \\ -2{,}759 \\ \hline 6{,}246 \\ 9{,}005 \end{array} \Bigg\downarrow \text{ add}$ Checks!

Ready to continue?

EXAMPLE 7:   Subtract: 4,983 − 789.

$$\begin{array}{r} 4,983 \\ -789 \\ \hline \end{array}$$

Borrow 1 from the 8 tens and add 10 to the digit 3.

$$\begin{array}{r} {}^{7}{}_{1} \\ 4,9\cancel{8}3 \\ -789 \\ \hline \end{array}$$

Subtract: $13 − 9 = 4$. Notice, we need to borrow again for the tens position.

$$\begin{array}{r} {}^{7}{}_{1} \\ 4,9\cancel{8}3 \\ -789 \\ \hline 4 \end{array}$$

Borrow 1 from the 9 hundreds and add 10 to the digit 7.

$$\begin{array}{r} {}^{8}{}^{7}{}_{1} \\ 4,\cancel{9}\cancel{8}3 \\ -789 \\ \hline 4 \end{array}$$

Subtract: $17 − 8 = 9$.

$$\begin{array}{r} {}^{8}{}^{7}{}_{1} \\ 4,\cancel{9}\cancel{8}3 \\ -789 \\ \hline 94 \end{array}$$

Subtract in the hundreds and thousands positions. No other borrowing is necessary.

$$\begin{array}{r} {}^{8}{}^{7}{}_{1} \\ 4,\cancel{9}\cancel{8}3 \\ -789 \\ \hline 4,194 \end{array}$$

Now practice by working the following problems. As before, solutions are in the right column, but try to work each *before* looking at the solution.

Examples	Solutions
Subtract: $596 − 372$.	$$\begin{array}{r} 596 \\ -372 \\ \hline 224 \end{array}$$ $6 − 2 = 4$ $9 − 7 = 2$ $5 − 3 = 2$
Subtract: $596 − 59$.	$$\begin{array}{r} 596 \\ -59 \\ \hline \end{array}$$ Need to borrow for ones position.

Examples	Solutions

	$\overset{8}{5}\overset{1}{9}6$ 9 becomes 8. 10 is added
	$-\ \ 59$ to 6.
	$\overline{537}$
	   $16-9=7$
	   $8-5=3$
	   $5-0=5$

Subtract and check: $169-53$.	$\begin{array}{r}169\\-\ \ 53\\\hline 116\\169\end{array}$ add
	  Answer.
	  Checks!

Subtract and check: 1,462
              $-$ 798
	$1,4\overset{5}{\cancel{6}}\overset{1}{2}$ Borrow. 6 becomes 5
	$-\ \ 798$ and 10 is added to 2.
	$\overline{\qquad\ 4}$ $12-8=4$.
	
	$1,\overset{3}{\cancel{4}}\overset{15}{\cancel{6}}\overset{1}{2}$ Borrow. 4 becomes 3
	$-\ \ 798$ and 10 is added to 5.
	$\overline{\quad\ 64}$ $15-9=6$.
	
	$\overset{0}{\cancel{1}},\overset{13}{\cancel{4}}\overset{15}{\cancel{6}}\overset{1}{2}$ Borrow. 1 becomes 0
	$-\ \ 798$ and 10 is added to 3.
	$\overline{\ 664}$ $13-7=6$.
	
	$\begin{array}{r}1,462\\-\ \ 798\\\hline 664\\1,462\end{array}$ add
	  Answer.
	  Checks!

Subtract and check: 7,527 $-$ 398	$7,\overset{4}{\cancel{5}}\overset{1}{2}\overset{1}{7}$ $-\ \ 398$ add $\overline{7,129}$ Answer. $7,527$ Checks!

Subtract: $10,002-99$.	Actual Estimation
	$\begin{array}{r}10,002\\-\ \ \ \ \ 99\end{array}$ Borrow from $\begin{array}{r}10,000\\-\ \ \ 100\\\hline 9,900\end{array}$
	        successive
	        positions.
	
	$\overset{0}{\cancel{1}}0,002$ 1 becomes 0 and 10 is
	$-\ \ \ \ \ 99$ added to 0.
	
	$\overset{0}{\cancel{1}}\overset{9}{\cancel{0}},002$ 10 becomes 9 and 10 is
	$-\ \ \ \ \ 99$ added to 0.

Examples	Solutions
	$\overset{0\,9\ \,9}{\cancel{1}\cancel{0},\cancel{0}02}$ $-\quad\ \ 99$ 10 becomes 9 and 10 is added to 0.
	$\overset{0\,9\ \,9\,9}{\cancel{1}\cancel{0},\cancel{0}\cancel{0}2}$ $-\quad\ \ 99$ $\overline{\quad 9,903}$ 10 becomes 9 and 10 is added to 2. Finally, subtract.
Subtract and check: $\begin{array}{r} 407 \\ -\ 38 \\ \hline \end{array}$	$\overset{3\,9}{\cancel{4}\cancel{0}\cancel{7}}$ $-\ 38$ ⎤ add $\overline{\ 369}$ ⎥ Answer! $\overline{\ 407}$ Checks!

Now try these—no hints given!

Subtract:	
(a) $593 - 251$	(a) 342
(b) $953 - 798$	(b) 155
(c) $804 - 48$	(c) 756

Please talk with an instructor or read this section again if you had difficulty with the last three problems.

Exercises Part I

Answers to these exercises can be found on page 131.

Find the indicated differences.

1.	$67 - 23$	2.	$54 - 37$	3.	$73 - 46$
4.	$70 - 14$	5.	$80 - 44$	6.	$198 - 63$
7.	$555 - 231$	8.	$425 - 157$	9.	$215 - 207$
10.	$802 - 408$	11.	$28,965 - 1,973$	12.	$6,914 - 883$
13.	$\begin{array}{r}28,701\\-10,808\\\hline\end{array}$	14.	$\begin{array}{r}7,165\\-\ \ 596\\\hline\end{array}$	15.	$\begin{array}{r}500,013\\-\ 34,726\\\hline\end{array}$

Exercises Part II

Find the indicated differences.

1.	$59 - 27$	2.	$65 - 49$	3.	$557 - 223$	4.	$452 - 87$
5.	$\begin{array}{r}25,018\\-\ 1,209\\\hline\end{array}$	6.	$\begin{array}{r}7,121\\-\ \ 987\\\hline\end{array}$				

BAL. FOR'D.	$ 560	00
DEPOSIT		
TOTAL		
THIS PAYMENT	− 95	00
OTHER DED.		
BAL. FOR'D.	$ 465	00

341

DATE ___Oct. 2___ 19 _86_ $ _95_ _____

TO___AB Automotive_____

FOR_____

341

___Oct. 2___19 _86_

PAY TO THE
ORDER OF___AB Automotive_____ $ | 95 |

___Ninety - five_____ DOLLARS

BANK OF TNCC

MEMO_____   _Jane Jones_____

⑆021244210562⑆ 402 100538 2⑈ 200096⑆

How much more?
How much did it increase or decrease?
How much remains?
What is the difference?

All these questions require a subtraction. The following examples are taken from situations at a bank.

EXAMPLE 1:   My average monthly balance for June was $325, and my average monthly balance for July was $379. By how much did my average balance increase?

$$\begin{array}{r} \text{Increase:} \quad 379 \\ -325 \\ \hline 54 \quad \text{dollars} \end{array}$$

EXAMPLE 2:   I have a balance of $362 in my checking account. How much remains if I write a check for $97?

$$\begin{array}{r} \text{How much remains:} \quad 362 \\ -\ 97 \\ \hline 265 \quad \text{dollars} \end{array}$$

My new balance is $265.

EXAMPLE 3: My goal is to have $500 in savings by December 1. If I currently have $312 in savings, how much more money do I need to reach my goal?

$$\begin{array}{r} \text{How much more:} \quad 500 \\ -312 \\ \hline 188 \quad \text{dollars} \end{array}$$

I need to deposit $188 more.

EXAMPLE 4: Starting with a balance of $361, I write checks for $5, $12, and $52. I also deposit $121. What is my new balance?

Starting balance:

$$\begin{array}{r} 361 \\ - \ \ 5 \quad \text{First check (subtract).} \\ \hline 356 \\ - \ 12 \quad \text{Second check.} \\ \hline 344 \\ - \ 52 \quad \text{Third check.} \\ \hline 292 \\ +121 \quad \text{Deposit (add).} \\ \hline 413 \end{array}$$

My new balance is $413.

EXAMPLE 5: Complete the stub entries for the check written.

The payment of $56 is subtracted from the previous balance to get the new balance forward.

BAL. FOR'D.	$ 503	00
DEPOSIT		
TOTAL		
THIS PAYMENT	56	00
OTHER DED.		
BAL. FOR'D.	$447	00

Now you try these problems. Check your solutions with mine, but try to work the problem first before looking at the answer.

Examples	Solutions
Manuel weighs 160 pounds; Elena weighs 135 pounds. What is the difference in their weights?	160 −135 ————— 25 pounds Manuel weighs 25 pounds more than Elena.
Anne has paid $204 to Ruth Willis for typing. If she owed a total of $354, how much more does she owe?	354 −204 ————— 150 dollars She owes $150 more.
Dan ran 30 miles the first week, but only 18 miles the second week. By how much did his distance decrease?	30 −18 ————— 12 miles He ran 12 fewer miles.
Monday the highest temperature reading was 52°. Tuesday the highest reading was 47°. What is the difference in these two temperatures?	52 −47 ————— 5 degrees There is a 5° difference.

No hints now.

You owe a total of $689 for carpeting. If you have paid $150 so far, how much more do you owe?	$539
Vera pays $45 a week for babysitting; Larry pays $29 a week. Who pays more and by how much?	Vera pays $16 more than Larry.
Your bank balance is $621. What is your new balance if you write checks for $29 and $15?	$577

If the last three problems gave you difficulty, please read this section again or talk with an instructor.

Exercises Part I

Answers to these exercises can be found on pages 131–132.

1. My goal is to save $623. If I currently have $217 in savings, how much more do I need to save?
2. Brenda may eat only 1,200 calories per day. If after breakfast and lunch she has eaten 987, how many calories may she eat for dinner?
3. Alaska has 566,432 square miles of land. Texas has 262,970 square miles. How much larger is Alaska?
4. The temperature at noon was 68°. At midnight the temperature was 39°. By how much did the temperature decrease?

5. The balance in a checking account was $529.
   (a) What is the new balance after a check for $73 is written?
   (b) What is the new balance after a subsequent deposit of $29 is made?

6. Stephanie owes Ray $1,653. She paid him $150 and $75. How much more does she owe him?

7. (a) A dress originally cost $58. If it now sells for $27 less, how much does it cost?
   (b) A shirt originally cost $18. If it now sells for $5 more, how much does it cost?

8. Starting with a balance of $514 in my checking account, I write three checks for $10, $52, and $13. I also deposit $60 in the account. What is the new balance?

9. Muriel read 56 pages the first day and 62 pages the second. How much more did she read the second day?

10. Alicia made 22 mint candies yesterday. If today she has a total of 57 pieces, how many did she make today?

11.

Find the actual cost of one jar of olives with this coupon if the store charges 92¢ for the jar.

12. Complete the deposit ticket for the deposit of two checks of $39 and $163 and $50 cash.

CASH		
TOTAL FROM OTHER SIDE		
TOTAL		
LESS CASH RECEIVED		
NET DEPOSIT		

13. Complete the following check stub for a check for $59.

BAL. FOR'D.	$ 102	00
DEPOSIT		
TOTAL		
THIS PAYMENT		
OTHER DED.		
BAL. FOR'D.		

1. Rosa wants to save $700 before Christmas. How much more does she need to save if she now has $573?

2. The temperature at noon was 72°. At midnight the temperature was 47°. By how much did the temperature drop?

3.

7½" SAW
SALE!
NOW ONLY
$40

Regular $65.

How much can be saved by buying the saw on sale?

4. Vera owes Dr. Payne $53. She paid him $10 and $15. How much more does she owe?

5. Starting with a balance of $323, I write checks for $57, $23, and $5. I also deposit $60 in the account. What is my new balance?

REVIEW MATERIAL

Review Skill Subtraction of Whole Numbers (Part I)

To subtract whole numbers, arrange the numbers in columns. Line up digits having the same place value. Borrow when necessary. Borrowing 1 from a digit to the left reduces the borrowing position by 1 and adds 10 to the original position.

$$58 - 39 = \qquad \overset{4}{\cancel{5}}\overset{1}{8}$$
$$\underline{-39}$$
$$19$$

Borrow 1 from the 5 tens, leaving 4. Add 10 to the digit 8.

$18 - 9 = 9$

$4 - 3 = 1$

Exercise A Answers to this exercise can be found on page 132.

Subtract.

1. 32 −21	**2.** 38 −29	**3.** 68 −25	**4.** 97 −37

5. 88 − 32	**6.** 79 − 30	**7.** 78 − 23	**8.** 89 − 21
9. 58 − 26	**10.** 92 − 29	**11.** 99 − 89	**12.** 38 − 29
13. 15 − 4	**14.** 27 − 9	**15.** 35 − 28	**16.** 40 − 6
17. 50 − 9	**18.** 60 − 45	**19.** 20 − 16	**20.** 25 − 10

Exercise B No answers are given.

Subtract.

1. 82 −51	**2.** 53 −31	**3.** 51 −50	**4.** 63 −59

5. $56 - 24$	6. $95 - 55$	7. $76 - 42$	8. $48 - 47$
9. $68 - 43$	10. $47 - 19$		

Review Skill Subtraction of Whole Numbers (Part II)

If a number contains the digit 0, it may be necessary to borrow from other than the position immediately to the left.

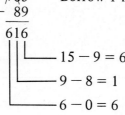

$$705 - 89 = $$

Borrow 1 from the 7 and add 10 to the 0.

Borrow 1 from the 10 and add 10 to the 5.

$15 - 9 = 6$

$9 - 8 = 1$

$6 - 0 = 6$

Exercise C   Answers to this exercise can be found on page 132.

Subtract.

1. $763 - 737$	2. $577 - 382$	3. $666 - 546$
4. $960 - 43$	5. $3,008 - 1,039$	6. $8,365 - 3,163$
7. $5,070 - 1,874$	8. $105 - 68$	9. $5,090 - 183$
10. $5,003 - 2,509$	11. $2,000 - 15$	12. $3,502 - 195$
13. $9,251 - 5,009$	14. $40,503 - 35,999$	15. $70,005 - 9,509$

Exercise D   No answers are given.

Subtract.

1. $849 - 787$	2. $575 - 471$	3. $626 - 213$
4. $910 - 39$	5. $9,004 - 8,428$	6. $8,625 - 8,172$
7. $5,030 - 2,144$	8. $7,003 - 4,705$	9. $305 - 49$
10. $1,035 - 928$		

REVIEW EXERCISES

Refer to the section listed if you have difficulty. Answers to these exercises can be found on page 132.

Section 3.2   Find the indicated differences.

1. $888 - 736$	2. $838 - 589$	3. $1,245 - 276$	4. $108 - 73$
5. $1,973 - 896$	6. $4,003 - 1,179$	7. $7,615 - 2,597$	

8. I have a balance of $659 in my checking account. What remains if I write a check for $93 and make a deposit of $25?

9. Maxine eats 1,800 calories per day; Nathan eats 1,560 calories per day. What is the difference in their caloric intake?

10. A book is 537 pages long. If I have read 259 pages, how many more pages remain to be read?

CHAPTER POST-TEST

This test should determine if you have mastered the topics in Chapter 3. Answers to this test can be found on page 132.

Find the indicated differences.

1. $785 - 344$	2. $8,407 - 606$	3. $2,085 - 716$
4. $71,575 - 2,819$	5. $300 - 145$	6. $7,000 - 1,368$
7. $81,284 - 2,784$	8. $6,005 - 847$	

9. At the beginning of the month there was $254 in the savings account. Find the balance if a withdrawal of $189 is made.

10. There is $525 in a checking account. If a check is written for $53 and a deposit of $75 is made, what is the new balance?

4

multiplication of whole numbers

SECTION 4.1
BASIC FACTS

$$2 \times 3 = \square$$

$$5 \times \square = 20$$

$$7 \times 9 = \square$$

This section explains and allows you to practice basic multiplication facts, such as

$$\begin{array}{r} 2 \\ \times\,3 \\ \hline 6 \end{array} \qquad \begin{array}{r} 5 \\ \times\,7 \\ \hline 35 \end{array} \qquad \begin{array}{r} 9 \\ \times\,5 \\ \hline 45 \end{array}$$

If you think you know these facts, turn to the exercises for practice. If you have difficulty, turn back and read this section completely.

Basic Facts   If a baby sitter is paid $2 per hour for working Saturday night, how much will he be paid for sitting 5 hours?

I could add $2 five times (once for each of the 5 hours).

$$\left.\begin{array}{r} 2 \\ 2 \\ 2 \\ 2 \\ +\ 2 \\ \hline \end{array}\right\} \text{ five 2's}$$
$$10$$

He would receive $10.

I could also multiply 5 and 2 to get the solution; that is

$$5 \times 2 = 10 \qquad \text{Read:} \quad 5 \text{ times 2 equals 10.}$$

factors   product

The 5 and 2 are called **factors**, and 10 is called the **product**.

Multiplication can be thought of as *repeated addition*.

$$5 \times 2 \quad \text{means} \quad \text{five 2's}$$
$$\text{means} \quad \underbrace{2 + 2 + 2 + 2 + 2}_{\text{five 2's}}$$
$$\text{means} \quad 10$$

Therefore, $5 \times 2 = 10$.

Likewise, 3×9 can be thought of as three 9's.

$$3 \times 9 \quad \text{means} \quad \text{three 9's}$$
$$\text{means} \quad \underbrace{9 + 9 + 9}_{\text{three 9's}}$$
$$\text{means} \quad 27$$

Therefore, $3 \times 9 = 27$.

You can multiply in either order and obtain the same product. Doing the same problem in the other order,

$$9 \times 3 \quad \text{means} \quad \text{nine 3's}$$
$$\text{means} \quad \underbrace{3 + 3 + 3 + 3 + 3 + 3 + 3 + 3 + 3}_{\text{nine 3's}}$$
$$\text{means} \quad 27$$

Therefore, both 3×9 and 9×3 equal 27.

If you forget a basic multiplication fact, you can always add to obtain the product; however, it is easier if you *memorize* the basic multiplication facts of all one-digit numbers.

 This chart contains the basic multiplication facts.

1	1	1	1	1	1	1	1	1
×1	×2	×3	×4	×5	×6	×7	×8	×9
1	2	3	4	5	6	7	8	9

	2	2	2	2	2	2	2	2
	×2	×3	×4	×5	×6	×7	×8	×9
	4	6	8	10	12	14	16	18

		3	3	3	3	3	3	3
		×3	×4	×5	×6	×7	×8	×9
		9	12	15	18	21	24	27

			4	4	4	4	4	4
			×4	×5	×6	×7	×8	×9
			16	20	24	28	32	36

				5	5	5	5	5
				×5	×6	×7	×8	×9
				25	30	35	40	45

					6	6	6	6
					×6	×7	×8	×9
					36	42	48	54

						7	7	7
						×7	×8	×9
						49	56	63

							8	8
							×8	×9
							64	72

								9
								×9
								81

One important application of multiplication is to count the total number of items when they are arranged in rows with the same number of items in each row.

5 × 3   means   five 3's   or   five rows of 3 items

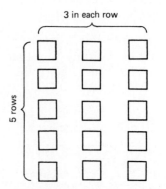

The total number of boxes is 15. That is, 5 × 3 = 15.

The product of 4 and 5 can be thought of as counting the number of boxes in 4 rows with 5 boxes in each row.

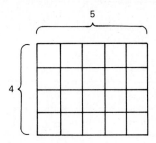

The total number of boxes is 4 × 5, or 20.

Multiplication by Zero What is the value of five 0's?

$$\text{five 0's} = 0 + 0 + 0 + 0 + 0$$
$$= 0$$

Therefore, 5 × 0 = 0.

> When any number is multiplied by zero, the product is zero.
>
> 5 × 0 = 0   and   0 × 9 = 0

Now try to multiply the following pairs of whole numbers. If you need to use repeated addition or think too long, practice this exercise over a period of days or weeks. Remember to cover the answers as you work.

Examples	Solutions
1. 2 × 1 = ?	2
2. 4 × 3 = ?	12
3. 6 × 5 = ?	30
4. 7 × 6 = ?	42
5. 9 × 5 = ?	45
6. 2 × 2 = ?	4
7. 4 × 4 = ?	16
8. 8 × 4 = ?	32
9. 9 × 6 = ?	54
10. 7 × 2 = ?	14
11. 7 × 5 = ?	35
12. 8 × 2 = ?	16
13. 5 × 5 = ?	25
14. 9 × 3 = ?	27

Examples	Solutions
15. 8 × 6 = ?	48
16. 3 × 2 = ?	6
17. 5 × 2 = ?	10
18. 5 × 4 = ?	20
19. 8 × 5 = ?	40
20. 8 × 3 = ?	24
21. 9 × 6 = ?	54
22. 7 × 7 = ?	49
23. 6 × 0 = ?	0
24. 6 × 3 = ?	18
25. 4 × 1 = ?	4
26. 1 × 1 = ?	1
27. 5 × 3 = ?	15
28. 7 × 4 = ?	28
29. 9 × 2 = ?	18
30. 9 × 8 = ?	72
31. 8 × 7 = ?	56
32. 9 × 4 = ?	36
33. 8 × 3 = ?	24
34. 6 × 4 = ?	24
35. 1 × 3 = ?	3
36. 4 × 2 = ?	8
37. 7 × 1 = ?	7
38. 1 × 6 = ?	6
39. 9 × 4 = ?	36
40. 6 × 6 = ?	36
41. 8 × 2 = ?	16
42. 3 × 0 = ?	0
43. 5 × 1 = ?	5
44. 7 × 3 = ?	21
45. 6 × 2 = ?	12

Practice will improve your speed and accuracy!

1. On a sheet of paper number from 1 to 45. Use this sheet to cover the right side of the Examples-Solutions section above. Time yourself! You should be able to complete the problems correctly in less than a minute and a half.

**Exercises
Part II**

1. Purchase a set of flash cards. Discover the facts that give you trouble by removing them from the deck. Practice these facts each day.

2. Multiplication Facts Test

5	8	3	6	5	4	2	0	9	2
×6	×8	×4	×7	×0	×2	×9	×8	×1	×5

7	3	7	2	8	9	4	1	0	1
×4	×9	×7	×1	×5	×3	×6	×1	×0	×0

6	5	0	3	8	5	9	7	9	0
×0	×8	×7	×8	×6	×3	×2	×2	×6	×4

4	6	4	3	4	1	2	9	7	2
×4	×9	×8	×6	×0	×4	×3	×9	×3	×7

1	6	8	3	0	4	5	6	6	7
×7	×8	×2	×7	×1	×5	×7	×2	×6	×0

4	8	0	9	7	1	3	2	5	2
×1	×7	×2	×8	×5	×5	×3	×0	×4	×8

1	9	7	2	0	5	6	1	4	3
×8	×7	×9	×2	×9	×5	×4	×3	×7	×0

0	3	5	8	7	5	3	3	1	6
×3	×2	×9	×4	×8	×2	×1	×5	×6	×5

7	9	8	4	2	8	4	8	5	1
×1	×4	×9	×3	×4	×0	×9	×3	×1	×2

8	2	9	0	7	6	6	1	0	9
×1	×6	×5	×6	×6	×1	×3	×9	×5	×0

3. Multiplication Facts Test

6	3	1	1	2	2	5	0	3	1
×1	×3	×6	×8	×2	×0	×1	×0	×2	×7

4	7	9	2	0	4	1	5	5	3
×1	×2	×2	×4	×4	×2	×9	×3	×2	×1

1	6	1	1	0	1	2	3	5	2
×0	×2	×1	×3	×3	×4	×7	×0	×5	×9

2	2	1	8	9	0	2	0	1	7
×5	×3	×2	×1	×1	×2	×1	×1	×5	×1

7	6	6	4	0	3	7	0	7	5
×0	×5	×3	×6	×8	×9	×6	×7	×5	×9

3	4	6	4	5	2	6	5	3	4
×8	×3	×0	×0	×4	×6	×4	×0	×6	×5

7	5	2	8	3	8	3	6	4	0
×3	×6	×8	×5	×7	×2	×5	×6	×4	×5

9	6	8	4	8	9	9	8	0	8
×8	×8	×0	×8	×9	×0	×4	×7	×9	×8

7	5	9	3	9	9	8	5	0	8
×4	×7	×5	×4	×3	×9	×3	×8	×6	×4

7	8	9	9	4	7	4	6	7	6
×7	×6	×7	×6	×9	×9	×7	×9	×8	×7

SECTION 4.2
MULTIPLICATION OF WHOLE NUMBERS

In problems dealing with electricity, voltage (E) is found by multiplying the amperes (I) times the ohms (R).

$$E = I \times R$$

This section explains multiplication of whole numbers using *shift* and *carry* procedures as illustrated by this example.

$$
\begin{array}{r}
\overset{1}{53} \\
\times\quad 34 \\
\hline
212 \\
1\,59 \\
\hline
1{,}802
\end{array}
$$

If you think you know these procedures, turn to the exercises and work *each* problem. If you have difficulty with the exercises, turn back and read this section completely.

Multiplication of Whole Numbers

EXAMPLE 1:   Consider 23 × 2.

First, write 23 and 2 using place names, then multiply by 2 ones. Remember that ones times ones equals ones and ones times tens equals tens.

$$
\begin{array}{r}
23 = \qquad \text{2 tens and 3 ones} \\
\times\ 2 = \qquad \times \underline{\qquad\qquad \text{2 ones}} \\
\text{4 tens and 6 ones} = 40 + 6 \\
= 46
\end{array}
$$

Look at the problem again, without the place names.

$$
\begin{array}{r}
2\,\fbox{3} \\
\times\ \fbox{2} \\
\hline
6
\end{array}
$$
Think: 2 ones times 3 ones equals 6 ones.
Write: Record 6 in ones position.

$$
\begin{array}{r}
\fbox{2}\,3 \\
\times\ \fbox{2} \\
\hline
4\ 6
\end{array}
$$
Think: 2 ones times 2 tens equals 4 tens.
Write: Record 4 in tens position.

Therefore, 23 × 2 = 46.

EXAMPLE 2: Consider 21 × 3.

$$
\begin{array}{r}
2\,\fbox{1} \\
\times\ \fbox{3} \\
\hline
3
\end{array}
$$
Think: 3 ones times 1 ones equals 3 ones.
Write: Record 3 in ones position.

$$
\begin{array}{r}
\fbox{2}\,1 \\
\times\ \fbox{3} \\
\hline
6\ 3
\end{array}
$$
Think: 3 ones times 2 tens equals 6 tens.
Write: Record 6 in tens position.

Therefore, 21 × 3 = 63.

Now you try these.

Examples	Solutions
Multiply: $\begin{array}{r}12\\ \times\ 3\\ \hline\end{array}$	$\begin{array}{r}1\,\fbox{2}\\ \times\ \fbox{3}\\ \hline 6\end{array}$ Think: 3 ones times 2 ones equals 6 ones. Write: Record 6 in ones position. $\begin{array}{r}\fbox{1}\,2\\ \times\ \fbox{3}\\ \hline 3\ 6\end{array}$ Think: 3 ones times 1 tens equals 3 tens. Write: Record 3 in tens position.
Multiply: $\begin{array}{r}42\\ \times\ 2\\ \hline\end{array}$	Multiply 2 times each digit. $\begin{array}{r}42\\ \times\ 2\\ \hline 84\end{array}$ 2 × 2 = 4 ones 2 × 4 = 8 tens
Multiply: $\begin{array}{r}123\\ \times\ 3\\ \hline\end{array}$	$\begin{array}{r}1\ 2\,\fbox{3}\\ \times\ \fbox{3}\\ \hline 9\end{array}$ Think: 3 ones times 3 ones equals 9 ones. Write: Record 9 in ones position.

Examples	Solutions
	$\begin{array}{r} 1\ 2\ 3 \\ \times\quad 3 \\ \hline 6\ 9 \end{array}$ Think: 3 ones times 2 tens equals 6 tens. Write: Record 6 in tens position.
	$\begin{array}{r} 1\ 2\ 3 \\ \times\quad 3 \\ \hline 3\ 6\ 9 \end{array}$ Think: 3 ones times 1 hundreds equals 3 hundreds. Write: Record 3 in hundreds position.

Multiply: $\begin{array}{r} 213 \\ \times\ \ 2 \\ \hline \end{array}$	Multiply 2 times each digit.
	 $\begin{array}{r} 213 \\ \times\ \ 2 \\ \hline 426 \end{array}$
	     2 X 3 = 6 ones
	     2 X 1 = 2 tens
	     2 X 2 = 4 hundreds

Ready to continue?

EXAMPLE 3:   Consider 26 X 2.

$\begin{array}{r} \overset{1}{2}\ 6 \\ \times\quad 2 \\ \hline 2 \end{array}$ Think: 2 X 6 = 12 ones.
                 Write: Record 2 and carry 1.

$\begin{array}{r} \overset{1}{2}\ 6 \\ \times\quad 2 \\ \hline 5\ 2 \end{array}$ Think: 2 X 2 = 4 tens; add carried value: 4 tens + 1 ten = 5 tens.
                 Write: Record 5.

Therefore, 26 X 2 = 52.

EXAMPLE 4:   Multiply 15 X 3.

$\begin{array}{r} \overset{1}{1}\ 5 \\ \times\quad 3 \\ \hline 5 \end{array}$ Think: 3 X 5 = 15.
                 Write: Record 5 and carry 1.

$\begin{array}{r} \overset{1}{1}\ 5 \\ \times\quad 3 \\ \hline 4\ 5 \end{array}$ Think: 3 X 1 = 3; add carried value: 3 + 1 = 4.
                 Write: Record 4.

Therefore, 15 X 3 = 45.

 When a two-digit answer occurs, carry the left digit and record the right digit. Remember to multiply *before* adding the carried value in the next position.

EXAMPLE 5:   Multiply: 153 × 5.

$$\begin{array}{r} 1\ \overset{1}{5}\ \boxed{3} \\ \times\ \boxed{5} \\ \hline 5 \end{array}$$

Think: 5 × 3 = 15.
Write: Record 5 and carry 1.

$$\begin{array}{r} \overset{2}{1}\ \overset{1}{5}\ 3 \\ \times\ 5 \\ \hline 6\ 5 \end{array}$$

Think: 5 × 5 = 25; add carried values: 25 + 1 = 26.
Write: Record 6 and carry 2.

$$\begin{array}{r} \overset{2}{1}\ \overset{1}{5}\ 3 \\ \times\ 5 \\ \hline 7\ 6\ 5 \end{array}$$

Think: 5 × 1 = 5; add carried value: 5 + 2 = 7.
Write: Record 7.

Therefore, 153 × 5 = 765.

EXAMPLE 6:   Find the product: 563 × 8.

$$\begin{array}{r} 5\ \overset{2}{6}\ \boxed{3} \\ \times\ \boxed{8} \\ \hline 4 \end{array}$$

Think: 8 × 3 = 24.
Write: Record 4, carry 2.

$$\begin{array}{r} \overset{5}{5}\ \overset{2}{6}\ 3 \\ \times\ 8 \\ \hline 0\ 4 \end{array}$$

Think: 8 × 6 = 48.
         48 + 2 = 50 (adding carried value).
Write: Record 0, carry 5.

$$\begin{array}{r} \overset{5}{5}\ \overset{2}{6}\ 3 \\ \times\ 8 \\ \hline 4,5\ 0\ 4 \end{array}$$

Think: 8 × 5 = 40.
         40 + 5 = 45.
Write: Record 5, carry 4 (in answer).

Therefore, 563 × 8 = 4,504.

Try these.

Examples	Solutions
Multiply: 685 × 9	$\begin{array}{r}\overset{4}{6}85\\ \times\ 9\\\hline 5\end{array}$ Think: 9 × 5 = 45. Write: Record 5 and carry 4. $\begin{array}{r}\overset{7}{6}\overset{4}{8}5\\ \times\ 9\\\hline 65\end{array}$ Think: 9 × 8 = 72, then add carried value: 72 + 4 = 76. Write: Record 6 and carry 7.

Examples	Solutions
	$$\overset{7\,4}{685}$$ $$\underline{\times\quad 9}$$ $$6{,}165$$ Think: $9 \times 6 = 54$, then add carried value: $54 + 7 = 61$. Write: Record 1 and carry 6 (in answer).
Multiply: 593 $\underline{\times\quad 7}$	$$\overset{2}{593}$$ $$\underline{\times\quad 7}$$ $$1$$ Think: $7 \times 3 = 21$. Write: Record 1 and carry 2.
	$$\overset{6\,2}{593}$$ $$\underline{\times\quad 7}$$ $$51$$ Think: $7 \times 9 = 63$, then add carried value: $63 + 2 = 65$. Write: Record 5 and carry 6.
	$$\overset{6\,2}{593}$$ $$\underline{\times\quad 7}$$ $$4{,}151$$ Think: $7 \times 5 = 35$, then add carried value: $35 + 6 = 41$. Write: Record 1 and carry 4 (in answer).
Multiply: 379 $\underline{\times\quad 8}$	$$\overset{6\,7}{379}$$ $$\underline{\times\quad 8}$$ $$3{,}032$$ $8 \times 9 = 72$ $8 \times 7 + 7 = 56 + 7 = 63$ $8 \times 3 + 6 = 24 + 6 = 30$

Be careful with zeros. For example,

$$\overset{\overset{2}{0}}{1\,0\,4}$$
$$\underline{\times\qquad 6}$$
$$4$$

   Think: $6 \times 4 = 24$.
Write: Record 4, carry 2.

$$\overset{2}{1\,0\,4}$$
$$\underline{\times\qquad 6}$$
$$2\,4$$

   Think: $6 \times 0 = 0$.
           $0 + 2 = 2$ (adding carried value).
Write: Record 2.

$$\begin{array}{r} \overset{2}{1 \ 0 \ 4} \\ \times \qquad 6 \\ \hline 6 \ 2 \ 4 \end{array}$$

Think: 6 × 1 = 6.
No carried value to add.
Write: Record 6.

Therefore, 104 × 6 = 624.

Try to work the following problems *before* you look at the solutions.

Examples	Solutions

Find the product.
 	
$\qquad\begin{array}{r} 29 \\ \times \ 3 \\ \hline \end{array}$
	$\begin{array}{r} \overset{2}{2} \ 9 \\ \times \quad 3 \\ \hline 7 \end{array}$ Think: 3 × 9 = 27.
	           Write: Record 7, carry 2.
	
	$\begin{array}{r} \overset{2}{2} \ 9 \\ \times \quad 3 \\ \hline 8 \ 7 \end{array}$ Think: 3 × 2 = 6.
	                6 + 2 = 8 (adding carried value).
	           Write: Record 8.
	
	Therefore, 29 × 3 = 87.

Find the product.
 	
$\qquad\begin{array}{r} 95 \\ \times \ 4 \\ \hline \end{array}$
	$\begin{array}{r} \overset{2}{9} 5 \\ \times \ 4 \\ \hline 0 \end{array}$ Think: 4 × 5 = 20.
	           Write: Record 0, carry 2.
	
	$\begin{array}{r} \overset{2}{9} 5 \\ \times \ 4 \\ \hline 380 \end{array}$ Think: 4 × 9 = 36.
	                36 + 2 = 38 (adding carried value).
	           Write: Record 8, carry 3 (in answer).
	
	Therefore, 95 × 4 = 380.

Find the product.
 	
$\qquad\begin{array}{r} 152 \\ \times \quad 3 \\ \hline \end{array}$
	$\begin{array}{r} 152 \\ \times \ \ 3 \\ \hline 6 \end{array}$ Think: 3 × 2 = 6.
	           Write: Record 6, no carried value!
	
	$\begin{array}{r} \overset{1}{152} \\ \times \ \ 3 \\ \hline 56 \end{array}$ Think: 3 × 5 = 15.
	                No carried value to add!
	           Write: Record 5, carry 1.
	
	$\begin{array}{r} \overset{1}{152} \\ \times \ \ 3 \\ \hline 456 \end{array}$ Think: 3 × 1 = 3.
	                3 + 1 = 4 (adding carried value).
	           Write: Record 4.
	
	Therefore, 152 × 3 = 456.

Examples	Solutions

Find the product.

$$857 \\ \times \quad 9$$

$$\overset{6}{857} \\ \times \quad 9 \\ \hline 3$$
Think: $9 \times 7 = 63$.
Write: Record 3, carry 6.

$$\overset{5\,6}{857} \\ \times \quad 9 \\ \hline 13$$
Think: $9 \times 5 = 45$. $45 + 6 = 51$.
Write: Record 1, carry 5.

$$\overset{5\,6}{857} \\ \times \quad 9 \\ \hline 7,713$$
Think: $9 \times 8 = 72$. $72 + 5 = 77$.
Write: Record 7, carry 7 (in answer).

Therefore, $857 \times 9 = 7,713$.

Find the product.

$$209 \\ \times \quad 5$$

$$\overset{4}{209} \\ \times \quad 5 \\ \hline 5$$
Think: $5 \times 9 = 45$.
Write: Record 5, carry 4.

$$\overset{4}{209} \\ \times \quad 5 \\ \hline 45$$
Think: $5 \times 0 = 0$. $0 + 4 = 4$.
Write: Record 4.

$$\overset{4}{209} \\ \times \quad 5 \\ \hline 1,045$$
Think: $5 \times 2 = 10$.
Write: Record 0, carry 1 (in answer).

Therefore, $209 \times 5 = 1,045$.

EXAMPLE 7:   Consider the multiplication 43×23.
First, multiply each digit of 43 by the 3 in the ones position.

$$\begin{array}{r} 4\ 3 \\ \times \quad 2③ \\ \hline 1\ 2\ 9 \end{array}$$

—— Start answer in ones position.

Next, multiply each digit of 43 by the 2 in the tens position. However, 2 tens $\times$ 3 ones = 6 *tens*. Therefore, record the 6 in the tens position—this is called a **shift**.

$$\begin{array}{r} 4\ 3 \\ \times \quad ②3 \\ \hline 1\ 2\ 9 \\ 6 \end{array}$$

—— Start answer in the tens position.

Continue by multiplying 2 times the next digit, 4.

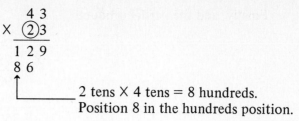

2 tens × 4 tens = 8 hundreds.
Position 8 in the hundreds position.

Finally, add the two partial products.

$$
\begin{array}{r}
43 \\
\times\ 23 \\
\hline
129 \\
86 \\
\hline
989
\end{array}
\ \right]\ \text{add}
$$

Therefore, 43 × 23 = 989.

The following example repeats the pattern for multiplication of whole numbers.

EXAMPLE 8:   Find the product of 53 × 34.

$$
\begin{array}{r}
\overset{1}{5}\,3 \\
\times\ 3\,④ \\
\hline
2\,1\,2
\end{array}
$$

Multiply 53 by the 4 ones first.
Record answer starting in the ones position.

$$
\begin{array}{r}
\overset{1}{5}\,3 \\
\times\ ③\,4 \\
\hline
2\,1\,2 \\
1\,5\,9
\end{array}
$$
← shift

Next, multiply 53 by the 3 tens.
Record answer starting in the tens position—this is a shift, which locates the answer directly under the multiplier 3.

$$
\begin{array}{r}
\overset{1}{5}\,3 \\
\times\ 3\,4 \\
\hline
2\,1\,2 \\
1\,5\,9 \\
\hline
1,8\,0\,2
\end{array}
\ \right]\ \text{add}
$$

Finally, add the partial products.

Therefore, 53 × 34 = 1,802.

 The shift lines up each partial product with its multiplier.

EXAMPLE 9:   Find the product of 98 and 75.
First, multiply by 5 ones.

$$
\begin{array}{r}
\overset{4}{9}8 \\
\times\ 75 \\
\hline
490
\end{array}
$$

Next, multiply by 7 tens. Shift to record the answer starting in the tens position, which lines it up under the 7.

$$
\begin{array}{r}
\overset{\substack{5 \\ 4}}{9}8 \\
\times\ 75 \\
\hline
490 \\
686
\end{array}
$$

You should either erase the previous carried value, or write the new one above it.

Finally, add the partial products.

$$\begin{array}{r} 98 \\ \times\ 75 \\ \hline 490 \\ 6\ 86 \\ \hline 7{,}350 \end{array}$$

Therefore, $98 \times 75 = 7{,}350$.

The shift technique can be used to multiply by any number of digits. As in addition, you begin the multiplication with the digit in the ones position.

EXAMPLE 10:   Find the product of 274 and 153.

$$\begin{array}{r} \overset{2}{2}\overset{1}{7}4 \\ \times 1\,5\,\textcircled{3} \\ \hline 8\,2\,2 \end{array}$$ Multiply 274 × 3 ones.

$$\begin{array}{r} \overset{3}{}\overset{2}{} \\ \overset{2}{2}\overset{1}{7}4 \\ \times\ \ \ 1\,\textcircled{5}\,3 \\ \hline 8\,2\,2 \\ 1\,3\,7\,0 \end{array}$$ Multiply 274 by 5 tens.
Shift to record the answer starting in the *tens* position. New carried values are shown above the previous ones.

$$\begin{array}{r} \overset{3}{}\overset{2}{} \\ \overset{2}{2}\overset{1}{7}4 \\ \times\ \ \ \textcircled{1}\,5\,3 \\ \hline 8\,2\,2 \\ 1\,3\,7\,0 \\ 2\,7\,4 \\ \hline 4\,1{,}9\,2\,2 \end{array}$$ Multiply 274 by 1 hundred.
Shift to record the answer starting in the *hundreds* position.

Add the partial products.

Therefore, $274 \times 153 = 41{,}922$.

EXAMPLE 11:   Find the product of 204 and 57.

$$\begin{array}{r} 2\overset{2}{0}4 \\ \times\ \ \ 5\,\textcircled{7} \\ \hline 1\,4\,2\,8 \end{array}$$ Multiply by 7 ones.

$7 \times 0 = 0$.
$0 + 2 = 2$ (adding carried value).

$$\begin{array}{r} 2\overset{2}{\overset{2}{0}}4 \\ \times\ \ \ \textcircled{5}\,7 \\ \hline 1\,4\,2\,8 \\ 1\,0\,2\,0 \\ \hline 1\,1{,}6\,2\,8 \end{array}$$ Multiply by 5 tens, shifting answer.

Therefore, $204 \times 57 = 11{,}628$.

EXAMPLE 12:   Find the product of 254 and 809.

$$
\begin{array}{r}
{}^{4\ 3}\\
2\ 5\ 4\\
\times\ \ 8\ 0\,\textcircled{9}\\
\hline
2\ 2\ 8\ 6
\end{array}
$$
Multiply by 9.

When you multiply by zero, you get zero.

$$
\begin{array}{r}
{}^{4\ 3}\\
2\ 5\ 4\\
\times\ \ 8\,\textcircled{0}\,9\\
\hline
2\ 2\ 8\ 6\\
0\ 0\ 0\longleftarrow
\end{array}
$$
Multiply by 0.
←—Shift!

$$
\begin{array}{r}
{}^{4\ 3}\\
{}^{4\ 3}\\
2\ 5\ 4\\
\times\ \ \ \ \textcircled{8}0\ 9\\
\hline
2\ 2\ 8\ 6\\
0\ 0\ 0\\
2\ 0\ 3\ 2\longleftarrow\\
\hline
2\ 0\ 5{,}4\ 8\ 6
\end{array}
$$
Multiply by 8.
←— Shift!

Note: Each shift lines up the partial product with its multiplier.

Therefore, 254 × 809 = 205,486.

EXAMPLE 13:   Find the product of 207 and 800.

$$
\begin{array}{r}
2\,0\,7\\
\times\,8\,0\,0\\
\hline
0\,0\,0
\end{array}
$$
Multiply by 0.

$$
\begin{array}{r}
2\,0\,7\\
\times\ \ 8\,0\,0\\
\hline
0\,0\,0\\
0\,0\,0
\end{array}
$$
Shift and multiply by 0.

$$
\begin{array}{r}
2\,0\,7\\
\times\ \ \ \ 8\,0\,0\\
\hline
0\,0\,0\\
0\ 0\,0\\
1\,6\,5\ 6\\
\hline
1\,6\,5{,}6\,0\,0
\end{array}
$$
Shift and multiply by 8.

Therefore, 207 × 800 = 165,600.

Now you try to work these problems *before* looking at the solutions.

Examples	Solutions
Find the product. $\begin{array}{r} 23 \\ \times\ 3 \\ \hline \end{array}$	$\begin{array}{r} 23 \\ \times\ 3 \\ \hline 69 \end{array}$ Think: 3 × 3 = 9 ones. 3 × 2 = 6 tens.

Examples	Solutions
Find the product. $\begin{array}{r} 54 \\ \times\ 6 \\ \hline \end{array}$	$\begin{array}{r} {\scriptstyle 2} \\ 54 \\ \times\ 6 \\ \hline 324 \end{array}$ Think: $6 \times 4 = 24$. Write: Record 4, carry 2. Think: $6 \times 5 = 30$. $\quad\quad 30 + 2 = 32$. Write: Record 2, carry 3 $\quad\quad$ in answer.
Find the product. $\begin{array}{r} 25 \\ \times 12 \\ \hline \end{array}$	$\begin{array}{r} {\scriptstyle 1} \\ 2\ 5 \\ \times\ 1②\\ \hline 5\ 0 \end{array}$ Multiply by 2 ones. $\begin{array}{r} {\scriptstyle 1} \\ 2\ 5 \\ \times\ ①2\\ \hline 5\ 0 \\ 2\ 5 \leftarrow \\ \hline 3\ 0\ 0 \end{array}$ Multiply by 1 ten, shifting answer. Add partial products.
Find the product. $\begin{array}{r} 153 \\ \times\ 29 \\ \hline \end{array}$	$\begin{array}{r} {\scriptstyle 1} \\ {\scriptstyle 4\,2} \\ 153 \\ \times\quad 29 \\ \hline 1\ 377 \\ 3\ 06\quad \\ \hline 4{,}437 \end{array}$
Find the product. $\begin{array}{r} 356 \\ \times 104 \\ \hline \end{array}$	To obtain a rough guess use estimation. Actual Estimation $\begin{array}{rr} {\scriptstyle 2\,2} & \\ 356 & 400 \\ \times\quad 104 & \times\quad 100 \\ \hline 1\ 424 & 000 \\ 0\ 00\quad & 0\ 00\quad \\ 35\ 6\quad\quad & 40\ 0\quad\quad \\ \hline 37{,}024 & 40{,}000 \end{array}$
Find the product. $\begin{array}{r} 603 \\ \times 279 \\ \hline \end{array}$	Actual Estimation $\begin{array}{rr} & {\scriptstyle 2} \\ & {\scriptstyle 2} \\ 603 & 600 \\ \times\quad 279 & \times\quad 300 \\ \hline 5\ 427 & 000 \\ 42\ 21\quad & 0\ 00\quad \\ 120\ 6\quad\quad & 180\ 0\quad\quad \\ \hline 168{,}237 & 180{,}000 \end{array}$

Now try these—no hints given!

Find the products.	
(a) 57×29	(a) 1,653
(b) 135×14	(b) 1,890
(c) 105×275	(c) 28,875

If you had difficulty with these problems, please see an instructor or read this section again.

Exercises Part I

Answers to these exercises can be found on page 132.

Find the indicated products.

1. 14×17 2. 35×47 3. 13×48
4. 185×94 5. 944×85 6. $2,084 \times 89$
7. 715×206 8. 715×790 9. $78,035 \times 284$
10. $4,635 \times 7,800$ 11. $\begin{array}{r} 21 \\ \times 35 \\ \hline \end{array}$ 12. $\begin{array}{r} 497 \\ \times 593 \\ \hline \end{array}$

13. $\begin{array}{r} 10,471 \\ \times \quad 507 \\ \hline \end{array}$ 14. $\begin{array}{r} 1,596 \\ \times \quad 700 \\ \hline \end{array}$ 15. $7,954 \times 213$

Exercises Part II

Find the indicated products.

1. 19×18 2. 27×30 3. 209×15 4. 725×189
5. $\begin{array}{r} 17 \\ \times 28 \\ \hline \end{array}$ 6. $\begin{array}{r} 507 \\ \times 173 \\ \hline \end{array}$ 7. $\begin{array}{r} 1,596 \\ \times \quad 500 \\ \hline \end{array}$

SECTION 4.3 EXPONENTS

$$5^4 = 5 \times 5 \times 5 \times 5 = 625$$

There is a simple notation for products in which the same number is repeated as a factor. For example,

$$
\begin{aligned}
5 \times 5 \times 5 \times 5 &= \underbrace{5 \times 5}_{} \times 5 \times 5 \\
&= \underbrace{25 \times 5}_{} \times 5 \qquad \text{Multiplying} \\
&= \underbrace{125 \times 5}_{} \qquad \text{two numbers} \\
&= 625 \qquad\qquad \text{at a time.}
\end{aligned}
$$

Instead of showing the factor 5 four times, the following notation is used.

$$5 \times 5 \times 5 \times 5 = 5^4$$

In this notation 5 is called the **base** and 4 is called the **exponent**. The entire symbol 5^4 is read: 5 to the fourth power.

The exponent counts the number of times the base is used as a factor.

$$5^4 = \underbrace{5 \times 5 \times 5 \times 5}_{\text{four factors of 5}} = 625$$

EXAMPLE 1:   Evaluate 4^3.

The expression 4^3 is read: 4 to the third power. It indicates the use of 4 as a factor 3 times.

$$4^3 = \underbrace{4 \times 4 \times 4}_{\text{3 factors of 4}} = \underbrace{4 \times 4 \times 4}_{}$$
$$= \underbrace{16 \times 4}_{}$$
$$= 64$$

Therefore, $4^3 = 64$.

EXAMPLE 2:   Write $3 \times 3 \times 3 \times 3 \times 3$, using an exponent.

$$3 \times 3 \times 3 \times 3 \times 3 = \underbrace{3 \times 3 \times 3 \times 3 \times 3}_{\text{5 factors of 3}} = 3^5$$

EXAMPLE 3:   Evaluate $4^2 \times 3^2$.

$$4^2 \times 3^2 = \underbrace{4 \times 4}_{\substack{\text{2 factors}\\\text{of 4}}} \times \underbrace{3 \times 3}_{\substack{\text{2 factors}\\\text{of 3}}} = \underbrace{4 \times 4 \times 3 \times 3}_{}$$
$$= \underbrace{16 \times 3 \times 3}_{}$$
$$= \underbrace{48 \times 3}_{}$$
$$= 144$$

EXAMPLE 4:   Write $2 \times 2 \times 2 \times 5 \times 5$, using exponents.

$$2 \times 2 \times 2 \times 5 \times 5 = \underbrace{2 \times 2 \times 2}_{\substack{\text{3 factors}\\\text{of 2}}} \times \underbrace{5 \times 5}_{\substack{\text{2 factors}\\\text{of 5}}}$$
$$= 2^3 \times 5^2$$

Try these!

Examples	Solutions
Evaluate 3^4.	3^4 means 3 used as a factor 4 times.
	$3^4 = \underbrace{3 \times 3 \times 3 \times 3}_{\text{4 times}}$
	$= \underbrace{3 \times 3 \times 3 \times 3}_{}$
	$= \underbrace{9 \times 3 \times 3}_{}$
	$= \underbrace{27 \times 3}_{}$
	$= 81$

Examples	Solutions
Evaluate $5^2 \times 7^2$.	$5^2 \times 7^2 = \underbrace{5 \times 5}_{\text{2 times}} \times \underbrace{7 \times 7}_{\text{2 times}}$ $= \underbrace{25 \times 7}_{} \times 7$ $= \underbrace{175 \times 7}_{}$ $= 1,225$
Write $3 \times 3 \times 3 \times 3 \times 3 \times 3$, using exponents.	$\underbrace{3 \times 3 \times 3 \times 3 \times 3 \times 3}_{\text{6 times}} = 3^6$
Write $3 \times 3 \times 5 \times 5 \times 5$, using exponents.	$\underbrace{3 \times 3}_{\text{2 times}} \times \underbrace{5 \times 5 \times 5}_{\text{3 times}} = 3^2 \times 5^3$
Write $2 \times 3 \times 3 \times 3 \times 5 \times 7 \times 7 \times 7 \times 7$, using exponents.	$2 \times \underbrace{3 \times 3 \times 3}_{\text{3 times}} \times 5 \times \underbrace{7 \times 7 \times 7 \times 7}_{\text{4 times}}$ $= 2 \times 3^3 \times 5 \times 7^4$

No hints this time!

Evaluate $2^3 \times 5^2$.	200
Write $2 \times 2 \times 7 \times 7 \times 7 \times 7$, using exponents.	$2^2 \times 7^4$

If you had difficulty, talk with an instructor or read this section again.

Exercises Part I

Answers to these exercises can be found on page 132.

Evaluate the following.

1. 5^3 2. 3^5 3. 2^5 4. 5^2
5. $5^2 \times 7$ 6. $2^4 \times 5^3$ 7. $3^2 \times 7^2$ 8. $2^2 \times 3^2 \times 5^2$

Write each expression using exponents.

9. $2 \times 2 \times 3 \times 3 \times 3$ 10. $3 \times 3 \times 5 \times 5 \times 5 \times 5 \times 7 \times 7 \times 7$
11. $2 \times 5 \times 5 \times 7$ 12. $3 \times 5 \times 5 \times 7 \times 7$
13. $2 \times 2 \times 2 \times 3 \times 7 \times 7$

Exercises Part II

Evaluate the following.

1. 2^4 2. 7^3 3. $2^3 \times 3 \times 5$ 4. $5^2 \times 2^2$

Write each expression using exponents.

5. $2 \times 3 \times 3 \times 3 \times 7 \times 7$ 6. $2 \times 2 \times 2 \times 2 \times 3 \times 3 \times 7 \times 7$

THIS CLASSROOM HAS 4 ROWS OF 9 STUDENT DESKS EACH, FOR A TOTAL OF 4 X 9 = 36 STUDENT DESKS.

EXAMPLE 1:   When Muriel drives her own car instead of the state car, she receives 18¢ for each mile driven. How much should she receive for driving her car 5 miles?

I could add 18¢ five times (once for each mile), but this repeated addition is the same as multiplication.

$$\text{five 18's} = 5 \times 18$$
$$= 90$$

She should receive 90¢.

EXAMPLE 2:   Muriel's old car gets only 15 miles per gallon of gas. How far can she drive on 9 gallons of gas?

The answer is 15 miles added 9 times, or nine 15's.

$$\text{nine 15's} = 9 \times 15$$
$$= 135$$

She can drive 135 miles.

EXAMPLE 3:   In her work, Muriel drives by a toll booth 6 times each day. If the toll is 10¢, how much does she pay each day in tolls?

The answer is 10¢ added 6 times, or six 10's.

$$\text{six 10's} = 6 \times 10$$
$$= 60$$

She pays 60¢ in tolls each day.

EXAMPLE 4:   Dan makes payments of $26 each month for his carpet. He owes a total of $395. After 14 payments, how much more does he owe?

I must work this problem in two steps.

Step 1:   Total paid:

26 paid 14 times, or fourteen 26's

fourteen 26's = 14 × 26

= 364

He has paid a total of $364.

Step 2:   Amount still owed:

$$395 \quad \text{Total.}$$
$$-364 \quad \text{Paid.}$$
$$\overline{31} \quad \text{Difference.}$$

He still owes $31.

Now it is your turn.

Examples	Solutions
A typewriter pad contains 70 sheets of paper. How many sheets are there in 28 pads?	The answer is 70 sheets added 28 times. twenty-eight 70's = 28 × 70 = 1,960 There are 1,960 sheets in 28 pads.
If each banana costs approximately 17¢, how much do 4 bananas cost?	The answer is 17¢ added 4 times, or four 17's. four 17's = 4 × 17 = 68 They cost 68¢.
Each bottle of correction fluid contains 6 ounces of liquid. How many ounces are there in 5 bottles?	The answer is 6 ounces added 5 times, or five 6's. five 6's = 5 × 6 = 30 There are 30 ounces in 5 bottles.

Now try these—no hints given.

Pat can type 53 words per minute. How many words can she type in 15 minutes?	795 words
A box of candy contains 35 assorted pieces. How many pieces are in 15 boxes?	525 pieces

If you had difficulty with these problems, read this section again or talk with an instructor.

**Exercises
Part I**

Answers to these exercises can be found on page 132.

1. A car gets 27 miles per gallon of gas. How far can it go on 7 gallons of gas?
2. A box of razor blades contains 5 separate blades. How many blades are contained in 48 boxes?
3. If one cookie costs 12¢, how much do 8 cookies cost?
4. If I read 50 pages each day, how many pages can I read in 20 days?
5. Each pencil costs 9¢. How much do 5 pencils cost?
6. A bag contains 95 balloons of assorted colors. How many balloons are contained in 16 bags?
7. Nine payments of $57 each have been made for a stereo. How much has been paid all together?
8. A total doctor bill is $562. How much more is owed if 8 payments of $61 each have been made?
9. For 3 weeks I paid $30 per week for daytime babysitting. For 5 weeks, I paid $35 per week. What is the total amount that I paid for 8 weeks?
10. One lollipop costs 10¢. One jawbreaker costs 8¢. What is the total bill for 3 lollipops and 4 jawbreakers?
11. How many chairs are in a classroom with 5 rows of 15 chairs each?
12. An apple grove has 35 rows of trees with 23 trees in each row. How many apple trees are in the grove?

**Exercises
Part II**

1. A car gets 32 miles per gallon of gas. How far can it go on 13 gallons of gas?
2. A package contains 19 markers. How many markers are contained in 18 packages?
3. The total bill is $753. How much more is owed if 7 payments of $91 each have been made?
4. One jar of rubber cement costs $2. One bottle of correction fluid costs $1. What is the total bill for 6 jars of rubber cement and 3 bottles of correction fluid?
5. How many chairs are in a classroom with 6 rows of 13 chairs each?

REVIEW MATERIAL

**Review Skill
Multiplication by
One Digit**

Multiply each position in turn. When the product is a two-digit answer, carry the left digit and record the right digit. Remember to multiply before adding the carried value in the next position.

$$
\begin{array}{r}
\overset{6}{5}\,7 \\
\times\ 9 \\
\hline
3
\end{array}
$$

Think: 9 × 7 = 63.
Write: Record 3 and carry 6.

Think: 9 × 5 = 45; then add carried value:
45 + 6 = 51.
Write: Record 1 and carry 5 in the answer.

Exercise A Answers to this exercise can be found on page 132.

Multiply.

1. $\;\;49$ $\times\;9$	2. $\;\;41$ $\times\;2$	3. $\;\;23$ $\times\;3$	4. $\;\;98$ $\times\;8$

5. 71 × 8 6. 49 × 9 7. 97 × 6 8. 51 × 5
9. 41 × 7 10. 84 × 7 11. 66 × 8 12. 62 × 7
13. 39 × 7 14. 89 × 8 15. 81 × 9 16. 762 × 6
17. 485 × 4 18. 654 × 7

Exercise B No answers are given.

Multiply.

1. 31 × 5 2. 63 × 8 3. 87 × 8 4. 82 × 8
5. 51 × 5 6. 99 × 7 7. 41 × 9 8. 63 × 9
9. 83 × 8 10. 24 × 9 11. 54 × 9 12. 83 × 7
13. 95 × 7 14. 68 × 7 15. 95 × 8 16. 278 × 9
17. 624 × 8 18. 527 × 6

Review Skill Multiplication of Whole Numbers Use the shift procedure to multiply by more than one digit.

$$
\begin{array}{r}
205 \\
\times\quad 123 \\
\hline
615 \longleftarrow 3 \times 205 \\
4\;10 \longleftarrow \text{Shift; } 2 \times 205 \\
20\;5 \longleftarrow \text{Shift; } 1 \times 205 \\
\hline
25{,}215
\end{array}
$$

Estimate to get a rough check for your answer.

Actual	Estimation
205	200
× 123	× 100
	000
	0 00
	20 0
	20,000

This tells you that the actual answer should be close to 20,000.

Exercise C Answers to this exercise can be found on page 132.

Multiply.

1. 23 × 22 2. 31 × 21 3. 34 × 23
4. 888 × 54 5. 675 × 35 6. 4,667 × 421
7. 803 × 5 8. 406 × 5 9. 2,004 × 3
10. 309 × 43 11. 712 × 304 12. 4,003 × 87
13. 4,030 × 7,040

Exercise D No answers are given.

Multiply.

1. 26×12
2. 33×32
3. 32×34
4. 139×36
5. 249×78
6. $6,996 \times 143$
7. 704×2
8. 509×3
9. $5,003 \times 9$
10. 708×24
11. 780×15
12. 432×203
13. $7,809 \times 9,060$

REVIEW EXERCISES

Refer to the section listed if you have difficulty. Answers to these exercises can be found on page 132.

Section 4.2 Find the indicated products.

1. 19×3
2. 27×15
3. 59×84
4. 987×93
5. $2,564 \times 48$
6. $\begin{array}{r} 204 \\ \times\ 27 \\ \hline \end{array}$
7. $\begin{array}{r} 195 \\ \times 207 \\ \hline \end{array}$
8. $\begin{array}{r} 489 \\ \times 600 \\ \hline \end{array}$

Section 4.3 Evaluate the following expressions.

9. 9^3
10. 4^2

Write each using exponents.

11. $2 \times 5 \times 5 \times 7 \times 7 \times 7$
12. $2 \times 2 \times 2 \times 5 \times 7 \times 7$

Section 4.4

13. Anne types 59 words per minute. How many words can she type in 16 minutes?
14. A box contains 8 crayons. How many crayons are in 16 boxes?
15. One pear costs 26¢. How much do 3 pears cost?
16. An orange grove has 34 rows of 45 trees each. How many trees are in the grove?

CHAPTER POST-TEST

This test should determine if you have mastered the topics in Chapter 4. Answers to this test can be found on page 132.

Find the indicated products.

1. 15×72
2. 158×96
3. $\begin{array}{r} 915 \\ \times\ 30 \\ \hline \end{array}$
4. $\begin{array}{r} 407 \\ \times 139 \\ \hline \end{array}$
5. $\begin{array}{r} 596 \\ \times 305 \\ \hline \end{array}$

6. Evaluate 5^3.
7. Write using exponents: $2 \times 2 \times 3 \times 3 \times 3 \times 5 \times 7 \times 7$
8. If I read 40 pages each day, how many pages can I read in 15 days?
9. One eraser costs 9¢. How much do 5 erasers cost?
10. A box of paper contains 500 sheets. How many sheets are in 73 boxes?

5

division of whole numbers

$36 \div 4 = \square$

$25 \div \square = 5$

$9 \div 3 = \square$

This section explains and allows you to practice basic division facts, such as

$$6 \overline{)54}^{\,9} \qquad 36 \div 9 = 4 \qquad 8 \overline{)32}^{\,4}$$

If you think you know these facts, turn to the exercises for practice. If you have difficulty, turn back and read this section completely.

Basic Facts   There are 15 people in the group. If 3 people make up a team, how many teams can be formed?

Let me subtract 3 people, repeatedly, keeping track of the number of subtractions.

$$
\text{Repeated subtraction:}
\begin{cases}
\begin{array}{rl}
15 & \\
-\ 3 & \text{1st team} \\
\hline
12 & \\
-\ 3 & \text{2nd team} \\
\hline
9 & \\
-\ 3 & \text{3rd team} \\
\hline
6 & \\
-\ 3 & \text{4th team} \\
\hline
3 & \\
-\ 3 & \text{5th team} \\
\hline
0 & \\
\end{array}
\end{cases}
$$

There are 5 teams, each containing 3 people—that is, 5 groups of 3.
This answer can also be found by division.

$$15 \div 3 = 5 \quad \underline{\underline{\text{or}}} \quad 3 \overline{)15}^{\,5}$$

Both of these are read: 15 divided by 3 is 5. The 3 is called the **divisor**. The 15 is called the **dividend**, and the 5 is called the **quotient**.

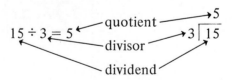

 Multiplication can be viewed as repeated addition, and division as repeated subtraction. Therefore, multiplication and division are considered *opposite* operations. Knowing this will help you learn the basic division facts.

For example, to answer the question,

$$12 \div 6 = ?$$

think the following multiplication question: Six times what (?) equals 12?

$$\overset{\text{times}}{\underset{6\,\overline{)12}}{\overset{?}{\big\lceil}}}$$

Because 6 × 2 = 12, 12 ÷ 6 = 2.

$$6 \overline{)12}^{\,2}$$

Here is another example to illustrate the point.

times
$$8 \overline{)56}^{\,?}$$

You should think: 8 times what equals 56? The answer is 7.

$$8 \overline{)56}^{\,7}$$

Finally,

$$9 \overline{)36}^{\,?}$$

The question is, 36 divided by 9 is what number? Think: 9 times what equals 36? The answer is 4.

$$9 \overline{)36}^{\,4} \quad \underline{\underline{\text{or}}} \quad 36 \div 9 = 4$$

Therefore, 36 divided by 9 is 4.

Division with Zero

$$5 \overline{)0}^{\,0} \quad \text{or} \quad 0 \div 5 = 0$$

because 0 × 5 = 0.

┌───┐
│ Zero divided by any nonzero number is zero. │
└───┘

However, division by zero is impossible.
  For example,

$$6 \div 0 = ?$$

What times 0 is 6?

$$0 \overline{)6}^{\,?}$$

No such number exists, because the product of a multiplication by zero is always zero. Thus, questions such as

$$9 \div 0 = ? \quad \text{and} \quad 5 \div 0 = ?$$

are impossible and have no answer.

┌─────────────────────────────────────┐
│ Division by zero is impossible. │
└─────────────────────────────────────┘

Now try these problems. By practice, you should be able to work these problems quickly.

Examples	Solutions
1. $12 \div 4 = ?$	*Remember:* $\div$ *means divided by.* $12 \div 4 = 3$, because $4 \times 3 = 12$.
2. $5\overline{)30}$	6, because $5 \times 6 = 30$.
3. $6\overline{)42}$	7
4. $45 \div 9 = ?$	5, because $9 \times 5 = 45$.
5. $2\overline{)4}$	2
6. $4\overline{)16}$	4
7. $8\overline{)32}$	4
8. $6\overline{)54}$	9
9. $2\overline{)14}$	7
10. $7\overline{)35}$	5
11. $8\overline{)16}$	2
12. $5\overline{)25}$	5
13. $9\overline{)27}$	3
14. $48 \div 8 = ?$	6
15. $6 \div 3 = ?$	2
16. $10 \div 5 = ?$	2
17. $20 \div 4 = ?$	5
18. $8\overline{)40}$	5
19. $3\overline{)24}$	8
20. $6\overline{)54}$	9
21. $7\overline{)49}$	7
22. $6\overline{)0}$	0
23. $3\overline{)18}$	6
24. $1\overline{)4}$	4
25. $1\overline{)1}$	1
26. $3\overline{)15}$	5
27. $4\overline{)28}$	7
28. $2\overline{)18}$	9

Chap. 5   Division of Whole Numbers

Examples	Solutions
29. 8⟌72	9
30. 7⟌56	8
31. 9⟌36	4
32. 8⟌24	3
33. 4⟌24	6
34. 3⟌3	1
35. 4⟌8	2
36. 7⟌7	1
37. 1⟌6	6
38. 4⟌36	9
39. 6⟌36	6
40. 8⟌16	2
41. 3⟌0	0
42. 1⟌5	5
43. 3⟌21	7
44. 6⟌12	2

Practice!

Exercises Part I

1. On a sheet of paper number from 1 to 44. Use this sheet to cover the right side of the Examples-Solutions section above. Time yourself! You should be able to complete the problems correctly in a minute.

Exercises Part II

1. Purchase a set of flash cards. Discover the facts that give you trouble by removing them from the deck. Practice these facts each day.

The average woman should consume 1,800 calories per day, or an average of 600 calories per meal.

This section explains the procedure for long division, as illustrated by the following example.

$$
\begin{array}{r}
215\ \text{R4} \\
41\overline{\smash{\big)}\,8{,}819} \\
\underline{8\ 2} \\
61 \\
\underline{41} \\
209 \\
\underline{205} \\
4
\end{array}
$$

If you think you know this procedure, turn to the exercises and work each problem. If you have difficulty, turn back and read this section completely.

Remainders   How many 7-foot sections can be cut from a board 18 feet long? That is, 7 feet can be subtracted from 18 feet how many times? Division gives the answer.

$$
\begin{array}{r}
? \\
7\overline{\smash{\big)}\,18}
\end{array}
$$

Two 7-foot sections = 2 X 7 = 14 feet

Three 7-foot sections = 3 X 7 = 21 feet

Seven does not divide 18 evenly, but it divides 18 two times, with 4 remaining.

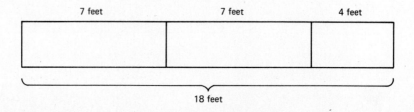

Here are the steps in the division.

```
  ?
7)18      Think:    7 times what equals 18?
```

The product closest to 18 (without going over) is 7 × 2, or 14.

```
  2
7)18      Write:    Place the 2 in the quotient.
```

```
  2
7)18                Multiply 7 times 2 and write the answer under the
 14                 dividend.
```

```
  2
7)18                Subtract to see what remains.
 14
 ──
  4
```

The answer to the division is 2 with a remainder of 4. This is indicated by writing 2 R4.

EXAMPLE 1:   Divide 29 by 6.

```
  ?
6)29      Think:    6 × ? = 29.
```

The product closest to 29 is 6 × 4, or 24.

```
  4
6)29      Write:    Record the 4.
```

```
  4
6)29                Multiply 6 times 4, write the answer under the divi-
 24                 dend, and subtract.
 ──
  5
```

```
  4 R5
6)29                The answer is 4 R5.
 24
 ──
  5
```

EXAMPLE 2:   Divide 9 by 4.

```
  2
4)9                 4 × 2, or 8 is the product closest to 9 (without going over).
  8
 ──
  1
```

The answer is 2 R1.

EXAMPLE 3:   Divide 4 by 9.

```
  ?
9)4       Think:    The product closest to 4 is 9 × 0, or 0.
```

9 divides 4 zero times with a remainder.

```
 0
9⟌4
 0
 ‾
 4
```

Write: Record 0 in the quotient.

Multiply and subtract as before.

```
 0 R4
9⟌4
 0
 ‾
 4
```

The answer is 0 R4.

  In division, the remainder is always smaller than the divisor.

```
   2 R4              4 R5              2 R1
7⟌18             6⟌29             4⟌9
  14                24                8
  ‾‾                ‾‾                ‾
   4  smaller than 7    5  smaller than 6    1  smaller than 4
```

Division is like repeated subtraction. After the divisor has been subtracted as many times as possible from the dividend, the number that remains is smaller than the divisor.

To check a division with remainder, multiply the divisior and quotient, and add the remainder. The answer should be the dividend.

```
   2 R4
7⟌18        Check:        7       Divisor times the quotient.
                       X  2
                       ‾‾‾
                         14
                       +  4       Add remainder.
                       ‾‾‾
                         18       The dividend!
```

```
   4 R5
6⟌29        Check:        6       Divisor times quotient.
                       X  4
                       ‾‾‾
                         24
                       +  5       Add remainder.
                       ‾‾‾
                         29       The dividend!
```

Try these problems *before* looking at the solutions! In each problem, divide and check.

Examples	Solutions

5⟌49	``` 9 R4```
	 ```5⟌49 Check: 5```
	 ``` 45 X 9```
	 ``` ‾‾ ‾‾‾```
	 ``` 4 45```
	 ``` + 4```
	 ``` ‾‾‾```
	 ``` 49```

Examples	Solutions

6)50	8 R2
	6)50
	  48
	   2 *Check:* 6
	              × 8
	              48
	            + 2
	              50

8)47	5 R7
	8)47
	  40
	   7 *Check:* 8
	              × 5
	              40
	            + 7
	              47

Please be sure you understand these before continuing!

Long Division

EXAMPLE 4: Divide 694 by 2.

2)[6]94

Think: First, notice that 2 divides the *hundreds position* exactly 3 times.

   3
2)694

Write: Record the 3 in the *hundreds position* over the dividend.

   3
2)694
  6
  0

Multiply and subtract.

$$2 \times 3 \text{ hundreds} = 6 \text{ hundreds}$$

$$6 \text{ hundreds} - 6 \text{ hundreds} = 0 \text{ hundreds}$$

Next, bring down the digit in the *tens position*.

   3
2)694
  6
  09

Think: 2 does not divide this digit evenly, but it will divide 4 times with remainder.

  34
2)694
  6
  09

Write: Record the 4 in the *tens position* above the dividend.

  34
2)694
  6
  09
   8
   1

Multiply 2 times the partial quotient 4 and subtract as before.

$$2 \times 4 \text{ tens} = 8 \text{ tens}$$

$$9 \text{ tens} - 8 \text{ tens} = 1 \text{ ten}$$

This remainder is in the tens position; when the *ones position* is considered, this gives

$$1 \text{ tens} + 4 \text{ ones} = 10 + 4$$
$$= 14$$

which is shown by bringing down the 4 in the ones position next to the remainder in the tens position, giving 14.

```
     34
  2 ⟌ 694
     6
    ‾‾
     09
      8
     ‾‾
     14
```

Think: Now 2 divides 14 ones evenly 7 times.

```
    347
  2 ⟌ 694
     6
    ‾‾
     09
      8
     ‾‾
     14
```

Write: Record this in the ones position above the dividend.

```
    347
  2 ⟌ 694
     6
    ‾‾
     09
      8
     ‾‾
     14
     14
     ‾‾
      0
```

Multiply 2 times the partial quotient 7 and subtract as before.

$$2 \times 7 \text{ ones} = 14 \text{ ones}$$
$$14 \text{ ones} - 14 \text{ ones} = 0 \text{ ones}$$

Therefore, 694 ÷ 2 is 347.

 If you keep the partial quotients and remainders lined up correctly, the digits of the final answer are in the correct positions.

Check by multiplication.

```
    327
  2 ⟌ 654
     6
    ‾‾
     05
      4
     ‾‾
     14
     14
     ‾‾
      0
```

Check:
```
    327
  X   2
  ‾‾‾‾‾
    654
  +   0        No remainder.
  ‾‾‾‾‾
    654        The dividend!
```

EXAMPLE 5:   Divide 563 by 9.

```
  9 ⟌ 56 3
```

Think: In this example, the 9 does not divide the 5 in the *hundreds position* (except zero times); therefore, I consider the first *two* digits, 56.

9 does not divide 56 evenly but will divide 6 times with remainder.

$$\begin{array}{r} 6 \\ 9\overline{)\smash{563}} \end{array}$$

Write: Because I was unable to divide in the hundreds position, I record this partial quotient in the *tens position*, over the 6.

$$\begin{array}{r} 6 \\ 9\overline{)\smash{563}} \\ 54 \\ \hline 2 \end{array}$$

Multiply and subtract.

$$9 \times 6 = 54$$
$$56 - 54 = 2$$

$$\begin{array}{r} 6 \\ 9\overline{)\smash{563}} \\ 54 \\ \hline 23 \end{array}$$

Bring down the 3, giving 23.

$$\begin{array}{r} 62 \\ 9\overline{)\smash{563}} \\ 54 \\ \hline 23 \end{array}$$

Think: 9 divides 23 twice with remainder.
Write: Record 2.

$$\begin{array}{r} 62 \\ 9\overline{)\smash{563}} \\ 54 \\ \hline 23 \\ 18 \\ \hline 5 \end{array}$$

Multiply and subtract.

$$9 \times 2 = 18$$
$$23 - 18 = 5$$

$$\begin{array}{r} 62\ R5 \\ 9\overline{)\smash{563}} \\ 54 \\ \hline 23 \\ 18 \\ \hline 5 \end{array}$$

5 is the remainder.

Check:
$$\begin{array}{r} 62 \\ \times\ \ 9 \\ \hline 558 \\ +\ \ 5 \\ \hline 563 \end{array}$$ The dividend!

EXAMPLE 6:   Divide 1,429 by 7.

$$7\overline{)\smash{\boxed{1,4}29}}$$

Think: 7 does not divide the digit in the *thousands position*, so consider the first two digits.

7 divides 14 evenly 2 times with 0 remainder.

$$\begin{array}{r} 2 \\ 7\overline{)\smash{1,429}} \end{array}$$

Write: Because the division begins in the *hundreds position*, record the 2 there. Multiply and subtract.

$$\begin{array}{r} 2 \\ 7\overline{\smash{)}1{,}429} \\ 1\,4 \\ \hline 0\,2 \end{array}$$

Bring down the 2 in the *tens position*.

$$\begin{array}{r} 20 \\ 7\overline{\smash{)}1{,}429} \\ 1\,4 \\ \hline 0\,2 \end{array}$$

Think: 7 does not divide 2 evenly; in fact, the partial quotient is 0.

Write: Record the 0.

$$\begin{array}{r} 20 \\ 7\overline{\smash{)}1{,}429} \\ 1\,4 \\ \hline 0\,2 \\ 0 \\ \hline \end{array}$$

Multiply by the partial quotient 0, and subtract as before.

$$7 \times 0 = 0$$
$$2 - 0 = 2$$

$$\begin{array}{r} 20 \\ 7\overline{\smash{)}1{,}429} \\ 1\,4 \\ \hline 0\,2 \\ 0 \\ \hline 29 \end{array}$$

Bring down the next digit, 9.

$$\begin{array}{r} 204 \\ 7\overline{\smash{)}1{,}429} \\ 1\,4 \\ \hline 0\,2 \\ 0 \\ \hline 29 \\ 28 \\ \hline 1 \end{array}$$

Think: 7 divides 29 four times with remainder.

Write: Record the 4.

Multiply by the partial quotient 4, and subtract as before.

$$7 \times 4 = 28$$
$$29 - 28 = 1$$

$$\begin{array}{r} 204\ \text{R}1 \\ 7\overline{\smash{)}1{,}429} \\ 1\,4 \\ \hline 0\,2 \\ 0 \\ \hline 29 \\ 28 \\ \hline 1 \end{array}$$

1 is the remainder.

Check:

$$\begin{array}{r} 204 \\ \times\ \ \ 7 \\ \hline 1{,}428 \\ +\ \ \ \ \ 1 \\ \hline 1{,}429 \end{array}$$ The dividend!

EXAMPLE 7:   Divide 6,153 by 5.

$$\begin{array}{r} 1 \\ 5\overline{\smash{)}6{,}153} \\ 5 \\ \hline 1 \end{array}$$

5 divides the 6 in the thousands position 1 time with remainder.

```
    1
5 ) 6,153        Bring down the next digit, giving 11.
    5
    1 1
```

```
    1 2
5 ) 6,153        5 divides 11 two times with remainder.
    5
    1 1
    1 0
      1
```

```
    1 2
5 ) 6,153        Bring down the next digit, giving 15.
    5
    1 1
    1 0
      15
```

```
    1 23
5 ) 6,153        5 divides 15 exactly 3 times.
    5
    1 1
    1 0
      15
      15
       0
```

```
    1 23
5 ) 6,153        Bring down the last digit, 3.
    5
    1 1
    1 0
      15
      15
      03
```

```
    1,230 R3
5 ) 6,153        5 divides 3 zero times with remainder.
    5
    1 1
    1 0
      15
      15
      03
       0                          5 × 0 = 0
       3     The remainder is 3.  3 − 0 = 3
```

Now you try these division problems. Try not to look at the solution until you have your answer.

Examples	Solutions
2)2,789	Divide the first position possible. Bring down the digit in the next position and divide the resulting number. Repeat this procedure for all the digits in the dividend. $$\begin{array}{r} 1{,}394 \text{ R1} \\ 2{\overline{\smash{\big)}\,2{,}789}} \\ \underline{2}\phantom{{,}789} \\ 0\,7 \\ \underline{6} \\ 18 \\ \underline{18} \\ 09 \\ \underline{8} \\ 1 \end{array}$$
4)2,957	$$\begin{array}{r} 739 \text{ R1} \\ 4{\overline{\smash{\big)}\,2{,}957}} \\ \underline{2\,8} \\ 15 \\ \underline{12} \\ 37 \\ \underline{36} \\ 1 \end{array}$$
9)3,659	$$\begin{array}{r} 406 \text{ R5} \\ 9{\overline{\smash{\big)}\,3{,}659}} \\ \underline{3\,6} \\ 05 \\ \underline{0} \\ 59 \\ \underline{54} \\ 5 \end{array}$$
8)595,701	$$\begin{array}{r} 74{,}462 \text{ R5} \\ 8{\overline{\smash{\big)}\,595{,}701}} \\ \underline{56}\phantom{5{,}701} \\ 35 \\ \underline{32} \\ 3\,7 \\ \underline{3\,2} \\ 50 \\ \underline{48} \\ 21 \\ \underline{16} \\ 5 \end{array}$$

Examples	Solutions
$7\overline{)2{,}456}$	$\begin{array}{r} 350 \text{ R}6 \\ 7\overline{)2{,}456} \\ \underline{2\ 1} \\ 35 \\ \underline{35} \\ 06 \\ \underline{0} \\ 6 \end{array}$

Please be sure you understand these before continuing.

EXAMPLE 8:   Divide 745 by 21.

$21\overline{)745}$

Think: 21 will not divide the first digit, so I consider the first two digits, 74. Therefore, the answer will begin in the *tens position.*

Because you do not know the multiplication facts for 21, I shall round 21 to 20. You know the basic facts for 2, which makes division by 20 easy also. 20 is called the **trial divisor.**

trial
divisor
(20) $\begin{array}{r} 3 \\ 21\overline{)745} \end{array}$

Think: 20 divides 74 three times with remainder because

but
$$20 \times 3 = 60$$
$$20 \times 4 = 80$$

$\begin{array}{r} 3 \\ 21\overline{)745} \\ \underline{63} \\ 11 \end{array}$

Write: which is larger than 74.
Record the 3, then multiply and subtract, but use the real divisor, 21.
$$21 \times 3 = 63$$
$$74 - 63 = 11$$

Notice that the remainder is smaller than the divisor, 21. If the remainder is larger than the divisor, your partial quotient is too small.

trial
divisor
(20) $\begin{array}{r} 3 \\ 21\overline{)745} \\ \underline{63}\downarrow \\ 115 \end{array}$

Write: Bring down 5, giving 115.

```
trial                35 R10
divisor      21 ) 745
 (20)              63
                  ─────
                   115
                   105
                  ─────
                    10
```

Think: 20 (the trial divisor) divides 115 five times with remainder because

$$20 \times 5 = 100$$

but

$$20 \times 6 = 120$$

which is too large.

Write: Record the 5, then multiply and subtract using the real divisor, 21. The remainder is 10.

EXAMPLE 9:   Divide 3,957 by 32.

```
trial
divisor      32 ) 3,957
 (30)
```

Think: The trial divisor is 30, which is 32 rounded to the leading digit.

The first digit of the quotient will be in the *hundreds position* because 32 does not divide the first digit, 3.

30 divides 39 once because

$$30 \times 1 = 30$$

but

$$30 \times 2 = 60$$

which is too large.

Using 30 instead of 32 makes it easier to find the partial quotient.

```
trial             1
divisor      32 ) 3,957
 (30)             3 2
                 ─────
                    7
```

Write: Record the 1, then multiply using 32, and subtract.

$$32 \times 1 = 32$$

$$39 - 32 = 7$$

```
trial             1
divisor      32 ) 3,957
 (30)             3 2 )
                 ─────
                   75
```

Write: Bring down the 5, giving 75.

Think: 30 divides into 75 twice because

$$30 \times 2 = 60$$

but

$$30 \times 3 = 90$$

which is too large.

```
trial             12
divisor      32 ) 3,957
 (30)             3 2
                 ─────
                   75
                   64
                 ─────
                   11
```

Write: Record the 2, then multiply and subtract using 32.

$$32 \times 2 = 64$$

$$75 - 64 = 11$$

Bring down the final digit, 7.

trial
divisor
(30)

$$12$$
$$32\overline{\smash{)}3{,}957}$$
$$\underline{3\,2}$$
$$75$$
$$\underline{64}$$
$$117$$

Think: 30 divides 117 three times with remainder because

$$30 \times 3 = 90$$
but
$$30 \times 4 = 120$$

which is too large.

trial
divisor
(30)

$$123\ R21$$
$$32\overline{\smash{)}3{,}957}$$
$$\underline{3\,2}$$
$$75$$
$$\underline{64}$$
$$117$$
$$\underline{96}$$
$$21$$

Write: Record the 3, then multiply and subtract.

The remainder is 21.

The trial divisor is used to simplify the problem of finding the partial quotients. It is a form of estimation. It can, however, give you the wrong result! The following example will show you how to adjust your answer.

EXAMPLE 10: Divide 65,058 by 185.

$$185\overline{\smash{)}\boxed{65{,}0}\,58}$$

Think: The division will begin in the *hundreds position* because 185 will not divide 6 or 65.

The first partial quotient will be written over the 0 in the dividend.

trial
divisor
(200)

$$185\overline{\smash{)}65{,}058}$$

Because, I do not know multiplications with 185, I think of the divisor as 200, which is 185 rounded to the leading digit.

200 divides 650 three times with remainder, because

$$200 \times 3 = 600$$
but
$$200 \times 4 = 800$$

which is too large.

trial
divisor
(200)

$$3$$
$$185\overline{\smash{)}65{,}058}$$
$$\underline{55\,5}$$
$$9\,5$$

Write: Record 3, then multiply using the real divisor and subtract.

$$185 \times 3 = 555$$

$$650 - 555 = 95$$

trial
divisor
(200)

$$3$$
$$185\overline{\smash{)}65{,}058}$$
$$55\,5$$
$$9\,55$$

Write: Bring down the next digit, giving 955.

```
                          34
trial
divisor        185 | 65,058
 (200)               55 5
                    ─────
                      9 55
                      7 40
                    ─────
                      2 15
```

Think: 200 divides 955 four times with remainder.

Write: Record 4, then multiply and subtract.

$$185 \times 4 = 740$$

$$955 - 740 = 215$$

The remainder, 215, is larger than the divisor, 185. This means that 185 divides more than 4 times. The trial divisor gave me a good guess, but it was not close enough. I shall erase the last step and replace the 4 with a 5.

```
                          35
trial
divisor        185 | 65,058
 (200)               55 5
                    ─────
                      9 55
                      9 25
                    ─────
                        30
```

Write: Record 5, then multiply by 5 and subtract.

$$185 \times 5 = 925$$

$$955 - 925 = 30$$

Now the remainder is smaller than 185. Continue by bringing down the next digit, 8, giving 308.

```
                          35
trial
divisor        185 | 65,058
 (200)               55 5
                    ─────
                      9 55
                      9 25
                    ─────
                       308
```

Think: 200 divides 308 once with remainder.

```
                       351 R123
trial
divisor        185 | 65,058
 (200)               55 5
                    ─────
                      9 55
                      9 25
                    ─────
                       308
                       185
                    ─────
                       123
```

Write: Record 1, then multiply and Subtract.

$$185 \times 1 = 185$$

$$308 - 185 = 123$$

The remainder is 123.

EXAMPLE 11:   Divide 8,819 by 41.

```
trial
divisor        41 | 8,819
 (40)
```

Think: The trial divisor is 40 and the division begins in the *hundreds position* because 41 will not divide 8. 40 divides 88 twice with remainder.

```
                      2
trial
divisor        41 | 8,819
 (40)               8 2
                   ────
                     61
```

Write: Multiply, subtract, and bring down next digit.

```
                     21
trial
divisor        41 | 8,819
 (40)               8 2
                   ────
                     61
                     41
                   ────
                    209
```

Think: 40 divides 61 one time with remainder.
Write: Multiply, subtract, and bring down next digit.

```
trial              215 R4
divisor      41 | 8,819        Think:  40 divides 209 five times with
  (40)              8 2                 remainder.
                    ──
                    61
                    41
                    ──
                    209
                    205
                    ───
                      4
```

Check by multiplying divisor and quotient and adding the remainder.

```
Check:        215
            ×  41
            ──────
              215
            8 60
            ──────
            8,815
          +     4
            ──────
            8,819      The dividend!
```

EXAMPLE 12:   Divide 2,805 by 52.

```
52 | 2,805
```
Think: The first digit of the quotient will be written
over the 0 in the dividend.
Use the trial divisor 50.

$$50 \times 5 = 250$$

but

$$50 \times 6 = 300$$

which is too large.
Therefore, use 5 as the first partial
quotient.

```
        5
52 | 2,805
    2 60
    ─────
     205
```
Write: Multiply 52 times 5 and subtract. Bring down
the 5.

```
       54
52 | 2,805
    2 60
    ─────
     205
     208
```
Think: Use 50 as the trial divisor again.

$$50 \times 4 = 200$$

but

$$50 \times 5 = 250$$

which is too large.
Therefore, use 4 as the next partial
quotient.

Write: Multiply.

I cannot subtract. This means that 52 does not divide 4 times after all. The
trial divisor gave me a good guess, but it was too large. I shall erase the last step
and replace the 4 with a 3.

```
      53 R49
52 | 2,805
    2 60
    ─────
     205
     156
     ───
      49
```
Write: Multiply and subtract.

The remainder is 49.

EXAMPLE 13:   Divide 306,123 by 30.

```
30 ⟌ 306,123          Think:   30 divides 30 once.
       1
30 ⟌ 306,123          Write:   Record 1, multiply, subtract, and bring down
       30                        the next digit.
       06

       10
30 ⟌ 306,123          Think:   30 divides 6 zero times with remainder.
       30              Write:   Record 0, multiply, subtract, and bring down
       06                        the next digit.
       00
        6 1

       10 2
30 ⟌ 306,123          Think:   30 divides 61 twice with remainder.
       30              Write:   Record 2, multiply, subtract, and bring down
       06                        the next digit.
       00
        6 1
        6 0
         12

       10 20
30 ⟌ 306,123          Think:   30 divides 12 zero times.
       30              Write:   Record 0, multiply, subtract, and bring down
       06                        the next digit.
       00
        6 1
        6 0
         12
         00
         123

       10,204 R3
30 ⟌ 306,123          Think:   Finally, 30 divides 123 four times with
       30                        remainder.
       06              Write:   Record 4, multiply, and subtract.
       00
        6 1                     The remainder is 3.
        6 0
         12
         00
         123
         120
           3
```

Now try the following division problems.

Examples	Solutions
23 $\overline{)744}$	trial divisor ⓵⓪(20) $\begin{array}{r} 32\ R8 \\ 23\overline{)744} \\ 69 \\ \hline 54 \\ 46 \\ \hline 8 \end{array}$
59 $\overline{)2{,}158}$	trial divisor (60) $\begin{array}{r} 36\ R34 \\ 59\overline{)2{,}158} \\ 1\ 77 \\ \hline 388 \\ 354 \\ \hline 34 \end{array}$
25 $\overline{)5{,}621}$	Trial divisor gives 1, but the remainder is too large. ↓ trial divisor (30) $\begin{array}{r} 224\ R21 \\ 25\overline{)5{,}621} \\ 5\ 0 \\ \hline 62 \\ 50 \\ \hline 121 \\ 100 \\ \hline 21 \end{array}$

Study the above examples carefully! Try these—no hints given.

21 $\overline{)5{,}961}$	283 R18
39 $\overline{)4{,}209}$	107 R36

If you had difficulty, please read this section again or talk with an instructor.

Exercises Part I Answers to these exercises can be found on page 132.

Divide and check.

1. $895 \div 11$
2. $1{,}574 \div 39$
3. $14\overline{)258}$
4. $15\overline{)500}$
5. $69\overline{)5{,}473}$
6. $32\overline{)2{,}357}$
7. $55\overline{)63{,}502}$
8. $67\overline{)5{,}215}$
9. $84\overline{)2{,}751}$
10. $42\overline{)9{,}776}$
11. $74\overline{)15{,}319}$
12. $96\overline{)288{,}770}$
13. $12\overline{)15{,}611}$
14. $121\overline{)59{,}632}$
15. $295\overline{)68{,}207}$
16. $355\overline{)72{,}956}$
17. $789\overline{)1{,}987{,}625}$
18. $217\overline{)956{,}321}$

Exercise Part II

1. $159 \div 12$
2. $1{,}569 \div 29$
3. $21\overline{)700}$
4. $25\overline{)1{,}983}$
5. $13\overline{)1{,}365}$
6. $222\overline{)56{,}932}$
7. $376\overline{)78{,}219}$
8. $115\overline{)235{,}753}$

The method called **short division** is used with one-digit divisors. The multiplication and subtraction steps are done mentally; only the partial quotients and remainders are recorded.

EXAMPLE 1:   Divide 654 by 2.

$$\begin{array}{r} 3 \\ 2\overline{)654} \end{array}$$

Think: 2 divides 6 *three* times with no remainder.
Write: Record 3.

$$\begin{array}{r} 32 \\ 2\overline{)65\overset{1}{4}} \end{array}$$

Think: 2 divides 5 *two* times with remainder.

$$2 \times 2 = 4$$

$$5 - 4 = 1$$

Write: Record the partial quotient 2 and write the remainder 1 next to the 4, giving 14.

$$\begin{array}{r} 327 \\ 2\overline{)65\overset{1}{4}} \end{array}$$

Think: 2 divides 14 *seven* times with no remainder.
Write: Record the partial quotient 7.

EXAMPLE 2:   Divide 74 by 3.

$$\begin{array}{r} 2 \\ 3\overline{)7\overset{1}{4}} \end{array}$$

Think: 3 divides 7 *two* times.

$$3 \times 2 = 6$$

$$7 - 6 = 1$$

Write: Record 2 and write the remainder to the left of the 4, giving 14.

$$
\begin{array}{r}
24\ \text{R}2 \\
3\overline{|\ \overset{1}{7}4}
\end{array}
$$

Think: 3 divides 14 *four* times.
$$3 \times 4 = 12$$
$$14 - 12 = 2$$

Write: Record 4.

The remainder is 2.

EXAMPLE 3: Divide 2,789 by 2.

$$
\begin{array}{r}
1 \\
2\overline{|\ 2,789}
\end{array}
$$

Think: 2 divides 2 *one* time with no remainder.
Write: Record 1.

$$
\begin{array}{r}
1\ 3 \\
2\overline{|\ 2,\overset{1}{7}89}
\end{array}
$$

Think: 2 divides 7 *three* times with remainder.
$$2 \times 3 = 6$$
$$7 - 6 = 1$$

Write: Record 3 and write 1 next to 8, giving 18.

$$
\begin{array}{r}
1\ 39 \\
2\overline{|\ 2,\overset{1}{7}89}
\end{array}
$$

Think: 2 divides 18 *nine* times with no remainder.
Write: Record 9.

$$
\begin{array}{r}
1\ 394\ \text{R}1 \\
2\overline{|\ 2,\overset{1}{7}89}
\end{array}
$$

Think: 2 divides 9 *four* times with remainder 1.
Write: Record 4.

The remainder is 1.

EXAMPLE 4: Divide 595,701 by 8.

$$
\begin{array}{r}
7 \\
8\overline{|\ 59\overset{3}{5},701}
\end{array}
$$

Think: 8 divides 59 *seven* times.
$$8 \times 7 = 56$$
$$59 - 56 = 3$$

Write: Record 7 and write 3 next to 5, giving 35.

$$
\begin{array}{r}
74 \\
8\overline{|\ 5\overset{3}{9}\overset{3}{5},701}
\end{array}
$$

Think: 8 divides 35 *four* times.
$$8 \times 4 = 32$$
$$35 - 32 = 3$$

Write: Record 4 and write 3 next to 7, giving 37.

$$
\begin{array}{r}
74\ 4 \\
8\overline{|\ 5\overset{3}{9}\overset{3}{5},\overset{5}{7}01}
\end{array}
$$

Think: 8 divides 37 *four* times.
$$8 \times 4 = 32$$
$$37 - 32 = 5$$

Write: Record 4 and write 5 next to 0, giving 50.

$$\begin{array}{r} 74\ 46 \\ \hline 8\,\lvert\,\overset{3\ \ 3\ 5\ 2}{595{,}701} \end{array}$$

Think: 8 divides 50 *six* times.

$$8 \times 6 = 48$$
$$50 - 48 = 2$$

Write: Record 6 and write 2 next to 1, giving 21.

$$\begin{array}{r} 74{,}462\ \text{R}5 \\ \hline 8\,\lvert\,\overset{3\ \ 3\ 5\ 2}{595{,}701} \end{array}$$

Think: 8 divides 21 *two* times.

$$8 \times 2 = 16$$
$$21 - 16 = 5$$

Write: Record 2.

The remainder is 5.

Try these problems using short division.

Examples	Solutions
$4\,\lvert\,\overline{265}$	$\begin{array}{r}66\ \text{R}1\\\hline 4\,\lvert\,\overset{2}{265}\end{array}$
$3\,\lvert\,\overline{513}$	$\begin{array}{r}171\\\hline 3\,\lvert\,\overset{2}{513}\end{array}$
$8\,\lvert\,\overline{1{,}235}$	$\begin{array}{r}154\ \text{R}3\\\hline 8\,\lvert\,1{,}2\overset{4\ 3}{35}\end{array}$

Try these—no hints given.

$5\,\lvert\,\overline{6{,}904}$	1,380 R4
$7\,\lvert\,\overline{2{,}793}$	399
$4\,\lvert\,\overline{1{,}084}$	271

If you had difficulty, talk with an instructor or read this section again.

Exercises Part I Answers to these exercises can be found on page 132.

Work each problem using short division.

1. $132 \div 5$ 2. $287 \div 2$ 3. $759 \div 7$ 4. $682 \div 8$
5. $1{,}305 \div 3$ 6. $2{,}475 \div 9$ 7. $8{,}562 \div 6$ 8. $4{,}893 \div 4$
9. $5{,}678 \div 8$ 10. $4{,}563 \div 7$

Exercises Part II Work each problem using short division.

1. $102 \div 2$ 2. $379 \div 6$ 3. $1{,}359 \div 8$ 4. $954 \div 9$
5. $8{,}576 \div 4$ 6. $2{,}756 \div 3$

CALORIE COST PER HOUR

Running	800 - 1,000
Swimming	300 - 650
Tennis	400 - 500
Walking	200 - 300

EXAMPLE 1: If I have to pay off a bill for $468 in 9 equal payments, how much will I have to pay each time?

9 times *what* equals 468? I shall use an open box to represent the missing number.

$$9 \times \square = 468$$

To find the missing factor, I divide.

$$\begin{array}{r} 52 \\ 9\overline{)468} \\ 45 \\ \hline 18 \\ 18 \\ \hline \end{array}$$

Therefore,

$$9 \times \boxed{52} = 468.$$

Nine 52's = 468; therefore, I have to make 9 payments of $52 each.

EXAMPLE 2: If you make a payment of $52 each week, how many weeks will it take to pay off a $468 bill?

How many 52's equal 468?

$$\square \times 52 = 468$$

The missing factor can be found by division.

$$\begin{array}{r} 9 \\ 52\overline{)468} \\ 468 \\ \hline \end{array}$$

Therefore,

$$\boxed{9} \times 52 = 468$$

It will take you 9 weeks to pay off the bill.

The previous examples illustrate two ways to consider division.

Division can be used to find:

1. The *size* of a fixed number of equal parts.

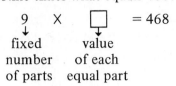

Nine times what equals 468?

$$9 \quad \times \quad \square = 468$$

fixed   value
number   of each
of parts equal part

2. The *number of equal parts* of a certain value.

How many 52's equal 468?

$$\square \quad \times \quad 52 \quad = 468$$

number   value
of equal   of each
parts   equal part

In each case, the product is divided by the known factor to find the missing factor.

1. known missing

$$9 \quad \times \quad \square = 468$$

divide product by
known factor

$$\begin{array}{r} 52 \\ 9\overline{)468} \\ 45 \\ \hline 18 \\ 18 \\ \hline 0 \end{array}$$
    52 is the missing factor.

2. $\square \times 52 \quad = \quad 468$

divide product by
known factor

$$\begin{array}{r} 9 \\ 52\overline{)468} \\ 468 \\ \hline 0 \end{array}$$
    9 is the missing factor.

EXAMPLE 3:   If Dan's salary is $15,600 for the year, how much does he make each month?

Because there are 12 months in a year,

Twelve times *what* is 15,600?

$$12 \quad \times \quad \square = 15,600$$

By division,

$$
\begin{array}{r}
1{,}300 \\
12\overline{\smash{\big)}\,15{,}600} \\
\underline{12\phantom{{,}600}} \\
3\,6 \\
\underline{3\,6} \\
00 \\
\underline{0} \\
00 \\
\underline{0} \\
0
\end{array}
$$

Therefore,

$$12 \times \boxed{1{,}300} = 15{,}600$$

He makes $1,300 per month.

EXAMPLE 4: If Dan saves $25 per week, how many weeks will it take him to save $500?

The question is,

How many 25's equal 500?

$$\Box \quad \times \quad 25 \quad = \quad 500$$

By division,

$$
\begin{array}{r}
20 \\
25\overline{\smash{\big)}\,500} \\
\underline{50} \\
00 \\
\underline{0} \\
0
\end{array}
$$

Therefore,

$$\boxed{20} \times 25 = 500$$

He will have to save for 20 weeks.

EXAMPLE 5: If 13 people share $587, how much does each get, and how much is left over?

The question is,

Thirteen times *what* is $587?

$$13 \quad \times \quad \Box \quad = \quad 587$$

By division,

$$
\begin{array}{r}
45 \\
13\overline{\smash{\big)}\,587} \\
\underline{52} \\
67 \\
\underline{65} \\
2
\end{array}
\qquad 13 \times \boxed{45} = 585
$$

Each gets $45, and $2 are left over.

Per Unit Another important application of division is reducing information to a *per unit* basis.

     EXAMPLE 6: If 5 pens cost 70¢, how much does each pen cost, or what is the price per pen?

The question is,

$$\text{Five times} \quad \textit{what} \quad \text{is 70?}$$

$$\downarrow$$

price for
each of the
five pens

$$5 \quad \times \quad \square \quad = 70$$

By division,

$$\begin{array}{r} 14 \\ 5\overline{)70} \\ \underline{5} \\ 20 \\ \underline{20} \end{array} \qquad 5 \times \boxed{14} = 70$$

Therefore, each pen costs 14¢.

    Clue: With per unit problems, the unit after the word *per* indicates the divisor. In this example,

$$\text{price per pen}$$
$$\uparrow$$
pen follows the word per

Therefore, you divide by the number of pens.

$$70 \div 5 = 14$$
$$\uparrow$$
price per pen

    EXAMPLE 7: Andy uses 15 gallons of gas for a 330-mile trip. How many miles does he drive on each gallon (miles per gallon)?

Because I am looking for miles per gallon, I divide by the number of *gallons* (gallon follows the word *per*).

$$330 \div 15 = ?$$
$$\uparrow$$
per gallon

Dividing,

$$\begin{array}{r} 22 \\ 15\overline{)330} \\ \underline{30} \\ 30 \\ \underline{30} \end{array}$$

He drives 22 miles on each gallon of gas.

    EXAMPLE 8: Anne types 252 words in 4 minutes. Karl types 156 words in 3 minutes. Who is the faster typist?

I first calculate their rates in words per minute.

        Anne: $252 \div 4 = ?$

Dividing,

$$
\begin{array}{r}
63 \\
4\overline{)252} \\
24 \\
\hline
12 \\
12 \\
\hline
\end{array}
$$

Anne types 63 words each minute.

Karl: $156 \div 3 = ?$

Dividing,

$$
\begin{array}{r}
52 \\
3\overline{)156} \\
15 \\
\hline
06 \\
6 \\
\hline
\end{array}
$$

Karl types 52 words each minute. Therefore, Anne is the faster typist.

EXAMPLE 9:   If it takes 6 minutes for a machine to make 3 paper logs, how long does it take to make 4 logs?

First, I calculate the number of minutes it takes to make one log, or minutes per log.

$$6 \div 3 = ?$$
per log

Dividing,

$$
\begin{array}{r}
2 \\
3\overline{)6} \\
6 \\
\hline
\end{array}
$$

Therefore, it takes 2 minutes per log. With 4 logs, it takes four 2's, or $4 \times 2 = 8$ minutes.

Average   The **average** of a group of numbers can be found by adding the quantities and dividing the sum by the number of quantities in the group.

For example, if a student has scores of 75 and 87, her average score is 81.

Average $= \underbrace{75 + 87}_{\substack{\text{sum of} \\ \text{quantities}}}$ divided by $\underbrace{2}_{\substack{\text{number of} \\ \text{quantities}}}$

$= 162 \div 2$

$= 81$

It is as if the student got 81 on each test, for a total of 162 points on 2 tests.

Actual scores: $75 + 87 = 162$ points

Average scores: $81 + 81 = 162$ points

EXAMPLE 10:   Jim drove 150 miles the first day, 269 miles the second day, and 331 miles the third day. How many miles did he average per day?

$$\text{Average} = \underbrace{150 + 269 + 331}_{\substack{\text{sum of}\\\text{quantities}}} \quad \text{divided by} \quad \underbrace{3}_{\substack{\text{number of}\\\text{quantities}}}$$

$$= 750 \div 3$$

$$= 250 \text{ miles}$$

It is as if he drove 250 miles each day, for a total of 750 miles.

Try to work the following problems before looking at the solutions.

Examples	Solutions
If you save $15 each week, how many weeks will it take to save $630?	*How many* 15's equal 630? ☐ X 15 = 630 Divide product by the known factor. $$\begin{array}{r} 42 \\ 15\overline{)630} \\ \underline{60} \\ 30 \\ \underline{30} \end{array}$$ It will take 42 weeks.
If 12 people share the cost of a party, how much does each pay for a party costing $756?	Twelve times *what* is 756? 12 X ☐ = 756 $$\begin{array}{r} 63 \\ 12\overline{)756} \\ \underline{72} \\ 36 \\ \underline{36} \end{array}$$ Each pays $63.
My car can go 273 miles on 13 gallons of gas. How many miles can I drive on each gallon (miles per gallon)?	$$273 \div 13 = ?$$ ↑ per gallon Dividing by the number of gallons, $$\begin{array}{r} 21 \\ 13\overline{)273} \\ \underline{26} \\ 13 \\ \underline{13} \end{array}$$ I get 21 miles per gallon of gas.

Examples	Solutions
Average 62 and 54.	Average = 62 + 54 divided by 2 = $116 \div 2$ = 58

Now try these—no hints given.

Benjamin types 312 words in 6 minutes. How many words does he type per minute?	He types 52 words per minute.
Denise gets a fixed salary each week for babysitting. If she makes $270 in 9 weeks, how much does she make each week?	$30 per week.
If you make payments of $63 each month, how many months will it take to pay off a bill for $1,449?	23 months.
Average 27, 29, 32, and 16.	26.

Please read this section again or talk with an instructor if you had difficulty with the last four problems.

Exercises Part I

Answers to these exercises can be found on page 133.

1. Jack earns $79 a day. How many days must he work to earn $9,717?
2. A set of tires lasts 24,000 miles. How many sets of tires will it take to drive 96,000 miles?
3. If Ruby can drive 360 miles on 15 gallons of gas, how many miles does she get per gallon?
4. Muriel must read 1,428 pages in 14 days. How many pages must she read each day?
5. A carpet costs $552. How much must you pay each month to pay it off in 24 months?
6. Which is cheaper—to get 5 pens for $1 or to pay 18¢ per pen?
7. Pedro's test scores are 84, 47, 88, 93, and 78. What is his average score?
8. Debbie's electric bills were $86, $64, and $72 for 3 months. What is her average electric bill per month?

Exercises Part II

1. Fran must work 1,515 problems in 15 days. How many problems must she work each day?
2. A carpet costs $552. How many $46 payments are necessary to pay for the carpet?
3. Which is cheaper—to get 12 boxes for $24 or to pay $3 per box?
4. Average these test scores: 96, 27, 83, and 74.

REVIEW MATERIAL

**Review Skill
One-Digit
Divisors**

```
              6,375 R1
        9 | 57,376
          54
          ──
           3 3
           2 7
           ───
             67
             63
             ──
              46
              45
              ──
               1
```

Exercise A   Answers to this exercise can be found on page 133.

Divide.

1. 4 | 56   2. 3 | 39   3. 2 | 108   4. 6 | 498
5. 4 | 1,012   6. 5 | 6,725   7. 7 | 9,802,471   8. 5 | 1,630
9. 8 | 9,872   10. 9 | 945

Exercise B   No answers are given.

Divide.

1. 7 | 84   2. 5 | 55   3. 8 | 256   4. 9 | 729
5. 8 | 3,408   6. 9 | 5,769   7. 6 | 4,876,518

**Review Skill
Division of Whole
Numbers**

Round the divisor to leading digit to use as a trial divisor. It helps estimate the correct quotient.

```
                       70 R12
trial
divisor     81 | 5,682
 (80)            5 67
                 ────
                   12
                   00
                   ──
                   12
```

Exercise C   Answers to this exercise can be found on page 133.

Divide.

1. 28 | 84   2. 54 | 795   3. 83 | 4,482   4. 78 | 93,894
5. 87 | 85,869   6. 97 | 9,703,889 7. 15 | 8,097   8. 37 | 2,954
9. 81 | 95,724   10. 135 | 67,904

Exercise D   No answers are given.

Divide.

1. $47\overline{)94}$ 2. $76\overline{)1,064}$ 3. $98\overline{)8,134}$ 4. $85\overline{)229,518}$
5. $77\overline{)72,457}$ 6. $92\overline{)8,299,119}$

REVIEW EXERCISES

Refer to the section listed if you have difficulty.   Answers to these exercises can be found on page 133.

Section 5.2   Divide.

1. $63,613 \div 27$ 2. $364,662 \div 81$ 3. $7,157 \div 125$

Section 5.3   Divide using short division.

4. $87,654 \div 3$ 5. $96,201 \div 8$

Section 5.4   6.   Each box contains 500 sheets of paper. How many boxes would I need to get 6,000 sheets of paper?
7.   Sue types 477 words in 9 minutes. How many words does she type per minute?
8.   To save $600 in 12 months, how much will I have to save each month?
9.   Average the following numbers: 15, 16, 19, and 10.
10.   Larry must read 64 pages in 4 hours. How many pages does he have to read each hour?

CHAPTER POST-TEST

This test should determine if you have mastered the topics in Chapter 5. Answers to this test can be found on page 133.

Divide.

1. $6\overline{)5,743}$
2. $32\overline{)3,417}$
3. $85\overline{)27,596}$
4. $123\overline{)59,632}$
5. $369\overline{)79,504}$
6.   If I must pay off a bill for $540 in 12 equal payments, how much will I have to pay each time?
7.   John saves $20 each week. How many weeks will it take him to save $1,040?
8.   Average 15, 24, and 36.

UNIT ONE WHOLE NUMBERS REVIEW AND POST-TEST

Read and write whole numbers in periods of three digits.

The number 5,508,271 is read five million, five hundred eight thousand, two hundred seventy-one.

One million, two hundred fifty-two thousand, fourteen is written 1,252,014.

To round whole numbers, look at the digit to the immediate right of the rounding position.

If this digit is 5 or more, round up.

$$8,\underline{5}74 \cong 8,600, \text{ to the nearest hundred.}$$

If this digit is less than 5, leave the rounding position unchanged.

$$8,5\underline{2}4 \cong 8,520, \text{ to the nearest ten.}$$

To add whole numbers arrange the numbers in columns. Line up digits having the same place value.

$$
128 + 9 + 57 =
\qquad
\begin{array}{r}
128 \\
9 \\
+\ 57 \\
\hline
194
\end{array}
\qquad
\begin{array}{r}
\text{Estimation} \\
100 \\
9 \\
+\ 60 \\
\hline
169
\end{array}
$$

The estimate (a good guess) is a quick way to see if the actual answer is reasonable.

To subtract whole numbers, arrange the numbers in vertical form as with addition. To borrow, reduce the borrowing position by 1 and add 10 to the original position.

$$
7,052 - 153 =
\qquad
\begin{array}{r}
7,052 \\
-\ 153 \\
\hline
6,899
\end{array}
\qquad
\begin{array}{r}
\text{Estimation} \\
7,000 \\
-\ 200 \\
\hline
6,800
\end{array}
$$

Use the shift procedure to multiply by more than one digit.

$$
\begin{array}{r}
213 \\
\times\quad 59 \\
\hline
1,917 \\
10\ 65 \\
\hline
12,567
\end{array}
\quad \longleftarrow \text{shift}
\qquad
\begin{array}{r}
\text{Estimation} \\
200 \\
\times\quad 60 \\
\hline
000 \\
12\ 00 \\
\hline
12,000
\end{array}
$$

To divide whole numbers, round the divisor to the first position to use a trial divisor.

```
trial              415
divisor    82 ) 34,030
 (80)          32 8
               ────
                1 23
                  82
                ────
                 410
                 410
                ────
```

Before working the Unit One Post-Test, work (or rework)

Chapter 1 Review Exercises, page 25
Chapter 2 Review Exercises, pages 45–46
Chapter 3 Review Exercises, pages 68–69
Chapter 4 Review Exercises, page 94
Chapter 5 Review Exercises, page 127

Study the sections for which you cannot work the sample problems.

WHOLE NUMBERS POST-TEST

Answers to this test can be found on page 133.

1. Write 6,053,153 as you would read it.
2. Write the numeral for three hundred fifty-seven thousand, sixty-two.
3. Round 45,693 to the nearest hundred.
4. Round 45,693 to the nearest ten.
5. Round 45,693 to the nearest ten thousand.
6. Perform the indicated operations.

   (a) 173   (b) $28,956 \div 53$ (c)   195
        245 -179
       +693

   (d)   143   (e)   5,006   (f)   4,063
       $\times\ 89$ $-1,273$ $\times\ \ 72$

   (g) $5,978 \div 29$ (h) $27 + 56 + 153$ (i) 496×400
   (j) $10 + 276 + 3$ (k) $796 - 399$ (l) $9,874 \div 202$

7. The beginning balance in a bank account is $536. What is the new balance after checks for $52 and $73 are written and a deposit of $69 is made?
8. If Stan saves $25 each month, how much will he have saved in 24 months?
9. Average the following numbers: 70, 86, 90, and 66.
10. Belinda paid the dentist $55 in January and $65 in February. If she still owes $175, how much was the total bill?
11. If I pay $26 per week, how many weeks will it take me to pay off a bill for $1,170?
12. Earlene types 56 words per minute. How many words can she type in 9 minutes?
13. The temperature at noon was 72°. At midnight the temperature was 53°. By how much did the temperature decrease?

ANSWERS

Unit One Pretest   **1.** (a) 0   (b) 2   (c) 4
2. one million, five hundred twenty-three thousand, five hundred seven.
3. 163,523   **4.** (a) 5,800   (b) 6,200   (c) 9,600
5. (a) 281   (b) 9,502   **6.** $85   **7.** (a) 938   (b) 5,215   **8.** 254
9. (a) 405   (b) 18,423   **10.** 64   **11.** 120
12. (a) 265 R12   (b) 47 R62   **13.** 61   **14.** 44

Chapter 1

Section 1.1   **1.** (a) hundreds   (b) tens   (c) ones
2. (a) ten-thousands   (b) hundreds   (c) tens
3. (a) 7   (b) 5   (c) 0   (d) 1   (e) 9   (f) 6   (g) 2
4. (a) 1   (b) 7   (c) 5

Section 1.2   **1.** five hundred twenty-three   **2.** five hundred two   **3.** fifty-seven
4. one thousand, five hundred twenty-one
5. twelve thousand, five hundred one
6. one hundred twenty-three thousand, one hundred twenty-three
7. one hundred fifty thousand, twenty-seven
8. fifteen million, four hundred sixty-nine thousand, five hundred twenty-three
9. two hundred fifty-seven million, three hundred five thousand, two hundred forty
10. one hundred five million, five thousand, four   **11.** 57,429   **12.** 172,052
13. 20,095,003   **14.** 7,035,063   **15.** 10,004,003
16. four hundred sixty-three   **17.** one thousand, two hundred five
18. five hundred two   **19.** 702   **20.** 1,053

Section 1.3   **1.** 200   **2.** 800   **3.** 900   **4.** 1,600   **5.** 2,600   **6.** 3,500   **7.** 3,900
8. 5,000   **9.** 4,000   **10.** 4,000   **11.** 2,000   **12.** 40,000

Exercise A   **1.** six hundred forty   **2.** fifty-six   **3.** four thousand, six hundred eighty-eight
4. forty-two thousand, two hundred seventy-three
5. Seven million, nine hundred fifty-one thousand, eight
6. thirty-one million, eight hundred fifty-eight thousand, five hundred twenty-one
7. forty-six   **8.** sixty-five   **9.** eight thousand, eight hundred sixty-four
10. thirty-seven thousand, two hundred twenty-four
11. eight million, one thousand, five hundred ninety-seven
12. one hundred twenty-five million, eight hundred fifty-eight thousand, three hundred ten
13. seven hundred forty-eight   **14.** seventy-seven thousand
15. fifty-nine thousand, one   **16.** 34   **17.** 19   **18.** 8,274   **19.** 626,603
20. 400,304

Exercise C   **1.** 380   **2.** 350   **3.** 5,000   **4.** 4,000   **5.** 500   **6.** 600   **7.** 630,000
8. 170,000   **9.** 5,800   **10.** 6,000   **11.** 29,640   **12.** 2,400   **13.** 5,000
14. 4,000   **15.** 17,000

Review Exercises   **1.** (a) thousands   (b) ones   (c) hundreds   (d) ten-thousands
2. (a) thousands   (b) tens   (c) millions   (d) ten-thousands
3. (a) 1   (b) 5   (c) 4   **4.** twenty-seven thousand, five hundred forty-two
5. one million, fifty-two thousand, two
6. thirteen thousand, five hundred twenty-four
7. ten thousand, five hundred three

8. fifteen million, four hundred two thousand, five hundred twenty-three
9. 162,549   10. 7,048,002   11. 240   12. 1,250   13. 2,580
14. 2,600   15. 263,000   16. 82,400   17. 96,200

Post-Test   1. (a) millions   (b) ten-thousands   (c) tens   (d) hundreds
(e) thousands   (f) ones   (g) ten-millions   (h) hundred-thousands
2. (a) 6   (b) 1   (c) 8
3. (a) five million, sixty-two thousand, one hundred three
(b) three million, one hundred two thousand, three hundred fifty-nine
4. (a) 50,227,043   (b) 105,007
5. (a) 44,000   (b) 48,000   (c) 150,000   (d) 42,000   (e) 149,000
(f) 56,000

Chapter 2

Section 2.2   1. 592   2. 23   3. 108   4. 99   5. 156   6. 1,200   7. 1,563
8. 638   9. 1,041   10. 1,393   11. 1,665   12. 1,173   13. 759
14. 113,338   15. 10,305   16.   16,083   17. 31,905   18. 1,599
19. 206   20. 58,118

Section 2.3   1. $33   2. $220   3. 18 feet   4. 49 feet   5. $240   6. 14 gallons
7. (a) $121   (b) $131   (c) $252   8. 33 minutes   9. 4 hours
10. 79,160 square miles   11. $494   12. $187

Exercise A   1. 56   2. 67   3. 93   4. 134   5. 120   6. 109   7. 101   8. 110
9. 84   10. 121   11. 36   12. 100   13. 30   14. 103   15. 21
16. 33   17. 105   18. 95   19. 60   20. 170

Exercise C   1. 1,500   2. 959   3. 1,212   4. 1,209   5. 4,654   6. 11,528
7. 7,301   8. 10,372   9. 6,508   10. 7,039   11. 32,135   12. 35,344
13. 37,102   14. 59,110   15. 52,410

Exercise E   1. 207   2. 175   3. 1,727   4. 2,345   5. 25,062   6. 20,881
7. 7,601   8. 48,394   9. 48,298   10. 3,702

Review Exercises   1. 81   2. 142   3. 363   4. 1,378   5. 4,701   6. 132   7. 1,916
8. 720   9. 16,793   10. 7,210   11. $73   12. 18 feet   13. $68,611

Post-Test   1. 71   2. 560   3. 10,040   4. 20,048   5. 7,730,353   6. $90
7. 23 feet   8. $21

Chapter 3

Section 3.2   1. 44   2. 17   3. 27   4. 56   5. 36   6. 135   7. 324   8. 268
9. 8   10. 394   11. 26,992   12. 6,031   13. 17,893   14. 6,569
15. 465,287

Section 3.3   1. $406   2. 213 calories   3. 303,462 square miles   4. 29°
5. (a) $456   (b) $427   6. $1,428   7. (a) $31   (b) $23   8. $499
9. 6 pages   10. 35 pieces   11. 77¢   12.   39
+ 169
―――
202   total
50   less cash received
―――
$152

13. $\begin{array}{r} 102\ 00 \\ \underline{59\ 00} \\ \$\ \ 43\ 00 \end{array}$ This payment.

Bal. forward.

Exercise A 1. 11 2. 9 3. 43 4. 60 5. 56 6. 49 7. 55 8. 68 9. 32 10. 63 11. 10 12. 9 13. 11 14. 18 15. 7 16. 34 17. 41 18. 15 19. 4 20. 15

Exercise C 1. 26 2. 195 3. 120 4. 917 5. 1,969 6. 5,202 7. 3,196 8. 37 9. 4,907 10. 2,494 11. 1,985 12. 3,307 13. 4,242 14. 4,504 15. 60,496

Review Exercises 1. 152 2. 249 3. 969 4. 35 5. 1,077 6. 2,824 7. 5,018 8. $591 9. 240 calories 10. 278 pages

Post-Test 1. 441 2. 7,801 3. 2,089 4. 68,756 5. 155 6. 5,632 7. 78,500 8. 5,158 9. $65 10. $547

Chapter 4

Section 4.2 1. 238 2. 1,645 3. 624 4. 17,390 5. 80,240 6. 185,476 7. 147,290 8. 564,850 9. 22,161,940 10. 36,153,000 11. 735 12. 294,721 13. 5,308,797 14. 1,117,200 15. 1,694,202

Section 4.3 1. 125 2. 243 3. 32 4. 25 5. 175 6. 2,000 7. 441 8. 900 9. $2^2 \times 3^3$ 10. $3^2 \times 5^4 \times 7^3$ 11. $2 \times 5^2 \times 7$ 12. $3 \times 5^2 \times 7^2$ 13. $2^3 \times 3 \times 7^2$

Section 4.4 1. 189 miles 2. 240 blades 3. 96¢ 4. 1,000 pages 5. 45¢ 6. 1,520 balloons 7. $513 8. $74 9. $265 10. 62¢ 11. 75 chairs 12. 805 trees

Exercise A 1. 441 2. 82 3. 69 4. 784 5. 568 6. 441 7. 582 8. 255 9. 287 10. 588 11. 528 12. 434 13. 273 14. 712 15. 729 16. 4,572 17. 1,940 18. 4,578

Exercise C 1. 506 2. 651 3. 782 4. 47,952 5. 23,625 6. 1,964,807 7. 4,015 8. 2,030 9. 6,012 10. 13,287 11. 216,448 12. 348,261 13. 28,371,200

Review Exercises 1. 57 2. 405 3. 4,956 4. 91,791 5. 123,072 6. 5,508 7. 40,365 8. 293,400 9. 729 10. 16 11. $2 \times 5^2 \times 7^3$ 12. $2^3 \times 5 \times 7^2$ 13. 944 words 14. 128 crayons 15. 78¢ 16. 1,530 trees

Post-Test 1. 1,080 2. 15,168 3. 27,450 4. 56,573 5. 181,780 6. 125 7. $2^2 \times 3^3 \times 5 \times 7^2$ 8. 600 pages 9. 45¢ 10. 36,500 sheets

Chapter 5

Section 5.2 1. 81 R4 2. 40 R14 3. 18 R6 4. 33 R5 5. 79 R22 6. 73 R21 7. 1,154 R32 8. 77 R56 9. 32 R63 10. 232 R32 11. 207 R1 12. 3,008 R2 13. 1,300 R11 14. 492 R100 15. 231 R62 16. 205 R181 17. 2,519 R134 18. 4,407 R2

Section 5.3   **1.** 26 R2   **2.** 143 R1   **3.** 108 R3   **4.** 85 R3   **5.** 435   **6.** 275
                   7. 1,427   **8.** 1,223 R1   **9.** 709 R6   **10.** 651 R6

Section 5.4   **1.** 123 days   **2.** 4 sets   **3.** 24 miles per gallon   **4.** 102 pages   **5.** $23
                   6. 18¢ per pen   **7.** 78   **8.** $74

Exercise A   **1.** 14   **2.** 13   **3.** 54   **4.** 83   **5.** 253   **6.** 1,345   **7.** 1,400,353
                 8. 326   **9.** 1,234   **10.** 105

Exercise C   **1.** 3   **2.** 14 R39   **3.** 54   **4.** 1,203 R60   **5.** 987   **6.** 100,040 R9
                 7. 539 R12   **8.** 79 R31   **9.** 1,181 R63   **10.** 502 R134

Review Exercises   **1.** 2,356 R1   **2.** 4,502   **3.** 57 R32   **4.** 29,218   **5.** 12,025 R1
                 6. 12 boxes   **7.** 53 words per minute   **8.** $50   **9.** 15   **10.** 16

Post-Test   **1.** 957 R1   **2.** 106 R25   **3.** 324 R56   **4.** 484 R100   **5.** 215 R169
                 6. $45   **7.** $52   **8.** 25

Unit One Post-Test   **1.** six million, fifty-three thousand, one hundred fifty-three   **2.** 357,062
                 3. 45,700   **4.** 45,700   **5.** 50,000
                 6. (a) 1,111   (b) 546 R18   (c) 16   (d) 12,727   (e) 3,733
                 (f) 292, 536   (g) 206 R4   (h) 236   (i) 198,400   (j) 289   (k) 397
                 (l) 48 R178
                 7. $480   **8.** $600   **9.** 78   **10.** $295   **11.** 45 weeks   **12.** 504 words
                 13. 19°

UNIT 2
fractions and mixed numbers

This test is given *before* you read the material to see what you already know. If you cannot work each sample problem, you need to read the indicated chapters carefully. Answers to this test can be found on page 277.

In high school, college, and the business world, the area of arithmetic with which most people have difficulty is fractions. You may omit reading those chapters for which you answer *each* sample problem correctly, but I recommend that you read *each* section of *each* chapter in this unit and work *every* exercise.

CHAPTER 6   Write a fraction that represents each shaded region.

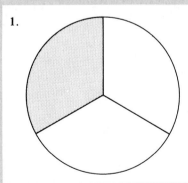   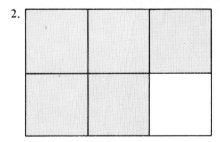

3. Write $\frac{29}{4}$ as a whole or mixed number.

4. Write $12\frac{2}{3}$ as an improper fraction.

5. Build the fraction $\frac{2}{9}$ to the denominator indicated. $\frac{2}{9} = \frac{}{108}$

6. Reduce $\dfrac{32}{240}$ to lowest terms.

7. Are $\dfrac{10}{15}$ and $\dfrac{19}{27}$ equal fractions?

CHAPTER 7

8. Fill in the box with $<$ or $>$ to indicate the correct order. $9 \,\square\, 2$

9. Find the prime factorization of 315.

10. Reduce $\dfrac{480}{540}$ to lowest terms.

11. Find the Least Common Multiple for 18 and 21.

12. Determine the greater fraction: $\dfrac{13}{18}$ and $\dfrac{7}{9}$.

CHAPTER 8

13. Add.

    (a) $\dfrac{7}{12} + \dfrac{7}{15}$ (b) $2\dfrac{5}{6} + 1\dfrac{3}{8}$

14. If it took $4\dfrac{1}{2}$ hours to go to work and $6\dfrac{3}{4}$ hours to return home, what was the total time spent traveling to and from work?

CHAPTER 9

15. Subtract.

    (a) $\dfrac{15}{16} - \dfrac{5}{24}$ (b) $8\dfrac{1}{2} - 3\dfrac{2}{3}$ (c) $11 - 4\dfrac{3}{8}$

CHAPTER 10

16. Multiply.

    (a) $\dfrac{5}{8} \times \dfrac{6}{15}$ (b) $2\dfrac{1}{2} \times 1\dfrac{1}{6}$

17. Divide.

    (a) $8 \div \dfrac{4}{31}$ (b) $\dfrac{\dfrac{5}{8}}{2\dfrac{1}{2}}$

CHAPTER 11

18. A book contains 136 pages. How many pages represent $\dfrac{1}{4}$ of the book?

19. If a car travels 55 miles per hour for $3\dfrac{1}{4}$ hours, how far has it gone?

20. If Sue types 390 words in $6\dfrac{1}{2}$ minutes, how many words can she type per minute?

21. Find the value of A when $A = \dfrac{1}{2}BH$, $B = 3$, and $H = 8$.

6
fractions and mixed numbers

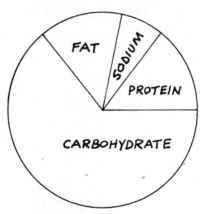

Nutrition Information for Cereal

PROTEIN $\frac{1}{7}$
CARBOHYDRATE $\frac{4}{7}$
FAT $\frac{1}{7}$
SODIUM $\frac{1}{14}$

This section introduces fractions. After reading this section, you will be able to answer questions such as these.

1. What fraction represents the shaded portion below?

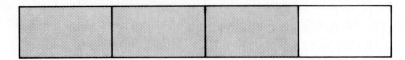

2. How can 7 people equally share 3 whole pies?

Division and Parts of Wholes

Situation One Suppose 3 people want to share 6 whole pickles equally. How many pickles should each person get? Three times *what number* is 6?
Dividing,

$$6 \div 3 = 2$$

Therefore, each person should receive 2 pickles.

Situation Two Suppose there are 2 pickles that are to be divided equally among 5 people. How many pickles should each person get? Five times *what number* is 2?
The answer should be given by division. But what does $2 \div 5$ mean? The answer lies in realizing that each person can receive only a *part of a whole* pickle. Therefore, the solution is not a whole number.
To see the answer, divide each whole pickle into 5 equal slices.

To share these slices equally, each of the 5 people should get 2 slices.

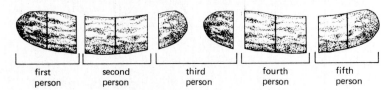

Each person receives 2 of the 5 equal parts of a whole pickle, as represented by the shaded slices.

Therefore, when you divide 2 by 5, the answer is a number idea that is not a whole number.

Fractions Fractions can represent parts of a whole. For example,

$$\frac{2}{5} \qquad \text{Read:} \quad \text{two-fifths}$$

is a fraction used to represent $2 \div 5$, or 2 of 5 equal parts of a whole.

The 2 is called the **numerator** and the 5 is called the **denominator.**

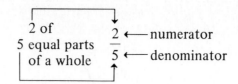

$$\begin{array}{l} \text{2 of} \\ \text{5 equal parts} \\ \text{of a whole} \end{array} \quad \dfrac{2}{5} \; \leftarrow \text{numerator} \\ \leftarrow \text{denominator}$$

 Fractions can be viewed in two ways:

1. Division of whole numbers.

$$\frac{2}{5} = 2 \div 5$$

2. Parts of a whole.

$$\frac{2}{5} = \begin{array}{l} \text{2 of} \\ \text{5 equal parts of a whole} \end{array}$$

Although a fraction can be seen as division of whole numbers, it is usually interpreted as a part of a whole.

EXAMPLE 1:   What fraction represents the shaded portion of the candy bar?

The whole candy bar has been divided into 6 equal pieces. The denominator indicates the number of parts into which the whole has been divided. Only 5 pieces are shaded. The numerator indicates the number of parts considered.

$$\begin{array}{l} \text{5 of} \\ \text{6 equal parts} \\ \text{of a whole} \end{array} \quad \dfrac{5}{6} \qquad \text{Read: five-sixths.}$$

Therefore, $\dfrac{5}{6}$ represents the shaded portion of the candy bar.

EXAMPLE 2:   If a pizza is divided into 12 equal slices, what fractional part remains if 5 slices are eaten?

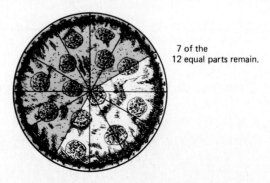

7 of the
12 equal parts remain.

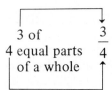

7 of
12 equal parts
of the whole $\dfrac{7}{12}$ Read: seven-twelfths.

Therefore, $\dfrac{7}{12}$ of the pizza remains.

EXAMPLE 3:   How can 3 oranges be equally divided among 4 children?
Arithmetically, 3 oranges divided equally among 4 children would be $3 \div 4$. As before,

$$3 \div 4 = \frac{3}{4}$$

To interpret this answer, think of $\dfrac{3}{4}$ as 3 of 4 equal parts of a whole orange.

3 of
4 equal parts
of a whole $\dfrac{3}{4}$ Read: three-fourths.

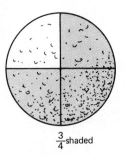

$\frac{3}{4}$ shaded

If each orange is divided into 4 equal pieces, and each child receives 3 of these pieces, the oranges have been shared equally.

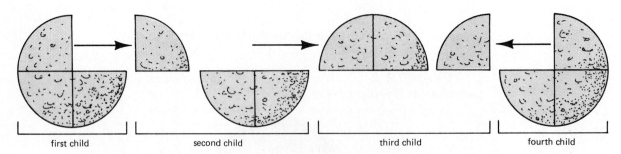

first child   second child   third child   fourth child

Therefore, each child should get $\dfrac{3}{4}$ of an orange.

Caution!   In this example there are 12 slices altogether. You might think that the denominator used to represent each child's portion is 12. I agree that there

are 12 pieces, but each *whole* is made up of 4 slices; therefore, I am working with fourths, not twelfths, of a whole! The denominator is 4.

EXAMPLE 4: Each of four whole candy bars is divided into 3 equal pieces. What fraction represents the shaded portion?

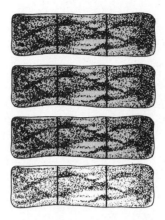

Because each whole is divided into 3 equal pieces, the denominator is 3. If you thought it should be 12, notice that 3 pieces make up each *whole*; therefore, these pieces are thirds, not twelfths.

Because 9 pieces are shaded, the numerator is 9.

$$\begin{array}{l} 9 \text{ of} \\ 3 \text{ equal parts} \\ \text{of a whole} \\ \text{candy bar} \end{array} \quad \frac{9}{3} \qquad \text{Read:} \quad \text{nine-thirds.}$$

Therefore, $\frac{9}{3}$ represents the shaded portion.

But looking at the candy bars, you can see that 3 whole bars are shaded. Therefore, $\frac{9}{3}$ should equal 3.

Viewing the fraction as division, the same conclusion is reached.

$$\frac{9}{3} = 9 \div 3$$
$$= 3$$

Try to work these problems before looking at the solution.

Examples	Solutions
What are the numerator and denominator of the fraction $\frac{5}{9}$?	$5 \longleftarrow$ numerator $9 \longleftarrow$ denominator

Examples	Solutions
What fraction represents the shaded portion?	3 of 5 equal parts of a whole → $\dfrac{3}{5}$ Therefore, $\dfrac{3}{5}$ represents the shaded portion.
If a tangerine is divided into 7 equal slices, how much remains if 4 slices are eaten?	Three-sevenths, or $\dfrac{3}{7}$, of the tangerine remains.
How can 7 people equally share 3 whole pies?	$3 \div 7 = \dfrac{3}{7}$ To interpret this answer, $\dfrac{3}{7}$ means 3 of 7 equal parts of a whole pie Therefore, each of the pies should be divided into 7 slices. Each person should get 3 of these slices.

Now try these—no hints given.

What fraction represents the shaded portion?	$\dfrac{5}{6}$ Read: five-sixths.
How can 3 people equally share 2 candy bars?	Each should get $\dfrac{2}{3}$ of a whole candy bar.

If you had difficulty with the last two problems, please talk with an instructor or read this section again.

Exercises
Part I

Answers to these exercises can be found on page 277.

Write a fraction that represents each of the shaded portions.

1.

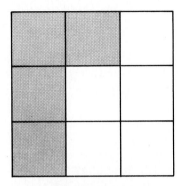

2.

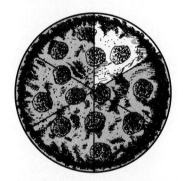

3.

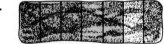

4. A book is divided into 8 equal sections. How much of the book remains to be read if a student has read 3 sections?

5. A book of stamps contains 20 stamps. If I have used 17 stamps, how much of the book is left?

6. How can 3 people equally share 1 peanut butter sandwich?

7. How can 7 people equally share 2 bananas?

Exercises
Part II

Write a fraction that represents each of the shaded portions.

1.

2.

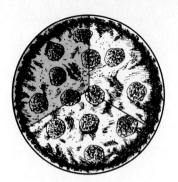

3. A book has 6 chapters. If a student has read 5 chapters, how much of the book remains to be read?
4. How can 5 people equally share 3 pizzas?

SECTION 6.2
PROPER FRACTIONS, IMPROPER FRACTIONS, AND MIXED NUMBERS

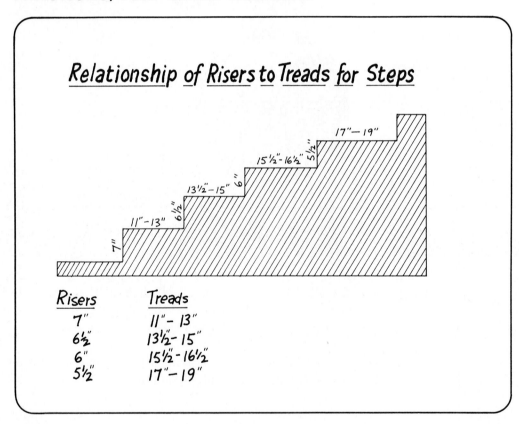

Relationship of Risers to Treads for Steps

Risers	Treads
7″	11″ – 13″
6½″	13½″– 15″
6″	15½″– 16½″
5½″	17″– 19″

This section explains two conversions that are necessary for working with fractions, as illustrated by these two examples.

Improper
Fraction to
Mixed Number

$$\frac{11}{9} = 11 \div 9$$

$$= 1\frac{2}{9}$$

Mixed Number
to Improper
Fraction

$$3\frac{2}{9} = \frac{27 + 2}{9}$$

$$= \frac{29}{9}$$

**Proper
and Improper
Fractions**

A **proper fraction** is a fraction whose numerator is less than its denominator. The following are proper fractions.

$$\frac{2}{9}, \quad \frac{5}{12}, \quad \frac{7}{8}, \quad \frac{19}{25}$$

An **improper fraction** is a fraction whose numerator is larger than or equal to its denominator. The following are improper fractions.

$$\frac{11}{9}, \quad \frac{17}{12}, \quad \frac{8}{8}, \quad \frac{39}{25}$$

Mixed Numbers What fraction represents the shaded parts of these pizzas?

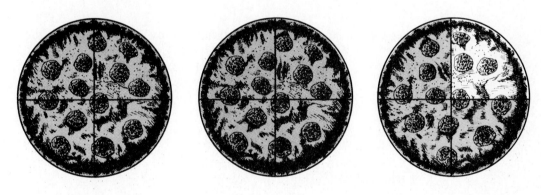

Each whole pizza has been divided into fourths. The shaded parts can be represented by the fraction $\frac{11}{4}$.

11 of
4 equal parts $\frac{11}{4}$
of a whole
pizza

I can also express the shaded parts as 2 wholes plus $\frac{3}{4}$ of a whole pizza, so $2 + \frac{3}{4}$ also represents the shaded portion.

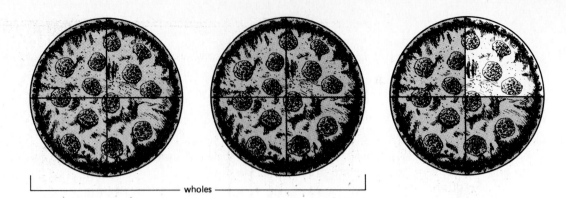

wholes

In arithmetic, $2 + \dfrac{3}{4}$ is written without the addition sign as $2\dfrac{3}{4}$. It is read and understood to mean two and three-fourths. A number like $2\dfrac{3}{4}$ is called a **mixed number** because it has both a whole number part, 2, and a fractional part, $\dfrac{3}{4}$.

In the example just discussed, there were two ways to express the shaded parts of the pizzas—as an improper fraction and as a mixed number. The expressions are equal because they represent the same parts. That is,

$$\frac{11}{4} = 2\frac{3}{4}$$

You can change $\dfrac{11}{4}$ to a mixed number by interpreting the fraction as division.

$$\frac{11}{4} = 11 \div 4 \qquad 4\overline{\big)\,11\,} \begin{array}{r} 2\ R3 \\ \hline 8 \\ \hline 3 \end{array}$$

This division gives the whole number 2. If you divide the remainder 3 by the denominator 4

$$3 \div 4$$

and express it as a fraction, the fractional part is $\dfrac{3}{4}$.

$$3 \div 4 = \frac{3}{4}$$

Therefore by division,

$$\frac{11}{4} = 2\frac{3}{4}$$

Any improper fraction can be changed to a mixed number by division.

> **Changing an Improper Fraction to a Mixed Number**
>
> 1. Divide the numerator by the denominator.
> 2. Write the quotient, which is the whole-number part of the mixed number.
> 3. Write the remainder over the denominator (divisor). This is the fractional part of the mixed number.

EXAMPLE 1: Change $\dfrac{11}{9}$ to a mixed number.

Divide numerator by denominator.

$$\frac{11}{9} = 11 \div 9 \qquad 9\overline{)\,11\,}\;\;{}^{1\,R2}$$
$$\frac{9}{2}$$

Therefore,

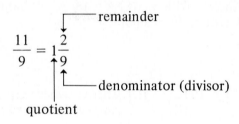

ninths of
two wholes

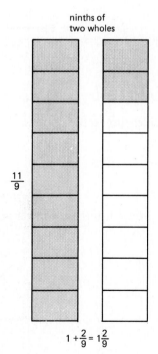

$$1 + \tfrac{2}{9} = 1\tfrac{2}{9}$$

EXAMPLE 2: Change $\dfrac{27}{12}$ to a mixed number.

Divide numerator by denominator.

$$\frac{29}{12} = 29 \div 12 \qquad 12\overline{)\,29\,}\ ^{2\ R5}$$
$$\frac{24}{5}$$

$$= \text{quotient}\ \frac{\text{remainder}}{\text{denominator}}$$

$$= 2\frac{5}{12}$$

EXAMPLE 3:   Change $\dfrac{53}{25}$ to a mixed number.

$$\frac{53}{25} = 53 \div 25$$

$$= 2\frac{3}{25}$$ A quotient of 2, with a remainder of 3, and a divisor of 25.

 Note: With a division of whole numbers, the remainder can be written as a fraction. For example,

$$12\overline{)\,29\,}\ ^{2\frac{5}{12}}$$
$$\frac{24}{5}$$

The remainder is placed over the divisor, giving a mixed number for the quotient.

Therefore, $29 \div 12 = 12\overline{)\,29\,} = 2\dfrac{5}{12}$.

EXAMPLE 4:   Change $\dfrac{8}{8}$ to a mixed number.

Divide numerator by denominator.

$$\frac{8}{8} = 8 \div 8$$

$$= 1$$

This is a whole number, *not* a mixed number.

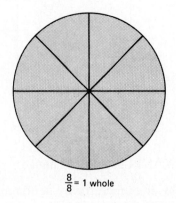

$\frac{8}{8}$ = 1 whole

EXAMPLE 5:   Change $\dfrac{10}{10}$ to a whole or mixed number.

$$\dfrac{10}{10} = 10 \div 10$$

$$= 1$$

EXAMPLE 6:   Change $\dfrac{18}{3}$ to a whole or mixed number.

$$\dfrac{18}{3} = 18 \div 3$$

$$= 6$$

Whole Numbers and Mixed Numbers to Fractions   By division,

$$\dfrac{4}{1} = 4 \div 1 = 4$$

$$\dfrac{6}{1} = 6 \div 1 = 6$$

$$\dfrac{29}{1} = 29 \div 1 = 29$$

Therefore,

$$4 = \dfrac{4}{1}$$

$$6 = \dfrac{6}{1}$$

$$29 = \dfrac{29}{1}$$

Any whole number can be written as a fraction with 1 as the denominator.

EXAMPLE 7:   Write 5 as a fraction.
$$5 = 5 \div 1$$
$$= \dfrac{5}{1}$$

EXAMPLE 8:   Write 69 as a fraction.
$$69 = 69 \div 1$$
$$= \dfrac{69}{1}$$

What mixed number represents the shaded region below?

The shaded area is $3\dfrac{2}{9}$.

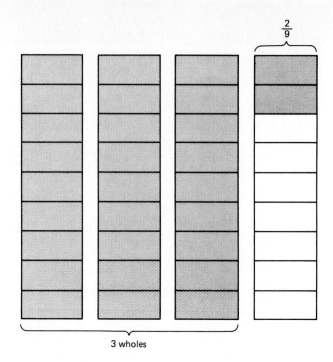

3 wholes

If you count ninths, you will see that there are $\frac{29}{9}$ shaded. Therefore, $3\frac{2}{9} = \frac{29}{9}$.

You can change any mixed number to an improper fraction. The procedure is similar to the one used for checking a division problem.

EXAMPLE 9: Write $3\frac{2}{9}$ as an improper fraction.

Remember that a mixed number was calculated from an improper fraction by division.

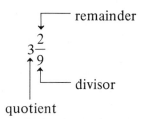

To change a mixed number to an improper fraction, reverse this procedure. *Multiply the divisor times the quotient and add the remainder.*

In this case,

| divisor × quotient + remainder |

$$3\frac{2}{9} = \frac{(9 \times 3) + 2}{9}$$

| denominator |

$$= \frac{27 + 2}{9} \qquad \text{Multiply } 9 \times 3.$$

$$= \frac{29}{9} \qquad \text{Add 2.}$$

EXAMPLE 10:   Write $4\frac{5}{9}$ as an improper fraction.

$$4\frac{5}{9} = \frac{(9 \times 4) + 5}{9}$$

$$= \frac{36 + 5}{9}$$

$$= \frac{41}{9}$$

Changing a Mixed Number to an Improper Fraction

1. Multiply the denominator of the fraction times the whole number.
2. To this product add the numerator of the fraction.
3. Write this result over the denominator of the fraction.

EXAMPLE 11:   Write $6\frac{2}{5}$ as an improper fraction.

$$6\frac{2}{5} = \frac{(5 \times 6) + 2}{5}$$

$$= \frac{30 + 2}{5}$$

$$= \frac{32}{5}$$

EXAMPLE 12:   Write $4\frac{2}{3}$ as an improper fraction, then change it back to a mixed number.

$$4\frac{2}{3} = \frac{(3 \times 4) + 2}{3}$$

$$= \frac{12 + 2}{3}$$

$$= \frac{14}{3}$$

$$\frac{14}{3} = 14 \div 3$$

$$= 4\frac{2}{3}$$ A quotient of 4, with a remainder of 2, and a divisor of 3.

Now it is your turn!

Examples	Solutions
Change $\frac{11}{9}$ to a whole or mixed number.	$\frac{11}{9} = 11 \div 9$ $= 1\frac{2}{9}$ A quotient of 1, with a remainder of 2, and a divisor of 9.

Examples	Solutions
Change $\dfrac{15}{3}$ to a whole or mixed number.	$\dfrac{15}{3} = 15 \div 3$ $\qquad = 5$
Change $\dfrac{9}{9}$ to a whole number.	$\dfrac{9}{9} = 9 \div 9$ $\qquad = 1$
Change $7\dfrac{3}{5}$ to an improper fraction.	$7\dfrac{3}{5} = \dfrac{(5 \times 7) + 3}{5}$ $\qquad = \dfrac{35 + 3}{5}$ $\qquad = \dfrac{38}{5}$
Change $11\dfrac{2}{3}$ to an improper fraction.	$11\dfrac{2}{3} = \dfrac{(3 \times 11) + 2}{3}$ $\qquad = \dfrac{33 + 2}{3}$ $\qquad = \dfrac{35}{3}$

Now try these—no hints given.

Change to a whole or mixed number. (a) $\dfrac{62}{5}$ (b) $\dfrac{35}{7}$ (c) $\dfrac{12}{12}$	(a) $12\dfrac{2}{5}$ (b) 5 (c) 1
Change $6\dfrac{7}{12}$ to an improper fraction.	$\dfrac{79}{12}$

If you had difficulty with the last four problems, talk with an instructor or read this section again.

Exercises
Part I

Answers to these exercises can be found on page 277.

Write each as a whole or mixed number.

1. $\dfrac{23}{4}$ 2. $\dfrac{78}{78}$ 3. $\dfrac{27}{3}$ 4. $\dfrac{23}{6}$

Write each as an improper fraction.

5. $9\frac{2}{5}$ 6. $12\frac{3}{8}$ 7. $11\frac{3}{5}$ 8. $13\frac{2}{7}$

Divide and write the quotient as a mixed number.

9. $345 \div 4$ 10. $125{,}839 \div 15$

Exercises Part II

Write each as a whole or mixed number.

1. $\frac{59}{4}$ 2. $\frac{39}{3}$

Write each as an improper fraction.

3. $4\frac{2}{7}$ 4. $8\frac{5}{9}$

Divide and write the quotient as a mixed number.

5. $956 \div 14$

**SECTION 6.3
BUILDING FRACTIONS**

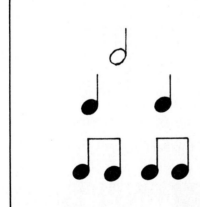

After reading this section, you will know how to build a fraction to an equivalent, or equal, fraction with a different denominator. For example,

$$\frac{2}{5} = \frac{2 \times 7}{5 \times 7}$$

$$= \frac{14}{35}$$

Building Fractions

Building fractions means writing an equivalent fraction with a greater numerator and denominator.

       EXAMPLE 1:   Suppose I cut a candy bar into 2 equal parts and give you 1 piece.

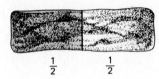

$$\frac{1}{2} \qquad \frac{1}{2}$$

I have given you $\frac{1}{2}$ of the bar.

       EXAMPLE 2:   Had I cut the bar into 4 equal parts, how much would I have to give you to equal $\frac{1}{2}$ of the candy bar?

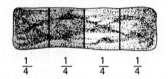

$$\frac{1}{4} \quad \frac{1}{4} \quad \frac{1}{4} \quad \frac{1}{4}$$

From the drawing, you can see that I would have to give you 2 of the 4 equal pieces of the whole bar, or $\frac{2}{4}$ of the bar. Therefore, $\frac{1}{2} = \frac{2}{4}$.

    To keep your portion of the candy the same when I double the number of equal parts, I also have to double the number of pieces I give you. This is indicated by multiplying the numerator and denominator by 2.

$$\frac{1}{2} = \frac{1 \times 2}{2 \times 2}$$

$$= \frac{2}{4}$$

       EXAMPLE 3:   Had I cut the bar into 6 equal parts, how much would I have to give you to equal $\frac{1}{2}$ of the bar?

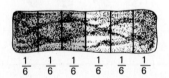

$$\frac{1}{6} \quad \frac{1}{6} \quad \frac{1}{6} \quad \frac{1}{6} \quad \frac{1}{6} \quad \frac{1}{6}$$

I would have to give you $\frac{3}{6}$ of the bar. Therefore, $\frac{1}{2} = \frac{3}{6}$.

    To keep your portion of the candy bar the same when I multiply the number of equal parts by 3, I also have to multiply the number of pieces I give you by 3.

$$\frac{1}{2} = \frac{1 \times 3}{2 \times 3}$$

$$= \frac{3}{6}$$

EXAMPLE 4:   Build $\dfrac{2}{3}$ by multiplying with 5.

$$\frac{2}{3} = \frac{2 \times 5}{3 \times 5}$$

Both numerator and denominator are multiplied by the same number.

$$= \frac{10}{15}$$

EXAMPLE 5:   Build $\dfrac{2}{3}$ by multiplying with 9.

$$\frac{2}{3} = \frac{2 \times 9}{3 \times 9}$$

$$= \frac{18}{27}$$

EXAMPLE 6:   Build $\dfrac{2}{5}$ to an equal fraction with a denominator of 35.

$$\frac{2}{5} = \frac{?}{35}$$

I need to multiply both the numerator and denominator by the same number and get 35 in the denominator.

$$\frac{2}{5} = \frac{2 \times \square}{5 \times \square}$$

$$= \frac{?}{35}$$

The number can be found by division of denominators.

$$\frac{2}{5} = \frac{?}{35}$$

5 divides 35 seven times.

Therefore, by multiplying both numerator and denominator by 7,

$$\frac{2}{5} = \frac{2 \times 7}{5 \times 7}$$

$$= \frac{14}{35}$$

EXAMPLE 7:   Build $\dfrac{3}{11}$ to an equal fraction with a denominator of 33.

$$\frac{3}{11} = \frac{?}{33}$$

Because 11 divides 33 three times, use 3 to build the fraction.

$$\frac{3}{11} = \frac{3 \times 3}{11 \times 3}$$

$$= \frac{9}{33}$$

EXAMPLE 8:   Build $\frac{5}{7}$ to an equal fraction with a denominator of 91.
Because $91 \div 7 = 13$,

$$\frac{5}{7} = \frac{5 \times 13}{7 \times 13}$$

$$= \frac{65}{91}$$

EXAMPLE 9:   A ream of paper contains 500 sheets of paper. How many sheets represent $\frac{2}{5}$ of the ream?

$$\frac{2}{5} = \frac{?}{500}$$ How many of the 500 sheets of the ream?

Because 5 divides 500 one hundred times, I use 100 to build.

$$\frac{2}{5} = \frac{2 \times 100}{5 \times 100}$$

$$= \frac{200}{500}$$

Therefore, 200 sheets make up $\frac{2}{5}$ of a ream.

EXAMPLE 10:   A book contains 124 pages. How many pages make up $\frac{1}{4}$ of the book?

$$\frac{1}{4} = \frac{?}{124}$$

Use 31 to build because $124 \div 4 = 31$.

$$\frac{1}{4} = \frac{1 \times 31}{4 \times 31}$$

$$= \frac{31}{124}$$

Therefore, 31 pages make up $\frac{1}{4}$ of the book.

Try to answer these problems before looking at the solutions.

Examples	Solutions
Build $\dfrac{5}{8}$ by multiplying with 6.	$\dfrac{5}{8} = \dfrac{5 \times 6}{8 \times 6}$ Both numerator and denominator multiplied by 6. $= \dfrac{30}{48}$
Build $\dfrac{2}{7}$ to an equal fraction with a denominator of 63.	$\dfrac{2}{7} = \dfrac{?}{63}$ Because $63 \div 7 = 9$, use 9 to build. $\dfrac{2}{7} = \dfrac{2 \times 9}{7 \times 9}$ $= \dfrac{18}{63}$
Build $\dfrac{5}{6}$ to an equal fraction with a denominator of 90.	$\dfrac{5}{6} = \dfrac{?}{90}$ Because $90 \div 6 = 15$, $\dfrac{5}{6} = \dfrac{5 \times 15}{6 \times 15}$ $= \dfrac{75}{90}$
An exercise section contains 27 problems. How many problems make up $\dfrac{2}{3}$ of the exercises?	I must find how many problems of the 27 equal $\dfrac{2}{3}$ of the exercises. $\dfrac{2}{3} = \dfrac{?}{27}$ Because $27 \div 3 = 9$, $\dfrac{2}{3} = \dfrac{2 \times 9}{3 \times 9}$ $= \dfrac{18}{27}$ Therefore, 18 problems make up $\dfrac{2}{3}$ of the exercises.

Now try these—no hints given.

Build $\dfrac{2}{9}$ to a denominator of 36.	$\dfrac{8}{36}$

Examples	Solutions
Build $\dfrac{3}{4}$ to a denominator of 52.	$\dfrac{39}{52}$
A half-ream package contains 250 sheets of paper. How many sheets represent $\dfrac{2}{5}$ of this package?	100 sheets

If you had difficulty with the last three problems, please talk with an instructor or read this section again.

Exercises Part I

Answers to these exercises can be found on page 277.

Build each fraction as indicated.

1. $\dfrac{4}{7}$ to the denominator 56

2. $\dfrac{2}{3}$ to the denominator 30

3. $\dfrac{2}{13}$ to the denominator 65

4. $\dfrac{2}{21}$ to the denominator 147

5. $\dfrac{3}{19}$ to the denominator 209

6. $\dfrac{12}{15}$ to the denominator 90

7. $\dfrac{7}{22}$ to the denominator 330

8. $\dfrac{4}{5}$ to the denominator 195

9. A book contains 126 pages. How many pages make up $\dfrac{1}{3}$ of the book?

10. A ream contains 500 sheets of paper. How many sheets represent $\dfrac{7}{10}$ of the ream?

Exercises Part II

Build each fraction as indicated.

1. $\dfrac{5}{6}$ to the denominator 36

2. $\dfrac{2}{7}$ to the denominator 63

3. $\dfrac{5}{9}$ to the denominator 117

4. $\dfrac{3}{23}$ to the denominator 184

5. A book contains 95 pages. How many pages represent $\dfrac{4}{5}$ of the book?

LAG BOLTS

DIAMETER (IN INCHES)	LENGTH (IN INCHES)
4/16 OR 1/4	1 — 6
5/16	1 — 10
6/16 OR 3/8	1 — 12
7/16	1 — 12
8/16 OR 1/2	1 — 12
10/16 OR 5/16	1½— 16
12/16 OR 3/4	1½— 16
14/16 OR 7/8	2 — 16
16/16 OR 1	2 — 16

This section describes the procedure for reducing fractions to lowest terms, as illustrated in this example.

$$\frac{24}{40} = \frac{\overset{\displaystyle 3}{\overset{\displaystyle \cancel{6}}{\cancel{24}}}}{\underset{\displaystyle 5}{\underset{\displaystyle \cancel{10}}{\cancel{40}}}}$$

$$= \frac{3}{5}$$

Reducing **Reducing** a fraction means writing an equivalent fraction with a smaller numerator and denominator. Reducing is the *opposite* of building, so you use the opposite operation.

Reducing Fractions

To reduce a fraction to a smaller numerator and denominator, divide both by the same nonzero number.

EXAMPLE 1:

	Building	Reducing

$$\frac{2}{9} = \frac{2 \times 5}{9 \times 5} \qquad \frac{10}{45} = \frac{10 \div 5}{45 \div 5}$$

$$= \frac{10}{45} \qquad\qquad = \frac{2}{9}$$

EXAMPLE 2:   Reduce $\dfrac{14}{21}$.

Because both 14 and 21 are divisible by 7,

$$\frac{14}{21} = \frac{14 \div 7}{21 \div 7}$$

$$= \frac{2}{3}$$

EXAMPLE 3:   Reduce $\dfrac{24}{40}$.

Both 24 and 40 are divisible by 4.

$$\frac{24}{40} = \frac{24 \div 4}{40 \div 4}$$

$$= \frac{6}{10}$$

But 6 and 10 are divisible by 2.

$$= \frac{6 \div 2}{10 \div 2}$$

$$= \frac{3}{5}$$

Therefore,

$$\frac{24}{40} = \frac{3}{5}$$

This reduction could have been done in one step if I had divided by 8.

$$\frac{24}{40} = \frac{24 \div 8}{40 \div 8}$$

$$= \frac{3}{5}$$

In this example, $\dfrac{24}{40}$ was reduced to both $\dfrac{6}{10}$ and $\dfrac{3}{5}$. In the first case, $\dfrac{6}{10}$ could be further reduced; but in the second case, no number (other than 1) divides both the 3 and 5 of $\dfrac{3}{5}$. The fraction $\dfrac{3}{5}$ is said to be in **lowest terms**.

Lowest Terms

A fraction is in lowest terms when it cannot be further reduced—that is, when the numerator and denominator have no common divisor other than 1.

EXAMPLE 4:   Reduce $\dfrac{12}{16}$ to lowest terms.

Dividing first by 2,

$$\frac{12}{16} = \frac{12 \div 2}{16 \div 2}$$

$$= \frac{6}{8}$$

Dividing by 2 again,

$$= \frac{6 \div 2}{8 \div 2}$$

$$= \frac{3}{4}$$

The fraction $\dfrac{3}{4}$ is in lowest terms because 3 and 4 have no common divisor other than 1.

The fraction $\dfrac{12}{16}$ could have been reduced to lowest terms in one step by dividing by 4.

$$\frac{12}{16} = \frac{12 \div 4}{16 \div 4}$$

$$= \frac{3}{4}$$

Which method is correct?   Both!   There is only *one* final, reduced answer, but there are many ways to get that answer.

 Caution!   Both numerator and denominator must be divided by the same number.

I have been showing the divisor used to reduce the fraction.   The division of the numerator and denominator by a common divisor is usually done mentally and only the quotients are recorded.   For example, to reduce $\dfrac{24}{40}$ I do the following.

Cross out the numerator and denominator and write the quotients after dividing both by 4.

$$\frac{24}{40} = \frac{\overset{6}{\cancel{24}}}{\underset{10}{\cancel{40}}} \qquad \begin{array}{l} 24 \div 4 \\ 40 \div 4 \end{array}$$

Divide by 2 mentally.

$$\frac{24}{40} = \frac{\overset{3}{\overset{\cancel{6}}{\cancel{24}}}}{\underset{5}{\underset{\cancel{10}}{\cancel{40}}}} \qquad \begin{array}{l} 6 \div 2 \\ 10 \div 2 \end{array}$$

$$= \frac{3}{5}$$

Here is Example 4 repeated using this method. Dividing by 2,

$$\frac{12}{16} = \frac{\overset{6}{\cancel{12}}}{\underset{8}{\cancel{16}}} \qquad \begin{array}{l} 12 \div 2 \\ 16 \div 2 \end{array}$$

then by 2 again.

$$\frac{12}{16} = \frac{\overset{3}{\overset{\cancel{6}}{\cancel{12}}}}{\underset{4}{\underset{\cancel{8}}{\cancel{16}}}} \qquad \begin{array}{l} 6 \div 2 \\ 8 \div 2 \end{array}$$

Therefore, $\dfrac{12}{16} = \dfrac{3}{4}$ in lowest terms.

EXAMPLE 5:   Reduce $\dfrac{30}{72}$ to lowest terms.

$$\frac{30}{72} = \frac{\overset{15}{\cancel{30}}}{\underset{36}{\cancel{72}}} \qquad \text{Dividing by 2.}$$

$$\frac{\overset{5}{\overset{\cancel{15}}{\cancel{30}}}}{\underset{12}{\underset{\cancel{36}}{\cancel{72}}}} \qquad \text{Dividing by 3.}$$

Therefore, $\dfrac{30}{72} = \dfrac{5}{12}$ in lowest terms.

EXAMPLE 6:   Reduce $\dfrac{28}{48}$ to lowest terms.

You might think to divide by 4.

$$\frac{28}{48} = \frac{\overset{7}{\cancel{28}}}{\underset{12}{\cancel{48}}}$$

Or you might divide by 2 twice.

$$\frac{28}{48} = \frac{\overset{\overset{\displaystyle 7}{\cancel{14}}}{\cancel{28}}}{\underset{\underset{\displaystyle 12}{\cancel{24}}}{\cancel{48}}}$$

In either case,

$$\frac{28}{48} = \frac{7}{12} \qquad \text{in lowest terms}$$

because 7 and 12 have no common divisor other than 1.

In the next chapter, I shall return to a discussion of reduction and explain a method to reduce fractions with very large numerators and denominators. But for now, consider three divisibility tests that can be used to save time when reducing fractions.

Divisibility Tests

Test One Any number is divisible (meaning no remainder) by 2 if its ones digit is 0, 2, 4, 6, or 8.

Test	Check
42 is divisible by 2 because the ones digit is 2:	$42 \div 2 = 21$
758 is divisible by 2 because the ones digit is 8:	$758 \div 2 = 379$
6,956 is divisible by 2 because the ones digit is 6:	$6{,}956 \div 2 = 3{,}478$

Test	Check
41 is *not* divisible by 2:	$41 \div 2 = 20 \text{ R1}$
757 is *not* divisible by 2:	$757 \div 2 = 378 \text{ R1}$
6,955 is *not* divisible by 2:	$6{,}955 \div 2 = 3{,}477 \text{ R1}$

Test Two Any number is divisible by 3 if the sum of its digits is divisible by 3.

57 is divisible by 3 because the sum of the digits 5 and 7 is 12, and 12 is divisible by 3

474 is divisible by 3 because $4 + 7 + 4 = 15$, and 15 is divisible by 3

5,955 is divisible by 3 because $5 + 9 + 5 + 5 = 24$, and 24 is divisible by 3

58 is *not* divisible by 3 because $5 + 8 = 13$, and 13 is *not* divisible by 3

217 is *not* divisible by 3 because $2 + 1 + 7 = 10$, and 10 is *not* divisible by 3

Use long or short division to check that these conclusions are correct. Please do not worry about why these tests work! Just consider them tricks of the (arithmetic) trade.

Test Three   Any number is divisible by 5 if its last digit is either 5 or 0.

250 is divisible by 5 because its last digit is 0

4,755 is divisible by 5 because its last digit is 5

73,493 is *not* divisible by 5 because its last digit is *not* 0 or 5

Again, you can check by division to see that these conclusions are indeed correct!

EXAMPLE 7:   Reduce $\dfrac{64}{480}$ to lowest terms.

If the last digits are 0, 2, 4, 6, or 8, the numbers are divisible by 2. Therefore, both 64 and 480 are divisible by 2.

$$\frac{64}{480} = \frac{\overset{32}{\cancel{64}}}{\underset{240}{\cancel{480}}} \qquad \begin{array}{l} 64 \div 2 = 32 \\ 480 \div 2 = 240 \end{array}$$

The numbers 32 and 240 have last digits 2 and 0. Therefore, each is divisible by 2.

$$\frac{64}{480} = \frac{\overset{16}{\cancel{\overset{32}{\cancel{64}}}}}{\underset{120}{\cancel{\underset{240}{\cancel{480}}}}} \qquad \begin{array}{l} 32 \div 2 = 16 \\ \\ 240 \div 2 = 120 \end{array}$$

Both 16 and 120 are divisible by 2.

$$\frac{64}{480} = \frac{\overset{8}{\cancel{\overset{16}{\cancel{\overset{32}{\cancel{64}}}}}}}{\underset{60}{\cancel{\underset{120}{\cancel{\underset{240}{\cancel{480}}}}}}} \qquad \begin{array}{l} 16 \div 2 = 8 \\ \\ \\ 120 \div 2 = 60 \end{array}$$

Again, these numbers are divisible by 2.

$$\frac{64}{480} = \frac{\overset{4}{\cancel{\overset{8}{\cancel{\overset{16}{\cancel{\overset{32}{\cancel{64}}}}}}}}}{\underset{30}{\cancel{\underset{60}{\cancel{\underset{120}{\cancel{\underset{240}{\cancel{480}}}}}}}}} \qquad \begin{array}{l} 8 \div 2 = 4 \\ \\ \\ \\ 60 \div 2 = 30 \end{array}$$

Dividing by 2 again,

$$\frac{64}{480} = \frac{\begin{array}{c} 2 \\ \cancel{4} \\ \cancel{8} \\ \cancel{16} \\ \cancel{32} \\ \cancel{64} \end{array}}{\begin{array}{c} \cancel{480} \\ \cancel{240} \\ \cancel{120} \\ \cancel{60} \\ \cancel{30} \\ 15 \end{array}}$$

$4 \div 2 = 2$

$30 \div 2 = 15$

Therefore, $\dfrac{64}{480} = \dfrac{2}{15}$ in lowest terms.

EXAMPLE 8:   Reduce $\dfrac{150}{360}$ to lowest terms.

Both 150 and 360 are divisible by 2.

$$\frac{150}{360} = \frac{\begin{array}{c} 75 \\ \cancel{150} \end{array}}{\begin{array}{c} \cancel{360} \\ 180 \end{array}}$$

The number 75 does *not* end in the digits 0, 2, 4, 6, or 8, so the numerator is *not* divisible by 2.

Next, sum the digits to check if both are divisible by 3.

$\boxed{7 \quad 5}$

$7 + 5 = 12,$ and 12 is divisible by 3.

$\boxed{1 \quad 8 \quad 0}$

$1 + 8 + 0 = 9,$ and 9 is divisible by 3.

Therefore, both 75 and 180 are divisible by 3.

$$\frac{150}{360} = \frac{\begin{array}{c} 25 \\ \cancel{75} \\ \cancel{150} \end{array}}{\begin{array}{c} \cancel{360} \\ \cancel{180} \\ 60 \end{array}}$$

The number 25 is *not* divisible by 3 because $2 + 5 = 7$, and 7 is *not* divisible by 3. But both 25 and 60 are divisible by 5 because the last digits are 5 and 0, respectively.

$$\frac{150}{360} = \frac{\begin{array}{c} 5 \\ \cancel{25} \\ \cancel{75} \\ \cancel{150} \end{array}}{\begin{array}{c} \cancel{360} \\ \cancel{180} \\ \cancel{60} \\ 12 \end{array}}$$

Therefore, $\dfrac{150}{360} = \dfrac{5}{12}$ in lowest terms because 5 and 12 have no common divisor other than 1.

Now you try to reduce the following fractions to lowest terms. The most common mistake is to forget that *both* numerator and denominator must be divided by the *same* number.

Examples	Solutions
$\dfrac{56}{63}$	Both 56 and 63 are divisible by 7. $\dfrac{56}{63} = \dfrac{\overset{8}{\cancel{56}}}{\underset{9}{\cancel{63}}}$ $= \dfrac{8}{9}$
$\dfrac{15}{35}$	Both 15 and 35 are divisible by 5. $\dfrac{15}{35} = \dfrac{\overset{3}{\cancel{15}}}{\underset{7}{\cancel{35}}}$ $= \dfrac{3}{7}$
$\dfrac{30}{72}$	Divide both numerator and denominator by 6, or by 2 and then 3. In either case, $\dfrac{30}{72} = \dfrac{5}{12}$
$\dfrac{45}{144}$	These numbers are harder to work with, so I shall use the divisibility tests. The number 45 does not have the last digit 0, 2, 4, 6, or 8. Therefore, I cannot use 2. $\boxed{4\ \ 5}\qquad\boxed{1\ \ 4\ \ 4}$ $4 + 5 = 9\qquad 1 + 4 + 4 = 9$ Because the sums of the digits are divisible by 3, so are 45 and 144. $\dfrac{45}{144} = \dfrac{\overset{15}{\cancel{45}}}{\underset{48}{\cancel{144}}}$

Examples	Solutions

<table>
<tr><td></td><td align="center">

┌─────┐   ┌─────┐
│ 1 5 │   │ 4 8 │
└─────┘   └─────┘
   ↓ ↓   ↓ ↓
1 + 5 = 6   4 + 8 = 12

Therefore, 3 divides both numerator and denominator again.

$$\frac{45}{144} = \frac{\overset{5}{\cancel{\underset{48}{\cancel{15}}}}\cancel{45}}{\underset{16}{\cancel{\underset{48}{\cancel{144}}}}}$$

Therefore, $\dfrac{45}{144} = \dfrac{5}{16}$ in lowest terms.
</td></tr>
<tr><td>

$$\frac{150}{225}$$
</td><td>

The number 225 does not end in 0, 2, 4, 6, or 8. Therefore, I cannot use 2.

Because the sum of the digits for both numbers is divisible by 3, 3 divides both 150 and 225 evenly.

$$\frac{150}{225} = \frac{\overset{50}{\cancel{150}}}{\underset{75}{\cancel{225}}}$$

Both 50 and 75 end in 0 or 5. Dividing by 5,

$$\frac{150}{225} = \frac{\overset{10}{\cancel{\overset{50}{\cancel{150}}}}}{\underset{15}{\cancel{\underset{75}{\cancel{225}}}}}$$

Dividing by 5 again,

$$\frac{150}{225} = \frac{\overset{2}{\cancel{\overset{10}{\cancel{\overset{50}{\cancel{150}}}}}}}{\underset{3}{\cancel{\underset{15}{\cancel{\underset{75}{\cancel{225}}}}}}}$$

$$= \frac{2}{3}$$

Therefore, $\dfrac{150}{225} = \dfrac{2}{3}$ in lowest terms.
</td></tr>
</table>

Examples	Solutions
Now try these—no hints given.	
Reduce to lowest terms.	
(a) $\dfrac{25}{35}$	(a) $\dfrac{5}{7}$
(b) $\dfrac{30}{90}$	(b) $\dfrac{1}{3}$
(c) $\dfrac{150}{375}$	(c) $\dfrac{2}{5}$

If you had difficulty with the last three problems, please talk with an instructor or read this section again.

Exercises Part I   Answers to these exercises can be found on page 278.

Reduce each to lowest terms.

1. $\dfrac{32}{56}$ 2. $\dfrac{20}{30}$ 3. $\dfrac{10}{65}$ 4. $\dfrac{72}{90}$ 5. $\dfrac{105}{330}$ 6. $\dfrac{150}{480}$

7. $\dfrac{12}{35}$ 8. $\dfrac{96}{210}$

Exercises Part II   Reduce each to lowest terms.

1. $\dfrac{20}{36}$ 2. $\dfrac{30}{45}$ 3. $\dfrac{240}{300}$ 4. $\dfrac{252}{1,512}$

SECTION 6.5
EQUALITY

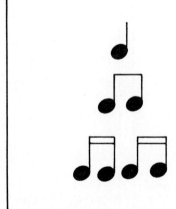

1 quarter note equals

2 eighth notes, or

4 sixteenth notes in duration.

This section gives an easy way to tell if two fractions are equal by comparing cross products.

$$\frac{2}{4} \diagdown\!\!\!\diagup \frac{3}{6} \longrightarrow 4 \times 3 = 12$$
$$\longrightarrow 2 \times 6 = 12$$

Equality   As we have already seen, by building or reducing, you obtain an **equal**, or **equivalent**, fraction. For example,

$$\frac{2}{4} = \frac{2 \div 2}{4 \div 2}$$

$$= \frac{1}{2}$$

Therefore, $\frac{2}{4} = \frac{1}{2}$. By building,

$$\frac{1}{2} = \frac{3}{6} = \frac{4}{8} = \frac{5}{10} = \frac{6}{12} = \cdots$$

The list is endless.

To check if two fractions are equal, see if their **cross products** are equal.

$$\frac{2}{4} \diagdown\!\!\!\diagup \frac{1}{2} \longrightarrow 4 \times 1 = 4 \qquad \frac{2}{4} \diagdown\!\!\!\diagup \frac{1}{2} \longrightarrow 2 \times 2 = 4$$

Equality

Fractions are equal when their cross products are equal.

EXAMPLE 1:   Is $\dfrac{6}{5} = \dfrac{30}{25}$?

$$\frac{6}{5} \diagdown\!\!\!\diagup \frac{30}{25} \longrightarrow 5 \times 30 = 150$$
$$\longrightarrow 6 \times 25 = 150$$

Yes, since their cross products are equal. The same conclusion is found by building.

$$\frac{6}{5} = \frac{6 \times 5}{5 \times 5}$$

$$= \frac{30}{25}$$

EXAMPLE 2:   Is $\dfrac{4}{6} = \dfrac{7}{9}$?

$$\frac{4}{6} \diagdown\!\!\!\diagup \frac{7}{9} \longrightarrow 6 \times 7 = 42$$
$$\longrightarrow 4 \times 9 = 36$$

No! The cross products are not equal. To indicate that the two fractions are not equal, draw a slash through the equality sign.

$$\frac{4}{6} \neq \frac{7}{9} \qquad \text{Read:} \qquad \frac{4}{6} \text{ is not equal to } \frac{7}{9}.$$

EXAMPLE 3:   Is $\dfrac{8}{12} = \dfrac{18}{27}$?

$$\frac{8}{12} \underset{\smile}{\overset{\frown}{\bowtie}} \frac{18}{27} \longrightarrow \begin{array}{l} 12 \times 18 = 216 \\ 8 \times 27 = 216 \end{array}$$

Yes, the fractions are equal.

EXAMPLE 4:   Is it the same to eat 6 slices of a pizza cut into 15 equal pieces as it is to eat 1 slice of the same size pizza cut into 5 equal pieces?

That is, is $\dfrac{6}{15} = \dfrac{1}{5}$? Check the cross products.

$$\frac{6}{15} \underset{\smile}{\overset{\frown}{\bowtie}} \frac{1}{5} \longrightarrow \begin{array}{l} 15 \times 1 = 15 \\ 6 \times 5 = 30 \end{array}$$

Therefore, $\dfrac{6}{15} \neq \dfrac{1}{5}$, and it is not the same.

EXAMPLE 5:   Are 3 slices of a pizza cut into 9 equal pieces the same as 5 slices of the same size pizza cut into 15 equal pieces?

That is, is $\dfrac{3}{9} = \dfrac{5}{15}$?

$$\frac{3}{9} \underset{\smile}{\overset{\frown}{\bowtie}} \frac{5}{15} \longrightarrow \begin{array}{l} 45 \\ 45 \end{array}$$

Yes! If you had reduced the fractions, you would have noted that both portions are equal to a third of the pizza.

Now you try these problems before looking at the solutions.

Examples	Solutions
Is $\dfrac{2}{15} = \dfrac{6}{45}$?	$\dfrac{2}{15} \underset{\smile}{\overset{\frown}{\bowtie}} \dfrac{6}{45} \longrightarrow \begin{array}{l} 15 \times 6 = 90 \\ 2 \times 45 = 90 \end{array}$ Yes, they are equal.
Is $\dfrac{10}{32} = \dfrac{150}{481}$?	Check the cross products. $32 \times 150 = 4,800$ $10 \times 481 = 4,810$ No; $\dfrac{10}{32} \neq \dfrac{150}{481}$.
Is $\dfrac{4}{6}$ of a pie the same as $\dfrac{10}{15}$ of a pie?	$\dfrac{4}{6} \underset{\smile}{\overset{\frown}{\bowtie}} \dfrac{10}{15} \longrightarrow \begin{array}{l} 6 \times 10 = 60 \\ 4 \times 15 = 60 \end{array}$ Yes; $\dfrac{4}{6} = \dfrac{10}{15}$.

Examples	Solutions
Now try these—no hints given.	
Is $\dfrac{5}{12} = \dfrac{30}{72}$?	Yes.
Is $\dfrac{3}{8} = \dfrac{27}{73}$?	No.
There are 500 sheets in a ream of paper. Is $\dfrac{3}{5}$ of a ream the same as 300 sheets of paper?	Yes.

If you had difficulty with the last three problems, talk with an instructor or read this section again.

Exercises Part I

Answers to these exercises can be found on page 278.

Determine if the two fractions are equal.

1. $\dfrac{4}{7}$ and $\dfrac{32}{55}$

2. $\dfrac{2}{3}$ and $\dfrac{20}{30}$

3. $\dfrac{12}{15}$ and $\dfrac{24}{90}$

4. $\dfrac{36}{42}$ and $\dfrac{96}{111}$

5. $\dfrac{12}{35}$ and $\dfrac{36}{105}$

6. Dan divided the novel into 25 equal sections and has read 10 sections. Larry divided the same novel into 20 equal sections and has read 8 sections. Have they read the same amount?

7. Mariko has $\dfrac{2}{3}$ of her business cards left. Warren had the same number in the beginning but now has only 335 cards left from the box of 500. Do they have the same number of cards remaining?

Exercises Part II

Determine if the two fractions are equal.

1. $\dfrac{3}{4}$ and $\dfrac{24}{32}$ 2. $\dfrac{2}{3}$ and $\dfrac{19}{36}$ 3. $\dfrac{14}{39}$ and $\dfrac{42}{125}$

4. Lupe has worked $\dfrac{2}{3}$ of the exercises. Tom has worked 30 of the 45 exercises. Have they worked the same amount?

REVIEW MATERIAL

Review Skill Building Fractions

To build a fraction, multiply both numerator and denominator by the same non-zero number.

$$\frac{5}{9} = \frac{?}{36}$$ Because $36 \div 9 = 4$, I shall build using 4.

$$\frac{5}{9} = \frac{5 \times 4}{9 \times 4} = \frac{20}{36}$$

Exercise A   Answers to this exercise can be found on page 278.

Build to the indicated denominator.

1. $\dfrac{1}{7} = \dfrac{}{49}$ 2. $\dfrac{3}{4} = \dfrac{}{72}$ 3. $\dfrac{8}{11} = \dfrac{}{143}$ 4. $\dfrac{3}{8} = \dfrac{}{32}$

5. $\dfrac{7}{12} = \dfrac{}{144}$ 6. $\dfrac{1}{7} = \dfrac{}{63}$ 7. $\dfrac{3}{4} = \dfrac{}{16}$ 8. $\dfrac{8}{11} = \dfrac{}{22}$

9. $\dfrac{3}{8} = \dfrac{}{112}$ 10. $\dfrac{7}{12} = \dfrac{}{180}$

Exercise B   No answers are given.

Build to the indicated denominator.

1. $\dfrac{1}{5} = \dfrac{}{35}$ 2. $\dfrac{5}{6} = \dfrac{}{36}$ 3. $\dfrac{7}{13} = \dfrac{}{52}$ 4. $\dfrac{3}{5} = \dfrac{}{100}$ 5. $\dfrac{7}{15} = \dfrac{}{75}$

Review Skill Reducing Fractions

To reduce a fraction, divide both numerator and denominator by the same non-zero number.

$$\frac{20}{36} = \frac{10}{18}$$ Dividing by 2.

$$= \frac{5}{9}$$ Dividing by 2.

Exercise C   Answers to this exercise can be found on page 278.

Reduce to lowest terms.

1. $\dfrac{7}{49}$ 2. $\dfrac{54}{72}$ 3. $\dfrac{104}{143}$ 4. $\dfrac{12}{32}$ 5. $\dfrac{84}{144}$

6. $\dfrac{9}{63}$ 7. $\dfrac{12}{16}$ 8. $\dfrac{16}{22}$ 9. $\dfrac{42}{112}$ 10. $\dfrac{105}{180}$

Exercise D   No answers are given.

Reduce to lowest terms.

1. $\dfrac{7}{35}$ 2. $\dfrac{30}{36}$ 3. $\dfrac{28}{52}$ 4. $\dfrac{60}{100}$ 5. $\dfrac{21}{75}$

Review Skill Mixed Numbers and Improper Fractions

Changing an Improper Fraction to a Mixed Number

Divide numerator by denominator.

$$\frac{59}{4} = 59 \div 4$$
$$= 14 \text{ R}3$$
$$= 14\frac{3}{4} \quad \begin{matrix} \leftarrow \text{ remainder} \\ \leftarrow \text{ divisor} \end{matrix}$$
$$\underset{\text{quotient}}{\uparrow}$$

Changing a Mixed Number to an Improper Fraction

Multiply the denominator by the whole number and add the numerator.

$$5\frac{2}{7} = \frac{(7 \times 5) + 2}{7}$$
$$= \frac{35 + 2}{7} = \frac{37}{7}$$

Exercise E Answers to this exercise can be found on page 278.

Change each to a mixed number.

1. $\dfrac{39}{7}$ 2. $\dfrac{52}{15}$ 3. $\dfrac{281}{18}$

Change each to an improper fraction.

4. $3\dfrac{3}{4}$ 5. $7\dfrac{3}{8}$ 6. $14\dfrac{8}{9}$

Exercise F No answers are given.

Change each to a mixed number.

1. $\dfrac{48}{5}$ 2. $\dfrac{124}{6}$ 3. $\dfrac{381}{4}$

Change each to an improper fraction.

4. $5\dfrac{5}{6}$ 5. $9\dfrac{2}{5}$ 6. $82\dfrac{5}{9}$

REVIEW EXERCISES

Refer to the section listed if you have difficulty. Answers to these exercises can be found on page 278.

Section 6.1   Write a fraction that represents each shaded region.

1.
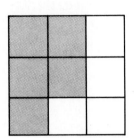

2.
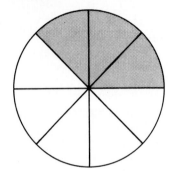

Section 6.2   Write each as a whole or mixed number.

3. $\dfrac{25}{4}$ 4. $\dfrac{24}{3}$ 5. $\dfrac{29}{5}$ 6. $\dfrac{36}{6}$

Write each as an improper fraction.

7. $9\dfrac{3}{5}$ 8. $15\dfrac{2}{3}$ 9. 12 10. $2\dfrac{7}{12}$

Section 6.3   Build each fraction to the denominator indicated.

11. $\dfrac{3}{5} = \dfrac{?}{45}$ 12. $\dfrac{7}{15} = \dfrac{?}{90}$ 13. $\dfrac{2}{3} = \dfrac{?}{69}$ 14. $\dfrac{5}{14} = \dfrac{?}{126}$

Section 6.4   Reduce each to lowest terms.

15. $\dfrac{36}{54}$ 16. $\dfrac{36}{90}$ 17. $\dfrac{160}{240}$ 18. $\dfrac{147}{168}$

Section 6.5   Determine if the two fractions are equal.

19. $\dfrac{9}{27}$ and $\dfrac{11}{32}$ 20. $\dfrac{10}{15}$ and $\dfrac{12}{18}$

This test should determine if you have mastered the topics in Chapter 6. Answers to this test can be found on page 278.

1. Write each as an improper fraction.

   (a) $6\dfrac{4}{5}$ (b) $25\dfrac{3}{4}$ (c) 27

2. Write each as a mixed number.

   (a) $\dfrac{14}{5}$ (b) $\dfrac{65}{9}$

3. Build each to the indicated denominator.

   (a) $\dfrac{2}{3} = \dfrac{?}{12}$ (b) $\dfrac{13}{15} = \dfrac{?}{135}$

4. Reduce each to lowest terms.

   (a) $\dfrac{21}{28}$ (b) $\dfrac{10}{15}$ (c) $\dfrac{120}{280}$ (d) $\dfrac{195}{255}$

5. Determine if the two fractions are equal.

   (a) $\dfrac{2}{3}$ and $\dfrac{5}{8}$ (b) $\dfrac{5}{16}$ and $\dfrac{11}{35}$

7

ordering fractions and equivalent forms

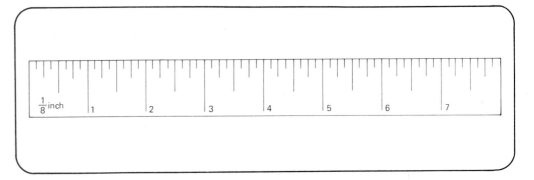

This section introduces the number line and the symbols used to indicate order.

Number Line Using a marking device similar to a ruler, I shall represent the whole numbers on a line.

First, draw a line.

Assume that the line extends infinitely in both directions, as indicated by the arrows.

Next, select any point on the line, mark the location, and label it with the whole number 0.

Now, mark off *equally spaced* points to the right. As you go, label these points with the whole numbers beginning with 1 and continuing consecutively.

176

A number line with marks 0 1 2 3 4 5 6 7

All of the whole numbers cannot be shown because these numbers continue without end.

Try these problems.

Examples	Solutions
What numbers are represented by the points A, B, and C? Number line with 0 marked, points A, B, C above	Labeling consecutively, Number line 0 1 2 3 4 5 6 with A above 2, B above 4, C above 5 A represents 2, B represents 4, and C represents 5.
What letter marks the position of 7? Number line with 0 marked, points A, B, C, D above	C, because the letter C is 7 equal spaces from 0.
What number is represented by the point A? Number line with A above first mark, 1 2 3 4 5	A is the starting point and represents 0.

Order The number line is a visual representation that also indicates the **ordering** of the whole numbers: Numbers become greater as you move to the right on the line.

EXAMPLE 1: Which number is greater, 5 or 2?

Of course, 5 is greater. Notice that the greater number is to the right of the smaller on the number line.

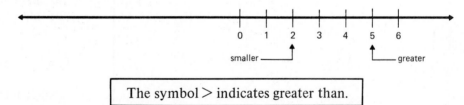

smaller ⎯ greater

The symbol $>$ indicates greater than.

I shall write

$$5 > 2$$

to mean that 5 is greater than 2.

EXAMPLE 2: Which is smaller, 1 or 3?

Of course, 1 is smaller. Again, notice that the greater number is to the right of the smaller on the number line.

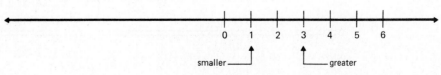

smaller ⎯ greater

The symbol $<$ indicates less than.

I shall write

$$1 < 3$$

to mean 1 is less than 3.

To remember how to read these symbols, notice that the greater number is always next to the large opening of the symbol.

greater $>$ smaller   Read: greater than.

smaller $<$ greater   Read: less than.

Now try these problems.

Examples	Solutions
Fill in the box to indicate the correct order. (a) 5 ☐ 7 (b) 2 ☐ 1 (c) 7 ☐ 3 (d) 3 ☐ 9	(a) 5 ☐ 7 smaller greater Large opening is next to greater number. $$5 < 7$$ Read: 5 is less than 7. (b) 2 ☐ 1 Greater number is first; therefore, use greater than. $$2 > 1$$ Read: 2 is greater than 1. (c) 7 ☐ 3 Greater number is first; therefore, use the symbol that has the large opening next to 7. $$7 > 3$$ Read: 7 is greater than 3. (d) 3 ☐ 9 Smaller number is first; therefore, use less than. $$3 < 9$$

Put a slash mark through each symbol that is not correct.
 	
(a) $5 < 6$
(b) $7 < 2$
(c) $2 > 5$
(d) $5 > 3$
	(a) smaller $<$ greater
	     True; 5 is less than 6.
	(b) $7 \not< 2$; 7 is *not* less than 2.
	(c) greater $>$ smaller
	     $2 \not> 5$; 2 is not greater than 5.
	(d) True; 5 is greater than 3.

Fractions on a Number Line

EXAMPLE 3:   How long is the pencil?

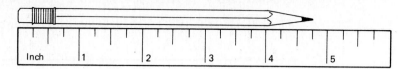

The pencil measures 4 inches plus part of the next inch. Because each inch is divided into 4 equal parts, the pencil measures $\frac{3}{4}$ of the next inch. The total length of the pencil is $4\frac{3}{4}$ inches.

This example illustrates the natural way fractions are represented on the number line.

EXAMPLE 4:   Represent $\frac{2}{5}$ on a number line.

First, divide each space between the whole numbers into 5 equal parts.

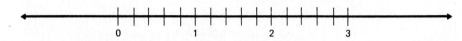

To locate $\frac{2}{5}$, count 2 spaces from 0 along the $\frac{1}{5}$ markings.

EXAMPLE 5:   Represent $\frac{7}{5}$ on a number line.

This time, count 7 spaces along the $\frac{1}{5}$ markings.

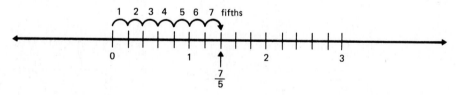

As a mixed number $\frac{7}{5} = 1\frac{2}{5}$. You can also locate $\frac{7}{5}$ by using the mixed-number form. Count 1 whole space plus 2 spaces along the $\frac{1}{5}$ markings.

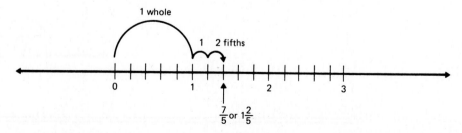

EXAMPLE 6:   Represent $\frac{3}{7}$ and $2\frac{5}{7}$ on a number line.

First, divide each space between the whole numbers into 7 equal parts.

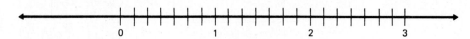

For $\frac{3}{7}$, count 3 sevenths. For $2\frac{5}{7}$, count 2 whole spaces and 5 sevenths.

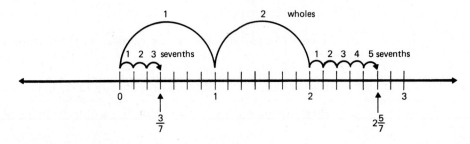

Try these before looking at the solutions.

Examples	Solutions
Represent the following fractions on a number line. (a) $\frac{2}{3}$ (b) $2\frac{1}{3}$ (c) $\frac{8}{3}$	Divide the spaces between the whole numbers into 3 equal parts.

(a)

(b)

(c)

Which fraction is greater, $\frac{2}{3}$ or $\frac{5}{8}$?

First, $\frac{2}{3} \neq \frac{5}{8}$ because their cross products are not equal.

A number line might help to answer this question, but the number of equal spaces is not the same (thirds and eighths) and is difficult to draw accurately.

An easy way to compare two fractions involves building each fraction to the *same denominator.*

$$\frac{2}{3} = \frac{2 \times 8}{3 \times 8} \qquad \frac{5}{8} = \frac{5 \times 3}{8 \times 3}$$

$$= \frac{16}{24} \qquad\qquad = \frac{15}{24}$$

Now I can compare them as numbers of 24 equal parts.

Because 16 is more than 15 of 24 equal parts,

$$\frac{2}{3} > \frac{5}{8}$$

In order to understand why I chose the denominator 24 from among many possibilities, you must study further. After you learn about prime factorization and least common multiples, you will be able to compare two fractions and determine which is greater or smaller. You will also be able to perform the arithmetic operations with fractions.

SECTION 7.2
FACTORS AND PRIME FACTORIZATIONS

Ways to arrange 12 chairs in a rectangular arrangement.

1 by 12 □□□□□□□□□□□□

2 by 6 □□□□□□   3 by 4 □□□□   12 by 1 □
       □□□□□□   □□□□   □
                          □□□□   □
4 by 3 □□□   □
       □□□   □
       □□□   □
       □□□   □
                                            □
6 by 2 □□   □
       □□   □
       □□   □
       □□   □
       □□   □
       □□   □

After reading this section, you will be able to write the prime factorization of numbers, as illustrated by this example.

$$
\begin{array}{r|r}
2 & 420 \\ \hline
2 & 210 \\ \hline
3 & 105 \\ \hline
5 & 35 \\ \hline
7 & 7 \\ \hline
& 1
\end{array}
$$

$$420 = 2 \times 2 \times 3 \times 5 \times 7$$
$$= 2^2 \times 3^1 \times 5^1 \times 7^1$$

Factors   Because 24 can be written as a product of 3 and 8,

$$24 = 3 \times 8$$

3 and 8 are called **factors** of 24.

$$\underset{\text{product}}{24} = \underset{\text{factors}}{3 \times 8}$$

Here is the list of all possible ways 24 can be written as a product of two whole-number factors. For each whole number, I tried to find a second factor that gave the product 24. Notice that in some cases this number could not be found.

$$
\begin{aligned}
24 = \quad & 1 \times 24 \\
& 2 \times 12 \\
& 3 \times 8 \\
& 4 \times 6 \\
& 5 \times \boxed{?} \\
& 6 \times 4 \\
& 7 \times \boxed{?} \\
& 8 \times 3 \\
& 9 \times \boxed{?} \\
& 10 \times \boxed{?} \\
& 11 \times \boxed{?} \\
& 12 \times 2 \\
& 13 \times \boxed{?} \\
& 14 \times \boxed{?} \\
& \vdots \quad \left.\right\} \text{ Continuing unsuccessfully.} \\
& 23 \times \boxed{?} \\
& 24 \times 1
\end{aligned}
$$

Here is the list again. This time it includes only the successful cases.

$$
\begin{aligned}
24 = \quad & 1 \times 24 \\
& 2 \times 12
\end{aligned}
$$

$$3 \times 8$$
$$4 \times 6$$
$$6 \times 4$$
$$8 \times 3$$
$$12 \times 2$$
$$24 \times 1$$

Notice that each multiplication expression has a *reversed* pair: for example, 2×12 and 12×2.

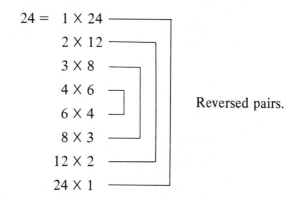

Reversed pairs.

Each of these expressions is called a **factorization** of 24. To **factor** a number is to express it as a product of other numbers.

The numbers 1, 2, 3, 4, 6, 8, 12, and 24 are called factors of 24. It is also correct to call these numbers **divisors** of 24 because each divides 24 without remainder.

$$24 = 1 \times 24$$
$$2 \times 12$$
$$3 \times 8$$
$$4 \times 6$$
$$6 \times 4$$
$$8 \times 3$$
$$12 \times 2$$
$$24 \times 1$$

→ Factorizations of 24.

1, 2, 3, 4, 6, 8, 12, 24   Factors, or divisors of 24.

EXAMPLE 1:   Find all possible factorizations of 48 (products of two whole-number factors), and list all the factors, or divisors, of 48.

$$48 = 1 \times 48$$
$$2 \times 24$$
$$3 \times 16$$
$$4 \times 12$$
$$5 \times \boxed{?}$$
$$6 \times 8 \leftarrow$$
$$7 \times \boxed{?}$$ Reversed pair.
$$8 \times 6$$

Once you have found a reversed pair, the remaining factorizations are just the reverse of the pairs above.

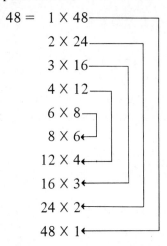

48 = 1 X 48
      2 X 24
      3 X 16
      4 X 12
      6 X 8
      8 X 6
     12 X 4
     16 X 3
     24 X 2
     48 X 1

Factorizations of 48.

1, 2, 3, 4, 6, 8, 12, 16, 24, 48   Factors, or divisors, of 48.

EXAMPLE 2:   Find all possible factorizations of 12 (products of two whole numbers), and list the factors of 12.

12 = 1 X 12
      2 X 6
      3 X 4
      4 X 3
      6 X 2
     12 X 1

Reversed pairs.

1, 2, 3, 4, 6, 12   Factors of 12.

EXAMPLE 3:   List the factors of 5.

5 = 1 X 5
    2 X ?
    3 X ?
    4 X ?
    5 X 1

The only factors of 5 are 1 and 5.

Prime and Composite Numbers

Some numbers have many factors or divisors; others have only themselves and 1 as factors.
From Examples 2 and 3,

1, 2, 3, 4, 6, 12   Factors of 12

1, 5   Factors of 5

> If a number has only 1 and itself as factors, then it is called **prime**. The number 5 is prime.
> If a number has factors other than 1 and itself, then it is called **composite**. The number 12 is composite.

EXAMPLE 4:   Is 36 prime or composite?

$$36 = \begin{array}{l} 1 \times 36 \\ 2 \times 18 \\ 3 \times 12 \\ 4 \times 9 \\ 5 \times \boxed{?} \\ 6 \times 6 \quad \text{Reverse of itself.} \\ 9 \times 4 \\ 12 \times 3 \\ 18 \times 2 \\ 36 \times 1 \end{array}$$

The numbers 36 has factors 1, 2, 3, 4, 6, 9, 12, 18, and 36. It is a composite number.

EXAMPLE 5:   Is 7 prime or composite?

The number 7 cannot be factored or broken down except by using the simple factors of 1 and itself.

$$7 = \begin{array}{l} 1 \times 7 \\ 7 \times 1 \end{array}$$

Therefore, 7 is a prime number.

It will be important later for you to recognize some of the smaller primes. Although the list of primes never ends, here are the first 10 primes.

$$2, 3, 5, 7, 11, 13, 17, 19, 23, 29$$

The only factors, or divisors, of these numbers are 1 and themselves.

Prime Factorization   Consider these factorizations of 12.

$$12 = 1 \times 12$$
$$12 = 4 \times 3$$
$$12 = 2 \times 2 \times 3$$

All these factorizations of 12 are products of two or more whole numbers. The last one is called a **prime factorization** of 12 because each factor used is a prime number.

$$12 = 2 \times 2 \times 3 \qquad \text{Each factor used is prime.}$$

There is a simple procedure for finding the prime factorization of a number using factors as divisors.

Finding the Prime Factorization of a Number

1. Divide the given number by the smallest prime that divides it evenly.
2. Continue dividing the resulting quotients by the smallest prime possible until you get the quotient 1.
3. Write the prime factorization of the number, which consists of all the prime divisors used.

EXAMPLE 6:   Find the prime factorization of 24.
First, divide by 2.

$$2 \mbig| \underline{24}$$
$$12$$ 2 divided into 24
equals 12.

Divide by 2 again.

$$2 \big| 24$$
$$2 \big| \underline{12}$$
$$6$$ 2 divided into 12
equals 6.

Divide the resulting quotient by the prime 2 again.

$$2 \big| 24$$
$$2 \big| 12$$
$$2 \big| \underline{6}$$
$$3$$ 2 divided into 6
equals 3.

Now 2 does not divide the quotient 3. The quotient 3 is divisible by the next prime, 3.

$$2 \big| 24$$
$$2 \big| 12$$
$$2 \big| 6$$
$$3 \big| \underline{3}$$
$$1$$ 3 divided into 3
equals 1.

The quotient 1 signals the end of the procedure.
The prime factorization consists of all the prime divisors used.

$$\begin{array}{c|c} 2 & 24 \\ 2 & 12 \\ 2 & 6 \\ 3 & 3 \\ \hline & 1 \end{array}$$

$$24 = 2 \times 2 \times 2 \times 3$$
$$= 2^3 \times 3$$

EXAMPLE 7:   Find the prime factorization of 420.
First, divide by 2,

$$2 \big| \underline{420}$$
$$210$$

and 2 again.

$$2 \big| 420$$
$$2 \big| \underline{210}$$
$$105$$

Next, try 3 because 105 is not divisible by 2.

$$2 \big| 420$$
$$2 \big| 210$$
$$3 \big| \underline{105}$$
$$35$$

Now divide by 5 because 3 does not divide into 35 evenly.

$$
\begin{array}{r|r}
2 & 420 \\
2 & 210 \\
3 & 105 \\
5 & 35 \\
\hline
& 7
\end{array}
$$

Finally, 7 can be divided only by 7.

$$
\begin{array}{r|r}
2 & 420 \\
2 & 210 \\
3 & 105 \\
5 & 35 \\
7 & 7 \\
\hline
& 1
\end{array}
$$

$$420 = 2 \times 2 \times 3 \times 5 \times 7$$
$$= 2^2 \times 3 \times 5 \times 7$$

It will make things easier later if you think of the exponent of 3, 5, and 7 in this example as 1.

$$3^1 = 3 \qquad \text{3 used as a factor once.}$$
$$5^1 = 5 \qquad \text{5 used as a factor once.}$$
$$7^1 = 7 \qquad \text{7 used as a factor once.}$$

Therefore, $420 = 2^2 \times 3^1 \times 5^1 \times 7^1$, using exponents.

For a larger number you might want to use the divisibility tests discussed on pages 163–164 to tell if the number is divisible by 2, 3, or 5.

EXAMPLE 8:   Find the prime factorization of 441.

Use each prime repeatedly until it no longer divides the quotient. Some primes may not divide at all.

Because 441 does not end in 0, 2, 4, 6, or 8, it is not divisible by 2. The sum of the digits of 441 is 9; therefore, the first divisor is 3.

$$
\begin{array}{r|r}
3 & 441 \\
\hline
& 147
\end{array}
$$

The sum of the digits of 147 is 12. Therefore, 3 divides 147.

$$
\begin{array}{r|r}
3 & 441 \\
3 & 147 \\
\hline
& 49
\end{array}
$$

Now 3 no longer divides; neither does the prime 5, because 49 does not end in 0 or 5. The next prime, 7, will divide,

$$
\begin{array}{r|r}
3 & 441 \\
3 & 147 \\
7 & 49 \\
\hline
& 7
\end{array}
$$

It can be used again.

$$
\begin{array}{r|r}
3 & 441 \\ \hline
3 & 147 \\ \hline
7 & 49 \\ \hline
7 & 7 \\ \hline
 & 1
\end{array}
$$

$$441 = 3 \times 3 \times 7 \times 7$$
$$= 3^2 \times 7^2$$

Now you try these.

Examples	Solutions
Find the prime factorization of 60.	$\begin{array}{r\|r} 2 & 60 \\ \hline 2 & 30 \\ \hline 3 & 15 \\ \hline 5 & 5 \\ \hline & 1 \end{array}$ $60 = 2 \times 2 \times 3 \times 5$ $= 2^2 \times 3^1 \times 5^1$
Find the prime factorization of 288.	$\begin{array}{r\|r} 2 & 288 \\ \hline 2 & 144 \\ \hline 2 & 72 \\ \hline 2 & 36 \\ \hline 2 & 18 \\ \hline 3 & 9 \\ \hline 3 & 3 \\ \hline & 1 \end{array}$ $288 = 2 \times 2 \times 2 \times 2 \times 2 \times 3 \times 3$ $= 2^5 \times 3^2$
Find the prime factorization of 770.	$\begin{array}{r\|r} 2 & 770 \\ \hline 5 & 385 \\ \hline 7 & 77 \\ \hline 11 & 11 \\ \hline & 1 \end{array}$ $770 = 2^1 \times 5^1 \times 7^1 \times 11^1$

Now try these—no hints given.

Find the prime factorization of the following numbers. (a) 18 (b) 182 (c) 980	 (a) $2^1 \times 3^2$ (b) $2^1 \times 7^1 \times 13^1$ (c) $2^2 \times 5^1 \times 7^2$

If you had difficulty with the last three problems, read this section again or talk with an instructor before continuing.

Prime Factorization and Reduction

When the numbers in a fraction are large, it is sometimes easier to reduce the fraction using prime factorization.

EXAMPLE 9: Reduce $\dfrac{150}{360}$.

First, find the prime factorization of each number.

$$
\begin{array}{r|r}
2 & 150 \\ \hline
3 & 75 \\ \hline
5 & 25 \\ \hline
5 & 5 \\ \hline
 & 1
\end{array}
\qquad
\begin{array}{r|r}
2 & 360 \\ \hline
2 & 180 \\ \hline
2 & 90 \\ \hline
3 & 45 \\ \hline
3 & 15 \\ \hline
5 & 5 \\ \hline
 & 1
\end{array}
$$

Next, rewrite the fraction in factored form.

$$\frac{150}{360} = \frac{2 \times 3 \times 5 \times 5}{2 \times 2 \times 2 \times 3 \times 3 \times 5}$$

Reduce by dividing both the numerator and denominator by the same number. This is easy in factored form because all the factors are shown.

Cross out a 2 in the numerator and denominator, indicating division by 2.

$$\frac{150}{360} = \frac{\overset{1}{\cancel{2}} \times 3 \times 5 \times 5}{\underset{1}{\cancel{2}} \times 2 \times 2 \times 3 \times 3 \times 5}$$

Eliminate one by one each prime factor common to both numerator and denominator by crossing it out. Thus each eliminated factor becomes a factor of 1. Continue by dividing by 3 and 5.

$$\frac{150}{360} = \frac{\overset{1}{\cancel{2}} \times \overset{1}{\cancel{3}} \times \overset{1}{\cancel{5}} \times 5}{\underset{1}{\cancel{2}} \times 2 \times 2 \times \underset{1}{\cancel{3}} \times 3 \times \underset{1}{\cancel{5}}}$$

$$= \frac{1 \times 1 \times 1 \times 5}{1 \times 2 \times 2 \times 1 \times 3 \times 1}$$

$$= \frac{5}{2 \times 2 \times 3}$$

$$= \frac{5}{12}$$

Multiply the remaining factors to get the reduced form.

EXAMPLE 10: Reduce $\dfrac{375}{750}$.

Find the prime factorization.

$$\frac{375}{750} = \frac{3 \times 5 \times 5 \times 5}{2 \times 3 \times 5 \times 5 \times 5}$$

Cancel common prime factors one for one in numerator and denominator.

$$\frac{375}{750} = \frac{\overset{1}{\cancel{3}} \times \overset{1}{\cancel{5}} \times \overset{1}{\cancel{5}} \times \overset{1}{\cancel{5}}}{2 \times \underset{1}{\cancel{3}} \times \underset{1}{\cancel{5}} \times \underset{1}{\cancel{5}} \times \underset{1}{\cancel{5}}}$$

All factors are crossed out in the numerator leaving 1's as factors.

$$\frac{375}{750} = \frac{\overset{1}{\cancel{3}} \times \overset{1}{\cancel{5}} \times \overset{1}{\cancel{5}} \times \overset{1}{\cancel{5}}}{2 \times \underset{1}{\cancel{3}} \times \underset{1}{\cancel{5}} \times \underset{1}{\cancel{5}} \times \underset{1}{\cancel{5}}}$$

$$= \frac{1}{2} \quad \text{In lowest terms.}$$

Practice this new technique with the following fractions.

Examples	Solutions
Reduce each by prime factorization.	Cancel common primes one for one in numerator and denominator.
(a) $\dfrac{25}{35}$	(a) $\dfrac{25}{35} = \dfrac{\overset{1}{\cancel{5}} \times 5}{\underset{1}{\cancel{5}} \times 7}$
(b) $\dfrac{30}{90}$	$= \dfrac{5}{7}$
(c) $\dfrac{45}{144}$	(b) $\dfrac{30}{90} = \dfrac{\overset{1}{\cancel{2}} \times \overset{1}{\cancel{3}} \times \overset{1}{\cancel{5}}}{\underset{1}{\cancel{2}} \times \underset{1}{\cancel{3}} \times 3 \times \underset{1}{\cancel{5}}}$
(d) $\dfrac{36}{6}$	$= \dfrac{1}{3}$
	(c) $\dfrac{45}{144} = \dfrac{\overset{1}{\cancel{3}} \times \overset{1}{\cancel{3}} \times 5}{2 \times 2 \times 2 \times 2 \times \underset{1}{\cancel{3}} \times \underset{1}{\cancel{3}}}$
	$= \dfrac{5}{16}$
	(d) $\dfrac{36}{6} = \dfrac{\overset{1}{\cancel{2}} \times 2 \times \overset{1}{\cancel{3}} \times 3}{\underset{1}{\cancel{2}} \times \underset{1}{\cancel{3}}}$
	$= \dfrac{6}{1}$
	$= 6$

Examples	Solutions

Now try these—no hints given.

Reduce each by prime factorization.

(a) $\dfrac{28}{48}$

(b) $\dfrac{300}{375}$

(a) $\dfrac{7}{12}$

(b) $\dfrac{4}{5}$

Read this section again or talk with an instructor if you had difficulty with the last two problems.

Exercises Part I

Answers to these exercises can be found on page 278.

Find the prime factorization of each number.

1. 42   2. 84   3. 256   4. 378   5. 1,078

Reduce each fraction to lowest terms.

6. $\dfrac{32}{56}$ 7. $\dfrac{20}{30}$ 8. $\dfrac{10}{65}$ 9. $\dfrac{72}{90}$ 10. $\dfrac{105}{330}$

11. $\dfrac{150}{480}$ 12. $\dfrac{96}{210}$

Exercises Part II

Find the prime factorization of each number.

1. 32   2. 76   3. 68

Reduce each fraction to lowest terms.

4. $\dfrac{42}{54}$ 5. $\dfrac{27}{63}$ 6. $\dfrac{36}{48}$

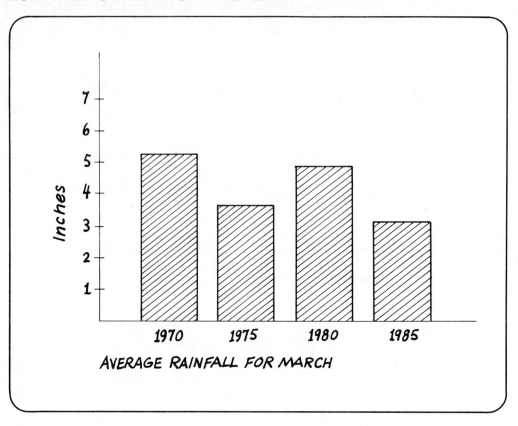

AVERAGE RAINFALL FOR MARCH

This section explains the procedure for finding the least common multiple of numbers and uses it to determine which of two fractions is greater.

Multiples   A **multiple** of 8 is any number that has 8 as a factor or divisor.

EXAMPLE 1:   List the multiples of 8.
The easiest way is to multiply 8 by each whole number starting with 1.

$$8 \times 1 = 8$$
$$8 \times 2 = 16$$
$$8 \times 3 = 24$$
$$8 \times 4 = 32$$
$$8 \times 5 = 40$$
$$\vdots$$

The multiples of 8 are

$$8, 16, 24, 32, 40, \ldots$$

There is, of course, no end to this list.

EXAMPLE 2:   List the multiples of 3.

$$3 \times 1 = 3$$
$$3 \times 2 = 6$$

$$3 \times 3 = 9$$
$$3 \times 4 = 12$$
$$3 \times 5 = 15$$
$$3 \times 6 = 18$$
$$3 \times 7 = 21$$
$$3 \times 8 = 24$$
$$3 \times 9 = 27$$
$$\vdots$$

The multiples of 3 are

$$3, 6, 9, 12, 15, 18, 21, 24, 27, \ldots$$

A **common multiple** of a group of numbers is a multiple of *each* number in the group.

EXAMPLE 3:   Find a common multiple of 8 and 3.
Comparing the lists,

Multiples of 8:   8, 16, (24), 32, 40, ...

Multiples of 3:   3, 6, 9, 12, 15, 18, 21, (24), 27, ...

I see that the multiple 24 appears in both lists. The number 24 is not the only common multiple of 8 and 3. Both numbers divide 48, 72, 96, and many more. The number 24 is called the **least common multiple** because, as the name implies, it is the smallest of the common multiples of 8 and 3.

Listing multiples takes too long. Fortunately there is a trick for finding the least common multiple (LCM) of a group of numbers, which involves prime factorization. It is an easy procedure; let us see why it works!

Finding the LCM

1. Write the prime factorization of each number, using exponents.
2. List each prime used in any of these factorizations.
3. For each prime listed, select the greatest exponent that appears with this prime number in any of the factorizations.
4. Find the product of all these primes, using the exponents selected.

This sounds difficult at first, but these examples show how easy it really is.

EXAMPLE 4:   Find the least common multiple, or LCM, of 18 and 84.

Step 1:   Find the prime factorizations of 18 and 84.

$$
\begin{array}{r|r}
2 & 18 \\ \hline
3 & 9 \\ \hline
3 & 3 \\ \hline
& 1
\end{array}
\qquad
\begin{array}{r|r}
2 & 84 \\ \hline
2 & 42 \\ \hline
3 & 21 \\ \hline
7 & 7 \\ \hline
& 1
\end{array}
$$

$$18 = 2^1 \times 3^2 \qquad 84 = 2^2 \times 3^1 \times 7^1$$

Step 2:   List the primes used in any of the factorizations.

$$2, 3, 7$$

Step 3: Select the largest exponent that appears with each prime. Consider the prime 2.

$18 = 2^1 \times 3^2$ → One factor of 2.

$84 = 2^2 \times 3^1 \times 7^1$ → Two factors of 2.

The greatest exponent of 2 is 2.
Consider the prime 3.

$18 = 2^1 \times 3^2$ → Two factors of 3.

$84 = 2^2 \times 3^1 \times 7^1$ → One factor of 3.

The greatest exponent of 3 is 2.
Finally, consider the prime 7.

$18 = 2^1 \times 3^2$ → No factor of 7.

$84 = 2^2 \times 3^1 \times 7^1$ → One factor of 7.

The greatest exponent of 7 is 1.

Step 4: Find the product of the primes, using these exponents.

$$\text{LCM of 18 and 84} = 2^2 \times 3^2 \times 7^1$$
$$= 4 \times 9 \times 7$$
$$= 36 \times 7$$
$$= 252$$

EXAMPLE 5: Find the LCM of 60 and 30.

Step 1: Factor first.

```
2 | 60          2 | 30
2 | 30          3 | 15
3 | 15          5 |  5
5 |  5              1
     1
```

$60 = 2^2 \times 3^1 \times 5^1$ $30 = 2^1 \times 3^1 \times 5^1$

Step 2: List the primes used in the factorizations.

$$2, 3, 5$$

Step 3: Select the greatest exponent that appears with each prime.

$$60 = 2^2 \times 3^1 \times 5^1$$
$$30 = 2^1 \times 3^1 \times 5^1$$

$$2^2$$
$$3^1$$
$$5^1$$

Step 4: Find the product of the primes, using the exponents selected above.

$$\text{LCM of 60 and 30} = 2^2 \times 3^1 \times 5^1$$
$$= 4 \times 3 \times 5$$
$$= 12 \times 5$$
$$= 60$$

Notice that the LCM for this particular pair is actually one of the given numbers.

EXAMPLE 6: Find the LCM of 14 and 15.
Factor.

$$\begin{array}{r|r} 2 & 14 \\ \hline 7 & 7 \\ \hline & 1 \end{array} \qquad \begin{array}{r|r} 3 & 15 \\ \hline 5 & 5 \\ \hline & 1 \end{array}$$

$$14 = 2^1 \times 7^1 \qquad 15 = 3^1 \times 5^1$$

List primes.

$$2, 3, 5, 7$$

Select exponents. The greatest exponent in each case is 1.

$$2^1, 3^1, 5^1, 7^1$$

Find the product of the primes, using these exponents.

$$\text{LCM of 14 and 15} = 2^1 \times 3^1 \times 5^1 \times 7^1$$
$$= 6 \times 5 \times 7$$
$$= 30 \times 7$$
$$= 210$$

EXAMPLE 7: Find the LCM of 12, 18, and 30.
Factor each number.

$$\begin{array}{r|r} 2 & 12 \\ \hline 2 & 6 \\ \hline 3 & 3 \\ \hline & 1 \end{array} \qquad \begin{array}{r|r} 2 & 18 \\ \hline 3 & 9 \\ \hline 3 & 3 \\ \hline & 1 \end{array} \qquad \begin{array}{r|r} 2 & 30 \\ \hline 3 & 15 \\ \hline 5 & 5 \\ \hline & 1 \end{array}$$

$$12 = 2^2 \times 3^1 \qquad 18 = 2^1 \times 3^2 \qquad 30 = 2^1 \times 3^1 \times 5^1$$

List all primes used above.

$$2, 3, 5$$

Look at *each* factorization and select the greatest exponent for each prime number.

$$12 = 2^2 \times 3^1$$
$$18 = 2^1 \times 3^2$$
$$30 = 2^1 \times 3^1 \times 5^1$$
$$2^2 \qquad 3^2 \qquad 5^1$$

Find the product of the primes, using the exponents selected.

$$\text{LCM of } 12, 18, 30 = 2^2 \times 3^2 \times 5$$
$$= 4 \times 9 \times 5$$
$$= 36 \times 5$$
$$= 180$$

Now try these *before* looking at the solution. In each problem, find the least common multiple of the numbers listed.

Examples	Solutions
12 and 15	$\begin{array}{r\|l} 2 & 12 \\ 2 & 6 \\ 3 & 3 \\ & 1 \end{array}$ $\qquad$ $\begin{array}{r\|l} 3 & 15 \\ 5 & 5 \\ & 1 \end{array}$ $12 = 2^2 \times 3^1 \qquad 15 = 3^1 \times 5^1$ $\text{LCM} = 2^2 \times 3^1 \times 5^1$ $\qquad = 4 \times 3 \times 5$ $\qquad = 12 \times 5$ $\qquad = 60$
24 and 36	$\begin{array}{r\|l} 2 & 24 \\ 2 & 12 \\ 2 & 6 \\ 3 & 3 \\ & 1 \end{array}$ $\qquad$ $\begin{array}{r\|l} 2 & 36 \\ 2 & 18 \\ 3 & 9 \\ 3 & 3 \\ & 1 \end{array}$ $24 = 2^3 \times 3^1 \qquad 36 = 2^2 \times 3^2$ $\text{LCM} = 2^3 \times 3^2$ $\qquad = 8 \times 9$ $\qquad = 72$
32 and 12	$\begin{array}{r\|l} 2 & 32 \\ 2 & 16 \\ 2 & 8 \\ 2 & 4 \\ 2 & 2 \\ & 1 \end{array}$ $\qquad$ $\begin{array}{r\|l} 2 & 12 \\ 2 & 6 \\ 3 & 3 \\ & 1 \end{array}$ $32 = 2^5 \qquad 12 = 2^2 \times 3^1$ $\text{LCM} = 2^5 \times 3^1$ $\qquad = 32 \times 3$ $\qquad = 96$

Examples	Solutions

24, 36, and 60

$$
\begin{array}{c|c}
2 & 24 \\
2 & 12 \\
2 & 6 \\
3 & 3 \\
\hline
& 1
\end{array}
\quad
\begin{array}{c|c}
2 & 36 \\
2 & 18 \\
3 & 9 \\
3 & 3 \\
\hline
& 1
\end{array}
\quad
\begin{array}{c|c}
2 & 60 \\
2 & 30 \\
3 & 15 \\
5 & 5 \\
\hline
& 1
\end{array}
$$

$$24 = 2^3 \times 3^1$$
$$36 = 2^2 \times 3^2$$
$$60 = 2^2 \times 3^1 \times 5^1$$
$$\text{LCM} = 2^3 \times 3^2 \times 5^1$$
$$= 8 \times 9 \times 5$$
$$= 72 \times 5$$
$$= 360$$

Now try these—no hints given.

21 and 15	105
12 and 9	36
18, 21, and 42	126

If you had difficulty with the last three problems, read this section again or talk with an instructor before continuing.

Ordering Fractions

EXAMPLE 8: Which is greater, $\dfrac{2}{3}$ or $\dfrac{5}{8}$?

To compare two fractions, first change each to the same number of equal parts, or to a common denominator. The smallest common denominator, called the least common denominator (LCD) of $\dfrac{2}{3}$ and $\dfrac{5}{8}$, will be the LCM of 3 and 8. Yes! The LCD of the fractions is the LCM of the denominators!

Here is the work necessary to find the LCD of $\dfrac{2}{3}$ and $\dfrac{5}{8}$.

$$3 = 3^1 \qquad 8 = 2 \times 2 \times 2 \qquad \text{Prime}$$
$$= 2^3 \qquad \text{factorization of denominators.}$$

$$\text{LCM of 3 and 8} = 2^3 \times 3^1$$
$$= 8 \times 3 \qquad \text{LCM of denominators.}$$
$$= 24$$

Next, build each fraction to the common denominator 24.

$$\frac{2}{3} = \frac{?}{24} \qquad \frac{5}{8} = \frac{?}{24}$$

Because 3 divides 24 eight times: Because 8 divides 24 three times:

$$\frac{2}{3} = \frac{2 \times 8}{3 \times 8} \qquad\qquad \frac{5}{8} = \frac{5 \times 3}{8 \times 3}$$

$$\frac{2}{3} = \frac{16}{24} \qquad\qquad \frac{5}{8} = \frac{15}{24}$$

Because *16 of 24 equal parts* is greater than *15 of 24 equal parts*,

$$\frac{16}{24} > \frac{15}{24}$$

Therefore,

$$\frac{2}{3} = \frac{16}{24} \quad \text{which is greater than} \quad \frac{15}{24} = \frac{5}{8}$$

that is,

$$\frac{2}{3} > \frac{5}{8}$$

The greater fraction is $\frac{2}{3}$.

Notice how all the pieces are fitting together—prime factorization, LCM, building, and parts of a whole. The picture will be even clearer in the next chapter.

EXAMPLE 9:   Which is greater, $\frac{7}{18}$ or $\frac{13}{24}$?

Prime factorization of denominators.

$$18 = 2^1 \times 3^2 \qquad 24 = 2^3 \times 3^1$$

Determine LCD.

$$\text{LCM of 18 and 24} = 2^3 \times 3^2$$
$$= 8 \times 9$$
$$= 72$$

Build fractions to LCD.

$$\frac{7}{18} = \frac{7 \times 4}{18 \times 4} \qquad\qquad \frac{13}{24} = \frac{13 \times 3}{24 \times 3}$$

$$\frac{7}{18} = \frac{28}{72} \qquad\qquad \frac{13}{24} = \frac{39}{72}$$

Because 39 of 72 equal parts is more than 28 of 72 equal parts,

$$\frac{39}{72} > \frac{28}{72} \quad \text{and thus} \quad \frac{13}{24} > \frac{7}{18}$$

Therefore, $\frac{13}{24}$ is greater.

EXAMPLE 10:   Which is smaller, $\frac{5}{12}$ or $\frac{11}{30}$?

Prime factorization of denominators.

$$12 = 2^2 \times 3^1 \qquad 30 = 2^1 \times 3^1 \times 5^1$$

Determine the LCD.

$$\text{LCM of } 12 \text{ and } 30 = 2^2 \times 3^1 \times 5^1$$
$$= 4 \times 3 \times 5$$
$$= 12 \times 5$$
$$= 60$$

Build fractions to LCD.

$$\frac{5}{12} = \frac{?}{60} \qquad\qquad \frac{11}{30} = \frac{?}{60}$$

$$= \frac{5 \times 5}{12 \times 5} \qquad\qquad = \frac{11 \times 2}{30 \times 2}$$

$$= \frac{25}{60} \qquad\qquad = \frac{22}{60}$$

Because $\frac{22}{60} < \frac{25}{60}, \frac{11}{30} < \frac{5}{12}$. Therefore, $\frac{11}{30}$ is smaller.

Now it is your turn!

Examples	Solutions	
Which is greater, $\frac{3}{28}$ or $\frac{10}{63}$?	$28 = 2^2 \times 7^1$ $63 = 3^2 \times 7^1$ $\text{LCD} = 2^2 \times 3^2 \times 7^1$ $= 252$	Prime factorization.
	$\frac{3}{28} = \frac{3 \times 9}{28 \times 9}$ $= \frac{27}{252}$	Building to LCD.
	$\frac{10}{63} = \frac{10 \times 4}{63 \times 4}$ $= \frac{40}{252}$	
	Therefore, $\frac{10}{63}$ is greater.	
Which is smaller, $\frac{1}{6}$ or $\frac{2}{9}$?	$6 = 2^1 \times 3^1$ $9 = 3^2$ $\text{LCD} = 2^1 \times 3^2$ $= 18$	
	$\frac{1}{6} = \frac{1 \times 3}{6 \times 3}$ $= \frac{3}{18}$	

Examples	Solutions
	$\dfrac{2}{9} = \dfrac{2 \times 2}{9 \times 2}$ $= \dfrac{4}{18}$ Therefore, $\dfrac{1}{6}$ is smaller.

Now try these—no hints given.

Find the LCM of 30 and 24.	120.
Which is greater, $\dfrac{7}{30}$ or $\dfrac{6}{25}$?	$\dfrac{6}{25}$

If you had difficulty with these two problems, read this section again or talk with an instructor.

Exercises Part I

Answers to these exercises can be found on page 278.

Find the LCM.

1. 9 and 12   2. 10 and 15   3. 4 and 14   4. 15 and 12
5. 18 and 28   6. 12 and 42   7. 8, 10, and 15   8. 15, 27, and 45

Determine the greater fraction.

9. $\dfrac{2}{7}$ and $\dfrac{1}{3}$ 10. $\dfrac{3}{4}$ and $\dfrac{7}{12}$ 11. $\dfrac{3}{5}$ and $\dfrac{2}{3}$ 12. $\dfrac{3}{18}$ and $\dfrac{7}{30}$

Exercises Part II

Find the LCM.

1. 8 and 12   2. 12 and 30   3. 3 and 18   4. 8, 27, and 15

Determine the greater fraction.

5. $\dfrac{7}{24}$ and $\dfrac{1}{4}$ 6. $\dfrac{4}{9}$ and $\dfrac{8}{15}$

REVIEW MATERIAL

Review Skill LCM

The least common multiple (LCM) of two or more numbers is found in four steps.

1. Write the prime factorization of each number, using exponents.
2. List primes used.
3. Select the greatest exponent that appears with each prime listed.
4. Find the product of all these primes, using the exponents selected.

LCM of 98 and 54:
$$98 = 2^1 \times 3^2 \times 5^1$$
$$54 = 2^1 \times 3^3$$
$$\overline{\text{LCM} = 2^1 \times 3^3 \times 5^1}$$
$$= 270$$

Exercise A Answers to this exercise can be found on page 278.

Find the LCM.

1. 8 and 12	2. 9 and 15	3. 24 and 32
4. 42 and 78	5. 8, 12, and 16	6. 63, 84, and 96
7. 9, 12, and 4	8. 4, 8, and 12	

Exercise B No answers are given.

Find the LCM.

1. 12 and 18	2. 8 and 4	3. 40 and 52
4. 63 and 96	5. 12, 18, and 27	6. 84, 54, and 56

REVIEW EXERCISES

Refer to the section listed if you have difficulty. Answers to these exercises can be found on page 278.

Section 7.1 Fill in the box with $<$ or $>$ to indicate the correct order.

1. 6 ☐ 8	2. 2 ☐ 1	3. 7 ☐ 8	4. 9 ☐ 12

Section 7.2 Find the prime factorization of each number.

5. 9	6. 10	7. 35	8. 56
9. 72	10. 480	11. 105	12. 330

Reduce each fraction to lowest terms.

13. $\dfrac{54}{156}$ 14. $\dfrac{63}{81}$ 15. $\dfrac{270}{480}$

Section 7.3 Find the LCM.

16. 6 and 9	17. 2 and 10	18. 21 and 35	19. 49 and 42
20. 56 and 72	21. 24 and 18	22. 18 and 21	23. 15 and 30
24. 32, 72, and 60			

Determine the greater fraction.

25. $\dfrac{10}{15}$ and $\dfrac{15}{21}$ 26. $\dfrac{17}{18}$ and $\dfrac{33}{36}$

CHAPTER POST-TEST

This test should determine if you have mastered the topics in Chapter 7. Answers to this test can be found on page 279.

Find the prime factorization of each number.

1. 48 **2.** 360 **3.** 305 **4.** 540 **5.** 375

Find the LCM.

6. 6 and 21 **7.** 15 and 70 **8.** 12 and 50 **9.** 15, 27, and 45

10. Which fraction is greater, $\dfrac{47}{56}$ or $\dfrac{57}{64}$?

8
addition of fractions and mixed numbers

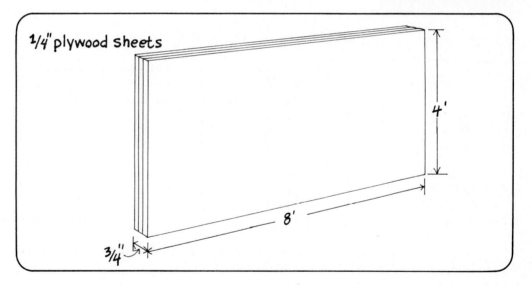

This section explains the rule for adding fractions with like denominators, as illustrated by this example.

$$\frac{1}{5} + \frac{2}{5} = \frac{1+2}{5}$$

$$= \frac{3}{5}$$

Addition with Like Denominators

Adding fractions is the same as counting the number of equal parts of a whole. For example,

$$\frac{1}{5} + \frac{2}{5} = \frac{1+2}{5}$$

$$= \frac{3}{5}$$

That is, counting fifths,

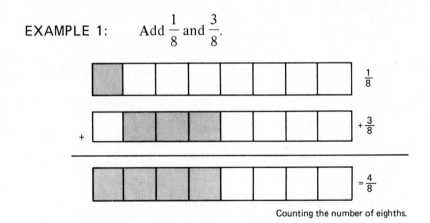

1 fifth

+ 2 fifths

= 3 fifths

Addition of Fractions with Like Denominators

To add fractions with the *same*, or *like*, *denominator*, add the numerators and write this result over the denominator. Reduce the sum to lowest terms.

EXAMPLE 1: Add $\frac{1}{8}$ and $\frac{3}{8}$.

$\frac{1}{8}$

+ $\frac{3}{8}$

= $\frac{4}{8}$

Counting the number of eighths.

Because each is divided into the same eight equal parts, add only the numerators.

$$\frac{1}{8}$$

$$+\frac{3}{8}$$

$$\frac{4}{8} = \frac{2}{4} \qquad \text{Reduce using division by 2.}$$

$$= \frac{1}{2} \qquad \text{Reduce using division by 2.}$$

Therefore, the sum is $\frac{1}{2}$.

EXAMPLE 2:   Add $\dfrac{7}{9}$ and $\dfrac{8}{9}$.

$$\dfrac{7}{9}$$

$$+\ \dfrac{8}{9}$$

$$\dfrac{15}{9} = \dfrac{5}{3} \qquad \text{Reduce using division by 3.}$$

This answer is reduced; however, most people prefer to change all improper fractions to mixed numbers.

$$\dfrac{5}{3} = 1\dfrac{2}{3}$$

Therefore, the sum is $1\dfrac{2}{3}$.

EXAMPLE 3:   The addition of three or more fractions with like denominators is treated in the same way.

$$\dfrac{3}{7} + \dfrac{2}{7} + \dfrac{6}{7} = \dfrac{3 + 2 + 6}{7}$$

$$= \dfrac{11}{7}$$

$$= 1\dfrac{4}{7}$$

Try these additions yourself.

Examples	Solutions
Add $\dfrac{1}{5}$ and $\dfrac{2}{5}$.	$\dfrac{1}{5} + \dfrac{2}{5} = \dfrac{1 + 2}{5}$ $= \dfrac{3}{5}$
$\dfrac{5}{6}$ $+\ \dfrac{1}{6}$ $?$	$\dfrac{5}{6}$ $+\ \dfrac{1}{6}$ $\dfrac{6}{6} = 1$
$\dfrac{5}{12}$ $+\ \dfrac{5}{12}$ $?$	$\dfrac{5}{12}$ $+\ \dfrac{5}{12}$ $\dfrac{10}{12} = \dfrac{5}{6}$

Examples	Solutions
Add $\dfrac{3}{8}$ and $\dfrac{7}{8}$.	$\dfrac{3}{8}$ $+\ \dfrac{7}{8}$ $\dfrac{10}{8} = \dfrac{5}{4}$ Reduce. $= 1\dfrac{1}{4}$ Mixed-number form.
$\dfrac{2}{5} + \dfrac{3}{5} + \dfrac{4}{5} = ?$	$\dfrac{2}{5} + \dfrac{3}{5} + \dfrac{4}{5} = \dfrac{2+3+4}{5}$ $= \dfrac{9}{5}$ $= 1\dfrac{4}{5}$

Now try these—no hints given.

$\dfrac{1}{4}$ $+\ \dfrac{2}{4}$ $\overline{}$?	$\dfrac{3}{4}$
$\dfrac{2}{9} + \dfrac{4}{9} = ?$	$\dfrac{2}{3}$
$\dfrac{6}{11} + \dfrac{8}{11} + \dfrac{10}{11} = ?$	$2\dfrac{2}{11}$

If you had difficulty with these last problems, please talk with an instructor or read this section again before working the exercises.

**Exercises
Part I**
Answers to these exercises can be found on page 279.

Perform the indicated additions.

1. $\dfrac{2}{8}$

      $+\dfrac{1}{8}$

2. $\dfrac{1}{6}$

      $+\dfrac{3}{6}$

3. $\dfrac{2}{3}$

      $+\dfrac{2}{3}$

4. $\dfrac{5}{9}$

    $+\dfrac{8}{9}$

5. $\dfrac{7}{12}$

    $+\dfrac{11}{12}$

6. $\dfrac{1}{9}+\dfrac{5}{9}+\dfrac{4}{9}$

7. $\dfrac{11}{24}+\dfrac{4}{24}+\dfrac{9}{24}$

8. $\dfrac{11}{16}+\dfrac{3}{16}+\dfrac{2}{16}$

Exercises Part II Perform the indicated additions.

1. $\dfrac{1}{3}+\dfrac{1}{3}$

2. $\dfrac{2}{9}+\dfrac{4}{9}$

3. $\dfrac{5}{11}+\dfrac{3}{11}+\dfrac{4}{11}$

4. $\dfrac{11}{18}+\dfrac{17}{18}+\dfrac{3}{18}$

SECTION 8.2
ADDITION OF FRACTIONS WITH UNLIKE DENOMINATORS

PECAN PIE

½ cup margarine   3 eggs
⅔ cup sugar   1 cup broken pecans
¾ cup white corn syrup

Melt margarine. Add sugar and syrup. Beat
mixture (1¹¹⁄₁₂ cups) well by hand.

After reading this section you will understand the procedure for adding fractions with unlike denominators, as illustrated by this example.

$$\dfrac{5}{24}=\dfrac{15}{72}$$

$$+\dfrac{7}{36}=\dfrac{14}{72}$$

$$\dfrac{29}{72}$$

Addition with Unlike Denominators Adding fractions is the same as counting the number of equal parts of a whole. But what happens when the fractions have unlike denominators?

$$\dfrac{1}{3}+\dfrac{1}{2}=?$$

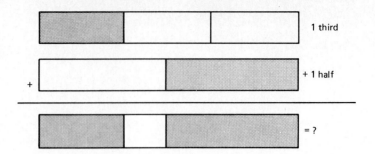

1 third

+ 1 half

= ?

Because the number of equal parts of the whole is different, or unlike, for the two fractions, you cannot simply count parts.

In Chapter 7, you learned how to change fractions to equivalent forms with a common denominator, or LCD. By thus changing each fraction, it is possible to add the equal parts of a whole. In this example the LCD is 6.

$$\frac{1}{3} = \frac{?}{6}$$

$$+ \frac{1}{2} = \frac{?}{6}$$

Build each fraction to this LCD.

$$\frac{1}{3} = \frac{1 \times 2}{3 \times 2} = \frac{2}{6}$$

$$+ \frac{1}{2} = \frac{1 \times 3}{2 \times 3} = \frac{3}{6}$$

$$\frac{5}{6} \qquad \text{Adding sixths.}$$

Addition of Fractions with Unlike Denominators

To add fractions with *different*, or *unlike*, *denominators*, build each fraction to an equivalent fraction with the LCD as the denominator. Add the fractions with like denominators, then reduce the sum to lowest terms.

EXAMPLE 1:   Add $\frac{7}{12}$ and $\frac{8}{15}$.

Step 1: Find prime factorizations of denominators.

$$
\begin{array}{r|r}
2 & 12 \\
\hline
2 & 6 \\
\hline
3 & 3 \\
\hline
 & 1
\end{array}
\qquad
\begin{array}{r|r}
3 & 15 \\
\hline
5 & 5 \\
\hline
 & 1
\end{array}
$$

$$12 = 2^2 \times 3^1 \qquad 15 = 3^1 \times 5^1$$

Step 2: Determine the LCD.

$$12 = 2^2 \times 3^1$$
$$15 = 3^1 \times 5^1$$
$$\overline{\text{LCD} = 2^2 \times 3^1 \times 5^1}$$
$$= 4 \times 3 \times 5$$
$$= 12 \times 5$$
$$= 60$$

Step 3: Build fractions to LCD and add.

$$\frac{7}{12} = \frac{7 \times 5}{12 \times 5} = \frac{35}{60} \qquad 35 \text{ sixtieths}$$

$$+\frac{8}{15} = \frac{8 \times 4}{15 \times 4} = \frac{32}{60} \qquad +32 \text{ sixtieths}$$

$$\frac{67}{60} \qquad = 67 \text{ sixtieths}$$

Although this fraction is in lowest terms, it is preferable to write the answer as a mixed number.

$$\frac{67}{60} = 1\frac{7}{60}$$

Therefore, $\dfrac{7}{12} + \dfrac{8}{15} = 1\dfrac{7}{60}$.

EXAMPLE 2:   Add $\dfrac{5}{24}$ and $\dfrac{7}{36}$.

Steps 1 and 2: Find prime factorizations and the LCD.

$$24 = 2^3 \times 3^1$$
$$36 = 2^2 \times 3^2$$
$$\overline{\text{LCD} = 2^3 \times 3^2}$$
$$= 8 \times 9$$
$$= 72$$

Step 3: Build and add.

$$\frac{5}{24} = \frac{5 \times 3}{24 \times 3} = \frac{15}{72}$$

$$+\frac{7}{36} = \frac{7 \times 2}{36 \times 2} = \frac{14}{72}$$

$$\frac{29}{72}$$

Now you can see why you had to learn all of the skills in Chapters 6 and 7.

EXAMPLE 3:   Add $\dfrac{1}{4}$, $\dfrac{1}{12}$, and $\dfrac{1}{2}$.

Steps 1 and 2: Find prime factorizations and the LCD.

$$4 = 2^2$$
$$12 = 2^2 \times 3^1$$
$$2 = 2^1$$
$$\overline{\text{LCD} = 2^2 \times 3^1}$$
$$= 12$$

Step 3: Build and add.

$$\frac{1}{4} = \frac{1 \times 3}{4 \times 3} = \frac{3}{12}$$

$$\frac{1}{12} = \frac{1 \times 1}{12 \times 1} = \frac{1}{12}$$

$$+\ \frac{1}{2} = \frac{1 \times 6}{2 \times 6} = \frac{6}{12}$$

$$\frac{10}{12} = \frac{5}{6} \qquad \text{In lowest terms.}$$

Now add these. Check your answers with the solutions.

Examples	Solutions
$\dfrac{14}{18} + \dfrac{12}{84} = ?$	$18 = 2^1 \times 3^2$ $84 = 2^2 \times 3^1 \times 7$ $\text{LCD} = 2^2 \times 3^2 \times 7$ $= 252$ $\dfrac{14}{18} = \dfrac{14 \times 14}{18 \times 14} = \dfrac{196}{252}$ $+\dfrac{12}{84} = \dfrac{12 \times 3}{84 \times 3} = \dfrac{36}{252}$ $\dfrac{232}{252}$ $= \dfrac{\overset{1}{\cancel{2}} \times \overset{1}{\cancel{2}} \times 2 \times 29}{\underset{1}{\cancel{2}} \times \underset{1}{\cancel{2}} \times 3 \times 3 \times 7}$ $= \dfrac{58}{63}$
$\dfrac{9}{14} + \dfrac{7}{15} = ?$	$14 = 2^1 \times 7^1$ $15 = 3^1 \times 5^1$ $\text{LCD} = 2^1 \times 3^1 \times 5^1 \times 7^1$ $= 210$ $\dfrac{9}{14} = \dfrac{9 \times 15}{14 \times 15} = \dfrac{135}{210}$ $+\dfrac{7}{15} = \dfrac{7 \times 14}{15 \times 14} = \dfrac{98}{210}$ $\dfrac{233}{210} = 1\dfrac{23}{210}$

Examples	Solutions
$\dfrac{5}{6} + \dfrac{3}{10} + \dfrac{4}{15} = ?$	$6 = 2^1 \times 3^1$ $10 = 2^1 \times 5^1$ $15 = 3^1 \times 5^1$ $\overline{\text{LCD} = 2^1 \times 3^1 \times 5^1}$ $= 30$ $\dfrac{5}{6} = \dfrac{5 \times 5}{6 \times 5} = \dfrac{25}{30}$ $\dfrac{3}{10} = \dfrac{3 \times 3}{10 \times 3} = \dfrac{9}{30}$ $+\dfrac{4}{15} = \dfrac{4 \times 2}{15 \times 2} = \dfrac{8}{30}$ $\dfrac{42}{30} = 1\dfrac{12}{30}$ $= 1\dfrac{2}{5}$ In lowest terms.

Now try these—no hints given.

$\dfrac{5}{12} + \dfrac{1}{9} = ?$	$\dfrac{19}{36}$
$\dfrac{17}{18} + \dfrac{11}{21} = ?$	$1\dfrac{59}{126}$

If you had difficulty with the last two problems, talk with an instructor or read this section again.

Exercises Part I

Answers to these exercises can be found on page 279.

Add.

1. $\dfrac{3}{4} + \dfrac{2}{3}$ 2. $\dfrac{5}{6} + \dfrac{4}{15}$ 3. $\dfrac{5}{12} + \dfrac{5}{24}$

4. $\dfrac{9}{30} + \dfrac{7}{24}$ 5. $\dfrac{1}{9} + \dfrac{11}{12}$ 6. $\dfrac{3}{4} + \dfrac{3}{14}$

7. $\dfrac{5}{18} + \dfrac{9}{28}$ 8. $\dfrac{7}{18} + \dfrac{11}{30}$ 9. $\dfrac{3}{8} + \dfrac{3}{10} + \dfrac{8}{15}$

10. $\dfrac{6}{15} + \dfrac{8}{27} + \dfrac{4}{45}$

11. Manuel bought $\dfrac{1}{8}$ pound of cashews and $\dfrac{3}{4}$ pound of peanuts. What is the total amount of nuts he purchased?

Add.

1. $\dfrac{2}{5} + \dfrac{4}{7}$ 2. $\dfrac{11}{12} + \dfrac{3}{4}$ 3. $\dfrac{3}{4} + \dfrac{5}{8} + \dfrac{7}{12}$ 4. $\dfrac{1}{6} + \dfrac{9}{10} + \dfrac{1}{5}$

5. Samantha bought $\dfrac{3}{4}$ pound of plain fudge and $\dfrac{1}{2}$ pound with nuts. What is the total amount of fudge she purchased?

SECTION 8.3
ADDITION OF MIXED NUMBERS

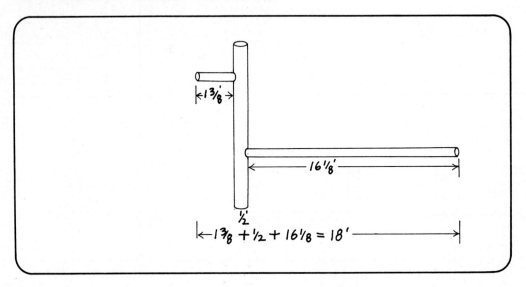

This section explains the addition of mixed numbers, as illustrated in this example.

$$2\dfrac{5}{6} = 2\dfrac{25}{30}$$

$$+\,5\dfrac{3}{10} = 5\dfrac{9}{30}$$

$$7\dfrac{34}{30} = 7 + 1\dfrac{4}{30}$$

$$= 8\dfrac{4}{30}$$

$$= 8\dfrac{2}{15} \qquad \text{In lowest terms.}$$

**Addition
of Mixed
Numbers**

EXAMPLE 1: Add $4\dfrac{1}{3}$ and $6\dfrac{1}{2}$.

First, build the fractional parts to the LCD 6.

$$4\dfrac{1}{3} = 4\dfrac{1 \times 2}{3 \times 2} = 4\dfrac{2}{6}$$

$$6\dfrac{1}{2} = 6\dfrac{1 \times 3}{2 \times 3} = 6\dfrac{3}{6}$$

Next, add the whole numbers and fractions separately.

$$4\frac{1}{3} = 4\frac{2}{6}$$

$$+ 6\frac{1}{2} = 6\frac{3}{6}$$

$$10\frac{5}{6}$$

Therefore, $4\frac{1}{3} + 6\frac{1}{2} = 10\frac{5}{6}$.

EXAMPLE 2:   Add $5\frac{1}{4}$, $6\frac{1}{12}$, and $2\frac{1}{2}$.

$$5\frac{1}{4} = 5\frac{1 \times 3}{4 \times 3} = 5\frac{3}{12}$$

$$6\frac{1}{12} = 6\frac{1 \times 1}{12 \times 1} = 6\frac{1}{12}$$

$$+ 2\frac{1}{2} = 2\frac{1 \times 6}{2 \times 6} = 2\frac{6}{12}$$

$$13\frac{10}{12}$$

Remember to reduce the fractional part.

$$13\frac{10}{12} = 13\frac{5}{6} \qquad \text{In lowest terms.}$$

Therefore, $5\frac{1}{4} + 6\frac{1}{12} + 2\frac{1}{2} = 13\frac{5}{6}$.

EXAMPLE 3:   Add $2\frac{5}{6}$ and $5\frac{3}{10}$.

$$2\frac{5}{6} = 2\frac{5 \times 5}{6 \times 5} = 2\frac{25}{30}$$

$$+ 5\frac{3}{10} = 5\frac{3 \times 3}{10 \times 3} = 5\frac{9}{30}$$

$$7\frac{34}{30}$$

In this example, the fractional part is an improper fraction. Change the improper fraction to a mixed number by division,

$$\frac{34}{30} = 34 \div 30 = 1\frac{4}{30}$$

and add the additional whole number to the digit 7.

$$7\frac{34}{30} = 7 + 1\frac{4}{30}$$

$$= 8\frac{4}{30}$$

$$= 8\frac{2}{15} \qquad \text{In lowest terms.}$$

Therefore, $2\frac{5}{6} + 5\frac{3}{10} = 8\frac{2}{15}$.

EXAMPLE 4:   Add $4\frac{5}{6}$ and $3\frac{7}{8}$.

$$4\frac{5}{6} = 4\frac{5 \times 4}{6 \times 4} = 4\frac{20}{24}$$

$$+\ 3\frac{7}{8} = 3\frac{7 \times 3}{8 \times 3} = 3\frac{21}{24}$$

$$7\frac{41}{24} = 7 + 1\frac{17}{24} \qquad \begin{array}{l}\text{Change improper fraction to}\\ \text{mixed number and add the}\\ \text{whole numbers.}\end{array}$$

$$= 8\frac{17}{24}$$

Now you try these.

Examples	Solutions
$2\frac{3}{16} + 1\frac{1}{6} = ?$	$2\frac{3 \times 3}{16 \times 3} = 2\frac{9}{48}$ $+\ 1\frac{1 \times 8}{6 \times 8} = 1\frac{8}{48}$ $3\frac{17}{48}$
$3\frac{1}{4} + 2\frac{1}{12} = ?$	$3\frac{1 \times 3}{4 \times 3} = 3\frac{3}{12}$ $+\ 2\frac{1 \times 1}{12 \times 1} = 2\frac{1}{12}$ $5\frac{4}{12} = 5\frac{1}{3}$ In lowest terms.

Examples	Solutions
$5\dfrac{3}{7} + 2\dfrac{4}{5} = ?$	$5\dfrac{3 \times 5}{7 \times 5} = 5\dfrac{15}{35}$
	$+\ 2\dfrac{4 \times 7}{5 \times 7} = 2\dfrac{28}{35}$
	$7\dfrac{43}{35} = 7 + 1\dfrac{8}{35}$
	$= 8\dfrac{8}{35}$

Now try these—no hints given.

$1\dfrac{7}{12} + 3\dfrac{3}{16} = ?$	$4\dfrac{37}{48}$
$3\dfrac{2}{15} + 7\dfrac{8}{9} = ?$	$11\dfrac{1}{45}$

Please talk with an instructor or read this section again if you had difficulty with the last two problems.

Exercises Part I

Answers to these exercises can be found on page 279.

Add.

1. $\dfrac{3}{5} + \dfrac{4}{5}$ 2. $\dfrac{7}{12} + \dfrac{3}{12}$ 3. $\dfrac{5}{15} + \dfrac{11}{15}$

4. $\dfrac{7}{8} + \dfrac{3}{4}$ 5. $\dfrac{5}{12} + \dfrac{3}{16}$ 6. $\dfrac{11}{48} + \dfrac{3}{64}$

7. $4\dfrac{1}{3} + 5\dfrac{2}{3}$ 8. $2\dfrac{3}{4} + 1\dfrac{5}{18}$ 9. $6\dfrac{5}{6} + 9\dfrac{4}{18}$

10. $6\dfrac{3}{4} + 5\dfrac{2}{3}$ 11. $2\dfrac{8}{9} + 1\dfrac{11}{12}$ 12. $3\dfrac{3}{8} + 1\dfrac{3}{10} + \dfrac{8}{15}$

13. $3\dfrac{1}{2} + 9 + 7\dfrac{2}{3}$ 14. $3\dfrac{7}{8} + 5$ 15. $6\dfrac{6}{15} + \dfrac{8}{27} + 5\dfrac{4}{45}$

16. A man drove $8\dfrac{1}{2}$ hours on the first day of a trip. His wife drove $5\dfrac{3}{4}$ hours the second day. What was their total driving time for the trip?

17. On the way to Washington, I stopped twice to buy gas. I bought $11\dfrac{3}{10}$ gallons and $12\dfrac{1}{2}$ gallons. How much gas did I buy on the trip?

Add.

1. $\dfrac{1}{6} + \dfrac{2}{6}$ 2. $\dfrac{3}{7} + \dfrac{5}{14}$ 3. $\dfrac{1}{4} + \dfrac{3}{50}$ 4. $2\dfrac{1}{3} + 3$ 5. $4\dfrac{4}{9} + 3\dfrac{7}{9}$

6. Brendan ran $2\dfrac{1}{3}$ miles in the morning and $5\dfrac{1}{8}$ miles in the afternoon. How many miles did he run that day?

REVIEW MATERIAL

Review Skill Addition of Fractions

To add fractions, build each to the least common denominator.

$$\dfrac{1}{2} = \dfrac{1 \times 5}{2 \times 5} = \dfrac{5}{10}$$
$$+\dfrac{4}{5} = \dfrac{4 \times 2}{5 \times 2} = \dfrac{8}{10}$$
$$\dfrac{13}{10} = 1\dfrac{3}{10}$$

Exercise A   Answers to this exercise can be found on page 279.

Add.

1. $\dfrac{3}{4} + \dfrac{7}{8}$ 2. $\dfrac{4}{5} + \dfrac{2}{3}$ 3. $\dfrac{7}{8} + \dfrac{15}{16}$ 4. $\dfrac{2}{3} + \dfrac{15}{16}$

5. $\dfrac{1}{4} + \dfrac{11}{12} + \dfrac{5}{16}$ 6. $\dfrac{1}{2} + \dfrac{4}{5} + \dfrac{2}{3}$ 7. $\dfrac{17}{16} + \dfrac{7}{24}$ 8. $\dfrac{17}{60} + \dfrac{9}{35}$

9. $\dfrac{5}{88} + \dfrac{11}{40}$ 10. $\dfrac{5}{18} + \dfrac{11}{42} + \dfrac{4}{63}$

Exercise B   No answers are given.

Add.

1. $\dfrac{5}{6} + \dfrac{15}{18}$ 2. $\dfrac{1}{4} + \dfrac{11}{12}$ 3. $\dfrac{5}{8} + \dfrac{1}{12}$ 4. $\dfrac{5}{12} + \dfrac{1}{9}$

5. $\dfrac{5}{6} + \dfrac{5}{8} + \dfrac{5}{9}$ 6. $\dfrac{5}{6} + \dfrac{5}{8} + \dfrac{1}{12}$ 7. $\dfrac{17}{16} + \dfrac{4}{35}$ 8. $\dfrac{11}{45} + \dfrac{3}{35}$

Review Skill Addition of Mixed Numbers

Add mixed numbers by adding the whole numbers and fractional parts separately.

$$6\frac{1}{3} = 6\frac{1 \times 2}{3 \times 2} = 6\frac{2}{6}$$

$$+ 5\frac{5}{6} = 5\frac{5 \times 1}{6 \times 1} = 5\frac{5}{6}$$

$$11\frac{7}{6} = 11 + 1\frac{1}{6}$$

$$= 12\frac{1}{6}$$

Exercise C Answers to this exercise can be found on page 279.

Add.

1. $9\frac{1}{2} + 7\frac{1}{3}$

2. $5\frac{3}{4} + 3\frac{5}{6}$

3. $9\frac{1}{2} + 10\frac{11}{14} + 6\frac{4}{7}$

4. $1\frac{2}{3} + 1\frac{1}{5}$

5. $2\frac{5}{8} + 1\frac{2}{3} + 2\frac{3}{4}$

6. $9 + 3\frac{5}{8}$

7. $14\frac{5}{9} + 1\frac{1}{9}$

8. $5\frac{1}{3} + 3\frac{1}{6}$

9. $4\frac{7}{20} + 6\frac{11}{15}$

10. $5\frac{1}{9} + 12\frac{4}{15}$

Exercise D No answers are given.

Add.

1. $7\frac{2}{3} + 5\frac{5}{6}$

2. $9\frac{1}{2} + 6\frac{1}{5}$

3. $10\frac{4}{9} + 2\frac{5}{6} + 4\frac{2}{3}$

4. $1\frac{1}{2} + 3\frac{2}{3}$

5. $4\frac{5}{8} + 2\frac{1}{4} + 1\frac{2}{5}$

REVIEW EXERCISES

Refer to the section listed if you have difficulty. Answers to these exercises can be found on page 279.

Section 8.1 Add.

1. $\frac{3}{5} + \frac{1}{5}$

2. $\frac{5}{6} + \frac{4}{6}$

3. $\frac{2}{9} + \frac{1}{9}$

Section 8.2 Add.

4. $\frac{3}{5} + \frac{2}{7}$

5. $\frac{5}{12} + \frac{7}{15}$

6. $\frac{5}{24} + \frac{3}{16} + \frac{5}{12}$

7. $2\dfrac{3}{4} + 5\dfrac{2}{3}$ 8. $7\dfrac{5}{6} + 1\dfrac{3}{8}$ 9. $2\dfrac{1}{14} + 3\dfrac{3}{16} + 1\dfrac{1}{7}$

10. A dress pattern calls for $2\dfrac{1}{4}$ yards of one fabric and $1\dfrac{1}{8}$ yards of another material. What is the total yardage needed to make the dress?

CHAPTER POST-TEST

This test should determine if you have mastered the topics in Chapter 8. Answers to this test can be found on page 279.

Add.

1. $\dfrac{4}{9} + \dfrac{8}{9}$ 2. $2\dfrac{1}{3} + 3\dfrac{2}{3}$ 3. $12\dfrac{5}{6} + 1\dfrac{7}{10}$

4. $\dfrac{3}{8} + \dfrac{1}{6}$ 5. $9\dfrac{4}{7} + 2\dfrac{1}{5}$ 6. $2\dfrac{8}{9} + 5\dfrac{7}{12}$

7. $\dfrac{10}{27} + 5\dfrac{7}{45}$

8. On a trip, I stopped to buy gas three times. I bought $6\dfrac{7}{10}$, $12\dfrac{9}{10}$, and $10\dfrac{3}{5}$ gallons. How much gas did I buy on the trip?

9

subtraction of fractions and mixed numbers

A healthy adult's temperature can differ from the normal $98\frac{6}{10}°$ by as much as $4\frac{°}{10}$ in either direction.

This section explains the procedures for subtraction of fractions, as illustrated by this example.

$$\frac{7}{12} = \frac{7 \times 5}{12 \times 5} = \frac{35}{60}$$

$$-\frac{8}{15} = \frac{8 \times 4}{15 \times 4} = \frac{32}{60}$$

$$\frac{3}{60} = \frac{1}{20}$$

Subtraction of Fractions To subtract fractions, build the denominators to a common denominator. The procedure is the same as for addition of fractions.

219

EXAMPLE 1:   Find the difference of $\dfrac{2}{5}$ and $\dfrac{1}{5}$.

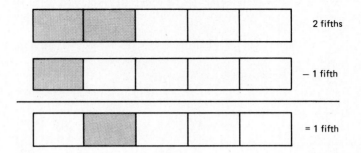

2 fifths

− 1 fifth

= 1 fifth

That is, $\dfrac{2}{5} - \dfrac{1}{5} = \dfrac{2-1}{5} = \dfrac{1}{5}$.

Subtraction of Fractions with Like Denominators

To subtract fractions with the *same*, or *like*, *denominator*, subtract the numerators and write this result over the denominator. Reduce the difference to lowest terms.

EXAMPLE 2:   Subtract $\dfrac{1}{8}$ from $\dfrac{3}{8}$.

$$\dfrac{3}{8}$$

$$-\dfrac{1}{8}$$

$$\dfrac{2}{8} = \dfrac{1}{4} \qquad \text{In lowest terms.}$$

Subtraction of Fractions with Unlike Denominators

To subtract fractions with *different*, or *unlike*, *denominators*, build each fraction to an equivalent fraction with the LCD as the denominator. Subtract the fractions with like denominators, then reduce the difference to lowest terms.

EXAMPLE 3:   Subtract $\dfrac{8}{15}$ from $\dfrac{7}{12}$.

Step 1:   Find the prime factorization of each denominator.

$$
\begin{array}{r|r} 3 & 15 \\ 5 & 5 \\ \hline & 1 \end{array}
\qquad\qquad
\begin{array}{r|r} 2 & 12 \\ 2 & 6 \\ 3 & 3 \\ \hline & 1 \end{array}
$$

$$15 = 3^1 \times 5^1 \qquad 12 = 2^2 \times 3^1$$

Step 2: Determine the LCD.

$$15 = 3^1 \times 5^1$$
$$12 = 2^2 \times 3^1$$
$$\overline{\text{LCD} = 2^2 \times 3^1 \times 5^1}$$
$$= 60$$

Step 3: Build fractions to LCD and subtract.

$$\frac{7}{12} = \frac{7 \times 5}{12 \times 5} = \frac{35}{60}$$
$$-\frac{8}{15} = \frac{8 \times 4}{15 \times 4} = \frac{32}{60}$$
$$\overline{ \frac{3}{60} = \frac{\overset{1}{\cancel{3}}}{2 \times 2 \times \underset{1}{\cancel{3}} \times 5}}$$
$$= \frac{1}{20}$$

You try these before looking at the solutions.

Examples	Solutions
Find the difference of $\dfrac{5}{24}$ and $\dfrac{7}{36}$.	$24 = 2^3 \times 3^1$ $36 = 2^2 \times 3^2$ $\overline{\text{LCD} = 2^3 \times 3^2}$ $= 72$ $\dfrac{5}{24} = \dfrac{5 \times 3}{24 \times 3} = \dfrac{15}{72}$ $-\dfrac{7}{36} = \dfrac{7 \times 2}{36 \times 2} = \dfrac{14}{72}$ $\overline{\dfrac{1}{72}}$
Subtract $\dfrac{1}{12}$ from $\dfrac{3}{4}$.	$4 = 2^2$ $12 = 2^2 \times 3^1$ $\overline{\text{LCD} = 2^2 \times 3^1}$ $= 12$ $\dfrac{3}{4} = \dfrac{3 \times 3}{4 \times 3} = \dfrac{9}{12}$ $-\dfrac{1}{12} = \dfrac{1 \times 1}{12 \times 1} = \dfrac{1}{12}$ $\overline{\dfrac{8}{12} = \dfrac{2}{3}}$ In lowest terms.

Examples	Solutions
Find the difference of $\dfrac{13}{18}$ and $\dfrac{5}{12}$.	$18 = 2^1 \times 3^2$ $12 = 2^2 \times 3^1$ $\overline{\text{LCD} = 2^2 \times 3^2}$ $= 36$ $\dfrac{13}{18} = \dfrac{13 \times 2}{18 \times 2} = \dfrac{26}{36}$ $-\dfrac{5}{12} = \dfrac{5 \times 3}{12 \times 3} = \dfrac{15}{36}$ $\dfrac{11}{36}$

Now try these—no hints given.

Subtract $\dfrac{1}{9}$ from $\dfrac{5}{12}$.	$\dfrac{11}{36}$
Find the difference of $\dfrac{17}{18}$ and $\dfrac{11}{21}$.	$\dfrac{53}{126}$

If you had difficulty with the last two problems, please talk with an instructor or read this section again.

**Exercises
Part I**

Answers to these exercises can be found on page 279.

Subtract.

1. $\dfrac{3}{4} - \dfrac{2}{3}$

2. $\dfrac{5}{6} - \dfrac{4}{15}$

3. $\dfrac{5}{12} - \dfrac{5}{24}$

4. $\dfrac{29}{30} - \dfrac{7}{24}$

5. $\dfrac{11}{12} - \dfrac{1}{9}$

6. $\dfrac{3}{4} - \dfrac{3}{14}$

7. $\dfrac{15}{27} - \dfrac{4}{45}$

8. $\dfrac{11}{18} - \dfrac{11}{30}$

9. $\dfrac{7}{15} - \dfrac{2}{27}$

10. A piece of fabric is $\dfrac{3}{4}$ yard long. How much fabric remains if $\dfrac{5}{8}$ yard is cut off?

**Exercises
Part II**

Subtract.

1. $\dfrac{1}{2} - \dfrac{1}{3}$

2. $\dfrac{8}{9} - \dfrac{2}{3}$

3. $\dfrac{7}{10} - \dfrac{4}{15}$

4. $\dfrac{7}{12} - \dfrac{1}{30}$

5. A piece of fabric is $\dfrac{3}{4}$ yard long. How much fabric remains if $\dfrac{1}{8}$ yard is cut off?

Wednesday stock AmCam closed at 3¾.
Today it closed at 4⅝, increasing by ⅞ point.

This section explains the procedure for subtraction of mixed numbers, as illustrated in this example.

$$5\frac{5}{6} = 5\frac{20}{24} = 4\frac{44}{24}$$

$$-3\frac{7}{8} = 3\frac{21}{24} = 3\frac{21}{24}$$

$$1\frac{23}{24}$$

Subtraction of Mixed Numbers When subtracting mixed numbers, subtract the whole numbers and fractions separately.

EXAMPLE 1:   Find the difference of $4\frac{1}{2}$ and $2\frac{1}{3}$.

First, build fractions to the LCD 6.

$$4\frac{1}{2} = 4\frac{1 \times 3}{2 \times 3} = 4\frac{3}{6}$$

$$-2\frac{1}{3} = 2\frac{1 \times 2}{3 \times 2} = 2\frac{2}{6}$$

Next, subtract the whole numbers and fractions separately.

$$4\frac{1}{2} = 4\frac{3}{6}$$

$$-2\frac{1}{3} = 2\frac{2}{6}$$

$$2\frac{1}{6}$$

Therefore, $4\frac{1}{2} - 2\frac{1}{3} = 2\frac{1}{6}$.

EXAMPLE 2:   Subtract $5\dfrac{3}{10}$ from $7\dfrac{5}{6}$.

$$7\dfrac{5}{6} = 7\dfrac{5 \times 5}{6 \times 5} = 7\dfrac{25}{30}$$

$$-5\dfrac{3}{10} = 5\dfrac{3 \times 3}{10 \times 3} = 5\dfrac{9}{30}$$

$$2\dfrac{16}{30} = 2\dfrac{8}{15} \qquad \text{In lowest terms.}$$

EXAMPLE 3:   Find the difference of $5\dfrac{5}{6}$ and $3\dfrac{7}{8}$.

$$5\dfrac{5}{6} = 5\dfrac{5 \times 4}{6 \times 4} = 5\dfrac{20}{24}$$

$$-3\dfrac{7}{8} = 3\dfrac{7 \times 3}{8 \times 3} = 3\dfrac{21}{24}$$

I cannot subtract $\dfrac{21}{24}$ from $\dfrac{20}{24}$; I must borrow from the whole number 5.

First, take 1 from the 5, leaving 4. Add the 1 to $\dfrac{20}{24}$.

$$5\dfrac{20}{24} = \overset{4}{\cancel{5}} + 1\dfrac{20}{24}$$

Next, change $1\dfrac{20}{24}$ to an improper fraction.

$$5\dfrac{20}{24} = \overset{4}{\cancel{5}} + 1\dfrac{20}{24} \qquad \left(1\dfrac{20}{24} = \dfrac{(24 \times 1) + 20}{24} \right.$$
$$= 4 \qquad \dfrac{44}{24} \qquad \qquad \left. = \dfrac{44}{24} \right)$$

This completes the borrowing. Now I can complete the subtraction. From the beginning,

$$\overset{\overbrace{\qquad\text{build}\qquad}}{}\overset{\overbrace{\qquad\text{borrow}\qquad}}{}$$

$$5\dfrac{5}{6} = 5\dfrac{20}{24} = \overset{4}{\cancel{5}} + 1\dfrac{20}{24} = 4\dfrac{44}{24}$$

$$-3\dfrac{7}{8} = 3\dfrac{7}{8} = \qquad 3\dfrac{21}{24} = 3\dfrac{21}{24}$$

$$1\dfrac{23}{24}$$

EXAMPLE 4:   Subtract $2\dfrac{5}{6}$ from $6\dfrac{3}{10}$.

$$6\dfrac{3}{10} = 6\dfrac{3 \times 3}{10 \times 3} = 6\dfrac{9}{30}$$

$$-2\dfrac{5}{6} = 2\dfrac{5 \times 5}{6 \times 5} = 2\dfrac{25}{30}$$

Again, I must borrow. Take 1 from the 6, and add it to the fractional part.

$$\overset{5}{\cancel{6}} + 1\frac{9}{30}$$

$$-2 \qquad \frac{25}{30}$$

Change $1\frac{9}{30}$ to an improper fraction and subtract.

$$\overset{5}{\cancel{6}} + 1\frac{9}{30} = 5\frac{39}{30}$$

$$-2 \quad \frac{25}{30} = 2\frac{25}{30}$$

$$3\frac{14}{30} = 3\frac{7}{15} \qquad \text{In lowest terms.}$$

EXAMPLE 5: Subtract $7\frac{1}{3}$ from 9.

$$9$$

$$-7\frac{1}{3}$$

I cannot subtract $\frac{1}{3}$ from no thirds, or $\frac{0}{3}$. Write $\frac{0}{3}$ next to the 9 and borrow as before.

$$9\frac{0}{3} = \overset{8}{\cancel{9}} + 1\frac{0}{3} = 8\frac{3}{3} \qquad \left(1\frac{0}{3} = \frac{(3 \times 1) + 0}{3}\right)$$

$$-7\frac{1}{3} = \qquad 7\frac{1}{3} = 7\frac{1}{3}$$

$$1\frac{2}{3}$$

EXAMPLE 6: Find the difference of 12 and $7\frac{3}{4}$.

$$12 \ = 12\frac{0}{4}$$

$$- \ 7\frac{3}{4} = \ 7\frac{3}{4}$$

Again, I cannot subtract $\frac{3}{4}$ from $\frac{0}{4}$. Borrow from the 12.

$$12\frac{0}{4} = \overset{11}{\cancel{12}} + 1\frac{0}{4} = 11\frac{4}{4}$$

$$- \ 7\frac{3}{4} = \qquad 7\frac{3}{4} = \ 7\frac{3}{4}$$

$$4\frac{1}{4}$$

EXAMPLE 7:   Subtract 2 from $3\frac{5}{8}$.

$$3\frac{5}{8} = 3\frac{5}{8}$$

$$-2 = 2\frac{0}{8}$$

$$\overline{ 1\frac{5}{8}}$$

In this example, I was able to subtract without borrowing.

Look carefully at Example 6 and Example 7! Can you tell when it is necessary to borrow? In Example 6, $\frac{3}{4}$ cannot be subtracted from $\frac{0}{4}$. In Example 7, $\frac{0}{8}$ can be subtracted from $\frac{5}{8}$ without borrowing.

Try to subtract these mixed numbers.

Examples	Solutions
$2\frac{3}{16} - 1\frac{1}{6} = ?$	$2\frac{3}{16} = 2\frac{3 \times 3}{16 \times 3} = 2\frac{9}{48}$ $-1\frac{1}{6} = 1\frac{1 \times 8}{6 \times 8} = 1\frac{8}{48}$ $\overline{1\frac{1}{48}}$
$3\frac{1}{12} - 2\frac{1}{4} = ?$	$3\frac{1}{12} = 3\frac{1 \times 1}{12 \times 1} = 3\frac{1}{12}$ $-2\frac{1}{4} = 2\frac{1 \times 3}{4 \times 3} = 2\frac{3}{12}$ It is necessary to borrow. The digit 3 becomes 2, and 1 is added to the fraction. $\overset{2}{\cancel{3}} + 1\frac{1}{12} = 2\frac{13}{12}$ $2\frac{3}{12} = 2\frac{3}{12}$ $\overline{\frac{10}{12} = \frac{5}{6}}$

Examples	Solutions
$5 - 3\dfrac{2}{9} = ?$	$5 \;= \overset{4}{\cancel{5}} + 1\dfrac{\overset{0}{9}}{9} = 4\dfrac{9}{9}$ $-3\dfrac{2}{9} = \qquad 3\dfrac{2}{9} = 3\dfrac{2}{9}$ $\rule{3cm}{0.4pt}$ $\hspace{3cm} 1\dfrac{7}{9}$
Now try these—no hints given.	
$3\dfrac{7}{12} - 3\dfrac{3}{16} = ?$	$\dfrac{19}{48}$
$7\dfrac{2}{15} - 3\dfrac{5}{9} = ?$	$3\dfrac{26}{45}$
$6 - 2\dfrac{4}{5} = ?$	$3\dfrac{1}{5}$

Please talk with an instructor or read this section again if you had difficulty with the last three problems.

**Exercises
Part I**

Answers to these exercises can be found on page 280.

Subtract.

1. $\dfrac{4}{5} - \dfrac{3}{5}$ 2. $\dfrac{7}{12} - \dfrac{3}{12}$ 3. $\dfrac{7}{8} - \dfrac{3}{4}$ 4. $5\dfrac{2}{3} - 4\dfrac{1}{3}$

5. $2\dfrac{3}{4} - 1\dfrac{5}{18}$ 6. $9\dfrac{5}{6} - 6\dfrac{4}{15}$ 7. $9\dfrac{2}{3} - 5$ 8. $6\dfrac{1}{4} - 2$

9. $6\dfrac{2}{3} - 4\dfrac{3}{4}$ 10. $3\dfrac{8}{9} - 1\dfrac{11}{12}$ 11. $7 - 5\dfrac{2}{5}$ 12. $8 - 7\dfrac{4}{9}$

13. $3\dfrac{3}{8} - \dfrac{8}{15}$ 14. $6\dfrac{8}{27} - \dfrac{4}{9}$

15. On a trip, a woman drove $8\dfrac{1}{2}$ hours, and her husband drove $5\dfrac{3}{4}$ hours. How much longer did the woman drive?

16. If $1\dfrac{7}{8}$ yards are cut from 6 yards of material, how much remains?

**Exercises
Part II**

Subtract.

1. $\dfrac{5}{12} - \dfrac{1}{12}$ 2. $\dfrac{9}{24} - \dfrac{1}{30}$ 3. $5\dfrac{1}{6} - 2\dfrac{1}{3}$ 4. $5 - 2\dfrac{4}{5}$ 5. $7\dfrac{2}{5} - 5$

REVIEW MATERIAL

**Review Skill
Subtraction
of Fractions**

To subtract fractions build each to a common denominator.

$$\frac{4}{5} = \frac{4 \times 2}{5 \times 2} = \frac{8}{10}$$

$$-\frac{1}{2} = \frac{1 \times 5}{2 \times 5} = \frac{5}{10}$$

$$\frac{3}{10}$$

Exercise A Answers to this exercise can be found on page 280.

Subtract.

1. $\dfrac{7}{8} - \dfrac{3}{4}$ 2. $\dfrac{4}{5} - \dfrac{2}{3}$ 3. $\dfrac{15}{16} - \dfrac{7}{8}$ 4. $\dfrac{15}{16} - \dfrac{2}{3}$ 5. $\dfrac{17}{16} - \dfrac{7}{24}$

6. $\dfrac{17}{60} - \dfrac{9}{35}$ 7. $\dfrac{9}{20} - \dfrac{1}{30}$ 8. $\dfrac{11}{20} - \dfrac{4}{45}$ 9. $\dfrac{17}{27} - \dfrac{5}{18}$ 10. $\dfrac{5}{63} - \dfrac{1}{18}$

Exercise B No answers are given.

Subtract.

1. $\dfrac{5}{6} - \dfrac{14}{18}$ 2. $\dfrac{11}{12} - \dfrac{1}{4}$ 3. $\dfrac{5}{8} - \dfrac{1}{12}$

4. $\dfrac{5}{12} - \dfrac{1}{9}$ 5. $\dfrac{17}{16} - \dfrac{4}{35}$ 6. $\dfrac{11}{45} - \dfrac{3}{35}$

**Review Skill
Subtraction of
Mixed Numbers**

Borrowing may be necessary to subtract mixed numbers.

$$18\frac{1}{5} = 18\frac{1 \times 2}{5 \times 2} = 18\frac{2}{10} = 17 + 1\frac{2}{10} = 17\frac{12}{10}$$

$$-\ 9\frac{1}{2} = \ 9\frac{1 \times 5}{2 \times 5} = \ 9\frac{5}{10} = \qquad 9\frac{5}{10} = \ 9\frac{5}{10}$$

$$8\frac{7}{10}$$

Exercise C Answers to this exercise can be found on page 280.

Subtract.

1. $9\dfrac{1}{2} - 7\dfrac{1}{3}$ 2. $5\dfrac{3}{4} - 3\dfrac{5}{6}$ 3. $10\dfrac{11}{14} - 6\dfrac{6}{7}$ 4. $3 - 1\dfrac{7}{8}$ 5. $4\dfrac{1}{7} - 3\dfrac{2}{3}$

Exercise D No answers are given.

Subtract.

1. $7\dfrac{2}{3} - 5\dfrac{5}{6}$ 2. $9\dfrac{1}{2} - 6\dfrac{1}{5}$ 3. $10\dfrac{4}{9} - 4\dfrac{2}{3}$ 4. $5 - 2\dfrac{2}{3}$ 5. $5\dfrac{1}{2} - 2\dfrac{1}{5}$

REVIEW EXERCISES

Refer to the section listed if you have difficulty. Answers to these exercises can be found on page 280.

Section 9.1 Subtract.

1. $\dfrac{2}{9} - \dfrac{1}{9}$ 2. $\dfrac{5}{6} - \dfrac{2}{6}$ 3. $\dfrac{3}{5} - \dfrac{2}{7}$

4. $\dfrac{3}{5} - \dfrac{1}{5}$ 5. $\dfrac{11}{15} - \dfrac{5}{12}$ 6. $\dfrac{13}{16} - \dfrac{5}{24}$

Section 9.2 Subtract.

7. $5\dfrac{3}{4} - 2\dfrac{2}{3}$ 8. $7\dfrac{3}{8} - 3\dfrac{5}{6}$ 9. $8\dfrac{1}{3} - 3$ 10. $11 - 4\dfrac{5}{8}$

CHAPTER POST-TEST

This test should determine if you have mastered the topics in Chapter 9. Answers to this test can be found on page 280.

Subtract.

1. $\dfrac{8}{13} - \dfrac{4}{13}$ 2. $3\dfrac{5}{12} - 2\dfrac{3}{12}$ 3. $12\dfrac{5}{6} - 7\dfrac{7}{10}$ 4. $9\dfrac{4}{7} - 2\dfrac{1}{5}$

5. $5\dfrac{7}{12} - 2\dfrac{8}{9}$ 6. $5\dfrac{7}{45} - \dfrac{10}{27}$ 7. $5\dfrac{2}{3} - 3$ 8. $6 - 5\dfrac{1}{9}$

9. A piece of wood $2\dfrac{3}{4}$ feet long is cut from an 8-foot board. How much remains?

10. Sue studied for $6\dfrac{1}{4}$ hours, whereas Steve studied for $5\dfrac{1}{2}$ hours. How much longer did Sue study?

10

multiplication and division of fractions and mixed numbers

MULTIPLICATION OF FRACTIONS AND MIXED NUMBERS

An object weighs ⅙ as much on the moon as it does on the earth.

Object	Earth	Moon
Man	180 pounds	30 pounds
Dog	70 pounds	11⅔ pounds

After studying this section you will be able to multiply fractions and mixed numbers, as shown in this example.

$$5 \times 1\frac{1}{2} \times \frac{2}{5} = \frac{5}{1} \times \frac{3}{2} \times \frac{2}{5}$$

$$= \frac{\overset{1}{\cancel{5}} \times 3 \times \overset{1}{\cancel{2}}}{1 \times \underset{1}{\cancel{2}} \times \underset{1}{\cancel{5}}}$$

$$= \frac{3}{1}$$

$$= 3$$

230

Multiplication of Fractions

EXAMPLE 1:   Dr. Newman, a psychologist, wishes to divide a group of 9 people into two separate groups. She wants $\frac{2}{3}$ of the people in the first group. How many people will be in the first group?

First, divide the whole group into 3 equal parts.

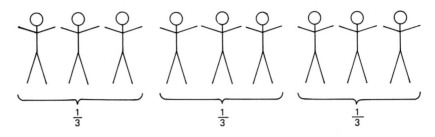

As the figure shows, $\frac{2}{3}$ of the group of 9 people is clearly 6 people. That is,

$$\frac{2}{3} \text{ of } 9 = 6$$

EXAMPLE 2:   A father decides to share a candy bar with his three children. He takes half, and shares the other half equally among the children. Each child receives $\frac{1}{3}$ of $\frac{1}{2}$ of a candy bar. What part of the whole bar does each child receive?

First, divide the bar in half.

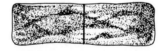

Next, divide each half into 3 equal pieces.

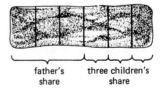

father's share   three children's share

Notice that each child receives $\frac{1}{6}$ of the whole bar.

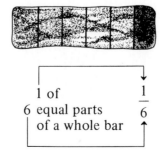

1 of 6 equal parts of a whole bar   → $\frac{1}{6}$

Therefore,

$$\frac{1}{3} \text{ of } \frac{1}{2} = \frac{1}{6}$$

Both of these problems can be solved by the procedure for multiplication of fractions.

Multiplication of Fractions

To find the product of two fractions, multiply the numerators to get the numerator of the product; multiply the denominators to get the denominator of the product. Reduce the resulting fraction to lowest terms, whenever possible.

EXAMPLE 3:   Multiply $\dfrac{2}{3}$ and $\dfrac{5}{9}$.

$$\frac{2}{3} \times \frac{5}{9} = \frac{2 \times 5}{3 \times 9} \qquad \text{Multiply numerators and multiply denominators.}$$

$$= \frac{10}{27}$$

EXAMPLE 4:   Any problem involved with finding a fractional part *of* a number can be solved by multiplication. Parts (a) and (b) show Examples 1 and 2 worked by multiplication.

(a) $\dfrac{2}{3}$ *of* $9 = \dfrac{2}{3} \times 9$

$\qquad\qquad = \dfrac{2}{3} \times \dfrac{9}{1} \qquad$ Write 9 as a fraction.

$\qquad\qquad = \dfrac{2 \times 9}{3 \times 1}$

$\qquad\qquad = \dfrac{18}{3}$

$\qquad\qquad = 18 \div 3$

$\qquad\qquad = 6$

(b) $\dfrac{1}{3}$ *of* $\dfrac{1}{2} = \dfrac{1}{3} \times \dfrac{1}{2}$

$\qquad\qquad = \dfrac{1 \times 1}{3 \times 2}$

$\qquad\qquad = \dfrac{1}{6}$

EXAMPLE 5:   Multiply $\dfrac{5}{9}$ and $\dfrac{6}{25}$.

$$\frac{5}{9} \times \frac{6}{25} = \frac{5 \times 6}{9 \times 25}$$

$$= \frac{30}{225}$$

$$= \frac{6}{45} \qquad \text{Reduce using division by 5.}$$

$$= \frac{2}{15} \qquad \text{Reduce using division by 3.}$$

It is more efficient to reduce *before* completing the calculations in the multiplication of fractions.

$$\frac{5}{9} \times \frac{6}{25} = \frac{5 \times 6}{9 \times 25}$$

Stop! Do not multiply! First, divide the factor 5 in the numerator and the factor 25 in the denominator by 5.

$$\frac{5}{9} \times \frac{6}{25} = \frac{\overset{1}{\cancel{5}} \times 6}{9 \times \underset{5}{\cancel{25}}} \qquad \begin{matrix} 5 \div 5 = 1 \\ 25 \div 5 = 5 \end{matrix}$$

Next, divide the factor 6 in the numerator and the factor 9 in the denominator by 3.

$$\frac{5}{9} \times \frac{6}{25} = \frac{\overset{1}{\cancel{5}} \times \overset{2}{\cancel{6}}}{\underset{3}{\cancel{9}} \times \underset{5}{\cancel{25}}} \qquad \begin{matrix} 6 \div 3 = 2 \\ 9 \div 3 = 3 \end{matrix}$$

Finally, multiply the remaining factors.

$$\frac{5}{9} \times \frac{6}{25} = \frac{\overset{1}{\cancel{5}} \times \overset{2}{\cancel{6}}}{\underset{3}{\cancel{9}} \times \underset{5}{\cancel{25}}}$$

$$= \frac{1 \times 2}{3 \times 5}$$

$$= \frac{2}{15}$$

As you can see, you get the same answer. However, it is easier to reduce in factored form because the numbers are smaller.

EXAMPLE 6: Multiply $\frac{14}{9}$ and $\frac{12}{21}$.

$$\frac{14}{9} \times \frac{12}{21} = \frac{14 \times 12}{9 \times 21}$$

Stop! Try to find a number that will divide into a factor of both the numerator and denominator.
Divide the factors 14 and 21 by 7.

$$= \frac{\overset{2}{\cancel{14}} \times 12}{9 \times \underset{3}{\cancel{21}}} \qquad \begin{matrix} 14 \div 7 = 2 \\ 21 \div 7 = 3 \end{matrix}$$

Divide the factors 12 and 9 by 3.

$$= \frac{\overset{2}{\cancel{14}} \times \overset{4 \leftarrow\!\!-12 \div 3 = 4}{\cancel{12}}}{\underset{3}{\cancel{9}} \times \underset{3}{\cancel{21}}} \quad 9 \div 3 = 3$$

Finally, multiply.

$$\frac{14}{9} \times \frac{12}{21} = \frac{\overset{2}{\cancel{14}} \times \overset{4}{\cancel{12}}}{\underset{3}{\cancel{9}} \times \underset{3}{\cancel{21}}}$$

$$= \frac{2 \times 4}{3 \times 3}$$

$$= \frac{8}{9}$$

The answer is already in lowest terms because reduction took place before the final calculations.

EXAMPLE 7:   Multiply $\frac{3}{7}, \frac{35}{36}$, and $\frac{8}{15}$.

$$\frac{3}{7} \times \frac{35}{36} \times \frac{8}{15} = \frac{3 \times 35 \times 8}{7 \times 36 \times 15}$$

Try to find all possible reductions before completing these multiplications.

As before, there is more than one *correct* way to get the final, reduced answer. Here is my approach.

I first consider the factor 3 in the numerator with each factor in the denominator.

$$\frac{3}{7} \times \frac{35}{36} \times \frac{8}{15} = \frac{\cancel{3} \times 35 \times 8}{7 \times \cancel{36} \times \cancel{15}}$$

The only factor 3 and 7 have in common is 1, but 3 and 36 are both divisible by 3. Dividing by 3,

$$= \frac{\overset{1 \leftarrow\!\!-\ 3 \div 3 = 1}{\cancel{3}} \times 35 \times 8}{7 \times \underset{12 \leftarrow\!\!-\ 36 \div 3 = 12}{\cancel{36}} \times 15}$$

Because the factor 3 is now 1, I next consider the factor 35 with each remaining factor in the denominator.

$$\frac{3}{7} \times \frac{35}{36} \times \frac{8}{15} = \frac{\overset{1}{\cancel{3}} \times \cancel{35} \times 8}{7 \times \underset{12}{\cancel{36}} \times \cancel{15}}$$

Both 35 and 7 are divisible by 7. Dividing by 7,

$$= \frac{\overset{1}{\cancel{3}} \times \overset{5 \leftarrow\!\!-\ 35 \div 7 = 5}{\cancel{35}} \times 8}{\underset{1}{\cancel{7}} \times \underset{12}{\cancel{36}} \times 15} \quad 7 \div 7 = 1$$

Because there is now a factor of 5 in the numerator, I continue to look at the factors in the denominator using the factor 5. The only common factor of 5 and 1 is 1, so I next consider the factor 12 in the denominator. Other than 1, nothing divides both 5 and 12 evenly. Finally, I considered 5 and 15. Both are divisible by 5.

$$= \frac{\overset{1}{\overset{\cancel{5}}{3} \times 35 \times 8}}{\underset{1 \quad 12 \quad 3}{7 \times 36 \times \cancel{15}}} \qquad \begin{matrix} \leftarrow 5 \div 5 = 1 \\ \\ \\ \leftarrow 15 \div 5 = 3 \end{matrix}$$

Next, I consider the factor 8 in the numerator with each of the remaining factors in the denominator.

$$\frac{3}{7} \times \frac{35}{36} \times \frac{8}{15} = \frac{\overset{1}{\cancel{3}} \times 35 \times \cancel{8}}{7 \times \cancel{36} \times \cancel{15}}$$

The only common factor of 8 and 1 is 1. The factors 8 and 12 are both divisible by 4.

$$= \frac{\overset{1}{\overset{\cancel{5}}{3}} \times 35 \times \overset{2}{\cancel{8}}}{\underset{1 \quad \underset{3}{\cancel{12}} \quad 3}{7 \times 36 \times 15}} \qquad \begin{matrix} \leftarrow 8 \div 4 = 2 \\ \\ \\ \leftarrow 12 \div 4 = 3 \end{matrix}$$

I continue the consideration with the 2 in the numerator; 2 and 3 have no divisors in common other than 1.

All reductions are complete. Now I multiply the remaining factors.

$$\frac{3}{7} \times \frac{35}{36} \times \frac{8}{15} = \frac{\overset{1}{\overset{\cancel{5}}{3}} \times 35 \times \overset{2}{\cancel{8}}}{\underset{1 \quad \underset{3}{\cancel{12}} \quad 3}{7 \times 36 \times 15}}$$

$$= \frac{1 \times 1 \times 2}{1 \times 3 \times 3}$$

$$= \frac{2}{9}$$

Practice finding all possible reductions by performing the following multiplications.

Examples	Solutions
$\dfrac{2}{7} \times \dfrac{3}{8} = ?$	$\dfrac{2}{7} \times \dfrac{3}{8} = \dfrac{2 \times 3}{7 \times 8}$

Examples	Solutions
	$1 \longleftarrow 2 \div 2 = 1$ $= \dfrac{\cancel{2} \times 3}{7 \times \cancel{8}}$ $\phantom{= \dfrac{2 \times 3}{7 \times}} 4 \longleftarrow 8 \div 2 = 4$ $= \dfrac{1 \times 3}{7 \times 4}$ $= \dfrac{3}{28}$
$\dfrac{35}{36} \times \dfrac{6}{25} = ?$	$\dfrac{35}{36} \times \dfrac{6}{25} = \dfrac{35 \times 6}{36 \times 25}$ $ 7 \longleftarrow 35 \div 5 = 7$ $= \dfrac{\cancel{35} \times 6}{36 \times \cancel{25}}$ $\phantom{= \dfrac{35 \times 6}{36 \times}} 5 \longleftarrow 25 \div 5 = 5$ $ 7 \quad 1 \longleftarrow 6 \div 6 = 1$ $= \dfrac{\cancel{35} \times \cancel{6}}{\cancel{36} \times \cancel{25}}$ $ 6 \quad 5 \qquad 36 \div 6 = 6$ $= \dfrac{7 \times 1}{6 \times 5}$ $= \dfrac{7}{30}$
$\dfrac{12}{13} \times \dfrac{5}{9} \times \dfrac{4}{15} = ?$	$\dfrac{12}{13} \times \dfrac{5}{9} \times \dfrac{4}{15} = \dfrac{12 \times 5 \times 4}{13 \times 9 \times 15}$ $ 4 \quad 1$ $= \dfrac{\cancel{12} \times \cancel{5} \times 4}{13 \times \cancel{9} \times \cancel{15}}$ $ 3 \quad 3$ $= \dfrac{16}{117}$ 12 and 9 were divided by 3; whereas 5 and 15 were divided by 5.

Multiplication with Mixed Numbers

> **Multiplication of Whole or Mixed Numbers**
>
> Change whole numbers or mixed numbers to improper fractions and then use the multiplication procedure for fractions.

EXAMPLE 8:   Multiply $\dfrac{2}{3}$ and 27.

$$\frac{2}{3} \times 27 = \frac{2}{3} \times \frac{27}{1}$$

$$= \frac{2 \times 27}{3 \times 1}$$

$$= \frac{2 \times \overset{9}{\cancel{27}}}{\underset{1}{\cancel{3}} \times 1}$$

$$= \frac{18}{1}$$

$$= 18$$

EXAMPLE 9:   Multiply $2\frac{1}{2}$ and $\frac{3}{5}$.

$$2\frac{1}{2} \times \frac{3}{5} = \frac{5}{2} \times \frac{3}{5} \qquad \text{Change } 2\frac{1}{2} \text{ to an improper fraction.}$$

$$= \frac{5 \times 3}{2 \times 5}$$

$$= \frac{\overset{1}{\cancel{5}} \times 3}{2 \times \underset{1}{\cancel{5}}}$$

$$= \frac{3}{2}$$

$$= 1\frac{1}{2}$$

EXAMPLE 10:   Multiply 5, $1\frac{1}{2}$, and $\frac{2}{5}$.

$$5 \times 1\frac{1}{2} \times \frac{2}{5} = \frac{5}{1} \times \frac{3}{2} \times \frac{2}{5}$$

$$= \frac{5 \times 3 \times 2}{1 \times 2 \times 5}$$

$$= \frac{\overset{1}{\cancel{5}} \times 3 \times \overset{1}{\cancel{2}}}{1 \times \underset{1}{\cancel{2}} \times \underset{1}{\cancel{5}}}$$

$$= \frac{3}{1}$$

$$= 3$$

EXAMPLE 11:   Multiply $6\frac{2}{5}$, $\frac{15}{16}$, and 4.

$$6\frac{2}{5} \times \frac{15}{16} \times 4 = \frac{32}{5} \times \frac{15}{16} \times \frac{4}{1}$$

$$= \frac{32 \times 15 \times 4}{5 \times 16 \times 1}$$

$$= \frac{\overset{2}{\overset{\cancel{4}}{\cancel{32}}} \times \overset{3}{\cancel{15}} \times 4}{\underset{1}{\cancel{5}} \times \underset{\underset{1}{\cancel{2}}}{\cancel{16}} \times 1}$$

I first divided 32 and 16 by 4; then I divided the resulting 4 and 2 by 2. Finally, I divided 15 and 5 by 5.

$$= \frac{2 \times 3 \times 4}{1 \times 1 \times 1}$$

$$= \frac{24}{1}$$

$$= 24$$

If you have difficulty finding all the reductions, particularly with larger numbers, you can use prime factorization to reduce.

EXAMPLE 12:   Multiply $\dfrac{65}{462}$ and $1\dfrac{31}{35}$.

$$\frac{65}{462} \times 1\frac{31}{35} = \frac{65}{462} \times \frac{66}{35}$$

$$= \frac{65 \times 66}{462 \times 35}$$

Because these factors are large, find the prime factorization of each (work not shown) and cancel any common primes.

$$\frac{65}{462} \times 1\frac{31}{35} = \frac{65 \times 66}{462 \times 35}$$

$$= \frac{5 \times 13 \times 2 \times 3 \times 11}{2 \times 3 \times 7 \times 11 \times 5 \times 7}$$

$$= \frac{\overset{1}{\cancel{5}} \times 13 \times \overset{1}{\cancel{2}} \times \overset{1}{\cancel{3}} \times \overset{1}{\cancel{11}}}{\underset{1}{\cancel{2}} \times \underset{1}{\cancel{3}} \times 7 \times \underset{1}{\cancel{11}} \times \underset{1}{\cancel{5}} \times 7}$$

$$= \frac{13}{49}$$

Now try these problems. If you cannot get the answer, talk with an instructor or read this section again.

Examples	Solutions
$\dfrac{7}{8} \times \dfrac{2}{14} = ?$	$\dfrac{1}{8}$
$\dfrac{3}{7} \times \dfrac{3}{8} = ?$	$\dfrac{9}{56}$
$\dfrac{5}{9} \times \dfrac{36}{25} = ?$	$\dfrac{4}{5}$
$\dfrac{7}{45} \times \dfrac{9}{14} \times \dfrac{5}{36} = ?$	$\dfrac{1}{72}$
$14 \times \dfrac{3}{4} = ?$	$\dfrac{21}{2} = 10\dfrac{1}{2}$
$\dfrac{4}{7} \times 24\dfrac{1}{2} = ?$	$\dfrac{14}{1} = 14$

**Exercises
Part I** Answers to these exercises can be found on page 280.

Find the indicated products.

1. $\dfrac{2}{3} \times \dfrac{3}{8}$ 2. $\dfrac{6}{8} \times \dfrac{4}{9}$ 3. $\dfrac{25}{28} \times \dfrac{21}{50}$ 4. $\dfrac{5}{7} \times \dfrac{3}{10} \times \dfrac{14}{15}$

5. $\dfrac{16}{17} \times \dfrac{3}{40} \times \dfrac{5}{6}$ 6. $\dfrac{7}{8} \times 1\dfrac{5}{7}$ 7. $9 \times 2\dfrac{2}{3}$ 8. $1\dfrac{11}{14} \times 2\dfrac{1}{10}$

9. $2\dfrac{6}{7} \times 4\dfrac{1}{5} \times \dfrac{2}{27}$ 10. $5\dfrac{1}{4} \times \dfrac{5}{18} \times 12$

**Exercises
Part II** 1. $\dfrac{3}{4} \times \dfrac{8}{15}$ 2. $\dfrac{3}{4} \times 36$ 3. $1\dfrac{1}{7} \times 4\dfrac{1}{5} \times 2\dfrac{1}{2}$ 4. $7 \times 3\dfrac{1}{14}$

5. $\dfrac{3}{8} \times 4 \times 1\dfrac{1}{7}$

SECTION 10.2
DIVISION OF FRACTIONS AND MIXED NUMBERS

RAINFALL FOR THE WEEK

MON	TUE	WED	THUR	FRI
½ inch	dry	⅖ inch	1/10 inch	dry

The average rainfall for the week: ⅕ inch

After reading this section you will understand the procedure for division of fractions as illustrated by this example.

$$\frac{2}{3} \div \frac{4}{9} = \frac{2}{3} \times \frac{9}{4}$$

$$= \frac{\overset{1}{\cancel{2}} \times \overset{3}{\cancel{9}}}{\underset{1}{\cancel{3}} \times \underset{2}{\cancel{4}}}$$

$$= \frac{3}{2}$$

$$= 1\frac{1}{2}$$

Division of Fractions and Mixed Numbers

The expression

$$\frac{2}{3} \div \frac{4}{9} = ?$$

is read: $\frac{2}{3}$ divided by $\frac{4}{9}$ equals what? Therefore, $\frac{4}{9}$ is the divisor.

Another way of writing the same statement is

$$\frac{\frac{2}{3}}{\frac{4}{9}} = ?$$

This expression is called a **complex fraction** because both the numerator and denominator are fractions. The complex denominator is the divisor.

$$\frac{\frac{2}{3}}{\frac{4}{9}} = \frac{2}{3} \div \frac{4}{9}$$

divisor

Inverting

The rule for division by fractions involves a procedure called **inverting**, which means that the numerator and denominator switch positions. When inverting whole numbers and mixed numbers, write them as improper fractions before inverting. The following are examples of inverting.

$$\frac{2}{5} \quad \text{becomes} \quad \frac{5}{2}$$

$$\frac{1}{5} \quad \text{becomes} \quad \frac{5}{1} = 5$$

$$7 \quad \left(\text{which equals } \frac{7}{1}\right) \quad \text{becomes} \quad \frac{1}{7}$$

$$2\frac{1}{3} \quad \left(\text{which equals } \frac{7}{3}\right) \quad \text{becomes} \quad \frac{3}{7}$$

Division by a Fraction

Here is the rule for dividing by a fraction.

> **Division by Fractions**
>
> To divide by a fraction, invert the divisor and multiply. Reduce the resulting fraction, whenever possible.

EXAMPLE 1: $\dfrac{2}{3} \div \dfrac{4}{9} = ?$

$$\frac{2}{3} \div \frac{4}{9} = \frac{2}{3} \times \frac{9}{4}$$

change to X

invert

$$= \frac{2 \times 9}{3 \times 4}$$

$$= \frac{\overset{1}{\cancel{2}} \times \overset{3}{\cancel{9}}}{\underset{1}{\cancel{3}} \times \underset{2}{\cancel{4}}}$$

$$= \frac{3}{2}$$

$$= 1\frac{1}{2}$$

EXAMPLE 2: Divide $\dfrac{7}{8}$ by $\dfrac{7}{16}$.

$$\frac{7}{8} \div \frac{7}{16} = \frac{7}{8} \times \frac{16}{7}$$

invert divisor and multiply

$$= \frac{7 \times 16}{8 \times 7}$$

$$= \frac{\overset{1}{\cancel{7}} \times \overset{2}{\cancel{16}}}{\underset{1}{\cancel{8}} \times \underset{1}{\cancel{7}}}$$

$$= \frac{2}{1}$$

$$= 2$$

Dividing by fractions is easy if you remember the rule and know which fraction is the divisor!

Here is an example with a whole number that shows that this division procedure, or rule, *really works*.

EXAMPLE 3:   If you divide 3 candy bars into halves, how many halves are there?

Arithmetically,

$$3 \div \frac{1}{2}$$

gives the number of halves in 3. Inverting the divisor and multiplying,

$$3 \div \frac{1}{2} = 3 \times \frac{2}{1}$$

$$= \frac{3}{1} \times \frac{2}{1}$$

$$= \frac{3 \times 2}{1 \times 1}$$

$$= \frac{6}{1}$$

$$= 6$$

Therefore, there should be 6 halves in 3 candy bars. The figure illustrates the result.

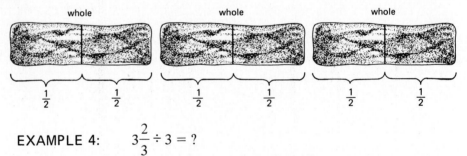

EXAMPLE 4: $3\frac{2}{3} \div 3 = ?$

First, change both the whole and the mixed number to improper fractions in order to invert the divisor and multiply.

$$3\frac{2}{3} \div 3 = \frac{11}{3} \div \frac{3}{1}$$

$$= \frac{11}{3} \times \frac{1}{3} \qquad \text{Invert the divisor (only) and multiply.}$$

$$= \frac{11 \times 1}{3 \times 3}$$

$$= \frac{11}{9}$$

$$= 1\frac{2}{9}$$

EXAMPLE 5: $\dfrac{2\dfrac{1}{3}}{\dfrac{3}{4}} = ?$

$$\frac{2\frac{1}{3}}{\frac{3}{4}} = 2\frac{1}{3} \div \frac{3}{4}$$

$$= \frac{7}{3} \div \frac{3}{4}$$

$$= \frac{7}{3} \times \frac{4}{3}$$

$$= \frac{7 \times 4}{3 \times 3}$$

$$= \frac{28}{9}$$

$$= 3\frac{1}{9}$$

EXAMPLE 6: $\dfrac{\dfrac{3}{4}}{2} = ?$

$$\frac{\frac{3}{4}}{2} = \frac{3}{4} \div 2$$

$$= \frac{3}{4} \div \frac{2}{1}$$

$$= \frac{3}{4} \times \frac{1}{2}$$

$$= \frac{3 \times 1}{4 \times 2}$$

$$= \frac{3}{8}$$

Now you try the following division problems.

Examples	Solutions
$\dfrac{5}{6} \div \dfrac{1}{2} = ?$	$\dfrac{5}{6} \div \dfrac{1}{2} = \dfrac{5}{6} \times \dfrac{2}{1}$ Invert divisor and multiply. $$= \dfrac{5 \times 2}{6 \times 1}$$ $$= \dfrac{5 \times \overset{1}{\cancel{2}}}{\underset{3}{\cancel{6}} \times 1}$$ $$= \dfrac{5}{3}$$ $$= 1\dfrac{2}{3}$$
$6 \div \dfrac{2}{3} = ?$	$6 \div \dfrac{2}{3} = \dfrac{6}{1} \div \dfrac{2}{3}$ $$= \dfrac{6}{1} \times \dfrac{3}{2}$$ $$= \dfrac{6 \times 3}{1 \times 2}$$ $$= \dfrac{\overset{3}{\cancel{6}} \times 3}{1 \times \underset{1}{\cancel{2}}}$$ $$= \dfrac{9}{1}$$ $$= 9$$
$\dfrac{1}{2} \div 6 = ?$	$\dfrac{1}{2} \div 6 = \dfrac{1}{2} \div \dfrac{6}{1}$ $$= \dfrac{1}{2} \times \dfrac{1}{6}$$ $$= \dfrac{1 \times 1}{2 \times 6}$$ $$= \dfrac{1}{12}$$
$4\dfrac{1}{2} \div 1\dfrac{3}{4} = ?$	$4\dfrac{1}{2} \div 1\dfrac{3}{4} = \dfrac{9}{2} \div \dfrac{7}{4}$ $$= \dfrac{9}{2} \times \dfrac{4}{7}$$

Examples	Solutions
	$$= \dfrac{9 \times \overset{2}{\cancel{4}}}{\underset{1}{\cancel{2}} \times 7}$$ $$= \dfrac{18}{7}$$ $$= 2\dfrac{4}{7}$$
$\dfrac{\frac{3}{4}}{\frac{9}{16}} = ?$	$$\dfrac{\frac{3}{4}}{\frac{9}{16}} = \dfrac{3}{4} \div \dfrac{9}{16}$$ $$= \dfrac{3}{4} \times \dfrac{16}{9}$$ $$= \dfrac{\overset{1}{\cancel{3}} \times \overset{4}{\cancel{16}}}{\underset{1}{\cancel{4}} \times \underset{3}{\cancel{9}}}$$ $$= \dfrac{4}{3}$$ $$= 1\dfrac{1}{3}$$
$\dfrac{12}{\frac{2}{3}} = ?$	$$\dfrac{12}{\frac{2}{3}} = 12 \div \dfrac{2}{3}$$ $$= \dfrac{12}{1} \div \dfrac{2}{3}$$ $$= \dfrac{12}{1} \times \dfrac{3}{2}$$ $$= \dfrac{\overset{6}{\cancel{12}} \times 3}{1 \times \underset{1}{\cancel{2}}}$$ $$= \dfrac{18}{1}$$ $$= 18$$

Now try these—no hints given.

$\dfrac{3}{14} \div \dfrac{5}{6} = ?$	$\dfrac{9}{35}$

Examples	Solutions
$3\dfrac{1}{3} \div 10 = ?$	$\dfrac{1}{3}$
$\dfrac{\dfrac{3}{4}}{2\dfrac{1}{3}} = ?$	$\dfrac{9}{28}$

If you had difficulty with the last three problems, talk with an instructor or read this section again.

Exercises Part I

Answers to these exercises can be found on page 280.

Perform the indicated divisions.

1. $\dfrac{5}{16} \div \dfrac{3}{8}$ 2. $\dfrac{5}{9} \div \dfrac{25}{36}$ 3. $\dfrac{1}{2} \div \dfrac{1}{4}$ 4. $\dfrac{2}{3} \div \dfrac{1}{6}$

5. $\dfrac{4}{5} \div 6$ 6. $6 \div \dfrac{1}{2}$ 7. $\dfrac{15}{28} \div \dfrac{6}{16}$ 8. $6\dfrac{2}{5} \div \dfrac{16}{25}$

9. $3\dfrac{3}{4} \div 2\dfrac{1}{2}$ 10. $1\dfrac{1}{4} \div 1\dfrac{7}{8}$ 11. $\dfrac{\dfrac{7}{40}}{\dfrac{21}{25}}$ 12. $\dfrac{15}{\dfrac{5}{9}}$

Exercises Part II

Divide.

1. $\dfrac{1}{2} \div \dfrac{2}{3}$ 2. $\dfrac{6}{7} \div \dfrac{13}{14}$ 3. $\dfrac{12}{27} \div \dfrac{10}{18}$ 4. $7\dfrac{5}{11} \div 4\dfrac{1}{10}$

5. $\dfrac{2\dfrac{1}{2}}{5}$ 6. $\dfrac{\dfrac{1}{3}}{2}$

REVIEW MATERIAL

Review Skill Multiplication of Fractions and Mixed Numbers

To multiply fractions and mixed numbers, change mixed numbers to improper fractions and place whole numbers over the denominator 1. Next, divide out any common factors from the numerators and denominators, then multiply numerators and denominators separately.

$$\frac{2}{35} \times \frac{15}{18} = \frac{\overset{1}{\cancel{2}} \times 15}{35 \times \underset{9}{\cancel{18}}} \qquad \text{Divide by 2.}$$

$$= \frac{\overset{1}{\cancel{2}} \times \overset{3}{\cancel{15}}}{\underset{7}{\cancel{35}} \times \underset{9}{\cancel{18}}}$$ Divide by 5.

$$= \frac{\overset{1}{\cancel{2}} \times \overset{\overset{1}{\cancel{3}}}{\cancel{15}}}{\underset{7}{\cancel{35}} \times \underset{\underset{3}{\cancel{9}}}{\cancel{18}}}$$ Divide by 3.

$$= \frac{1}{21}$$ Multiply numerators and denominators.

Exercise A Answers to this exercise can be found on page 280.

Multiply.

1. $\frac{4}{7} \times \frac{7}{4}$
2. $\frac{12}{35} \times \frac{7}{24}$
3. $\frac{2}{5} \times \frac{5}{22}$
4. $\frac{2}{3} \times \frac{3}{7}$

5. $\frac{1}{4} \times 40$
6. $1\frac{1}{2} \times \frac{2}{5}$
7. $3\frac{3}{8} \times 1\frac{7}{9}$
8. $\frac{2}{9} \times \frac{12}{8} \times \frac{4}{5}$

9. $4 \times 2\frac{1}{3} \times 1\frac{1}{8}$
10. $2\frac{1}{2} \times 1\frac{3}{4} \times \frac{8}{35}$

Exercise B No answers are given.

Multiply.

1. $\frac{2}{3} \times \frac{3}{2}$
2. $\frac{8}{100} \times \frac{10}{16}$
3. $\frac{3}{4} \times \frac{8}{9}$
4. $\frac{2}{3} \times \frac{6}{7}$

5. $\frac{3}{5} \times 20$
6. $3\frac{5}{6} \times \frac{3}{10}$
7. $10\frac{5}{6} \times 3\frac{3}{10}$

**Review Skill
Division
of Fractions and
Mixed Numbers** Divide fractions and mixed numbers by inverting the divisor and multiplying.

$$\frac{3}{4} \div 2\frac{5}{8} = \frac{3}{4} \div \frac{21}{8}$$

$$= \frac{3}{4} \times \frac{8}{21}$$ Invert and multiply.

$$= \frac{\overset{1}{\cancel{3}} \times \overset{2}{\cancel{8}}}{\underset{1}{\cancel{4}} \times \underset{7}{\cancel{21}}}$$

$$= \frac{2}{7}$$

Exercise C Answers to this exercise can be found on page 280.

Divide.

1. $\dfrac{3}{4} \div \dfrac{3}{7}$ 2. $6 \div \dfrac{2}{3}$ 3. $\dfrac{3}{14} \div \dfrac{6}{5}$ 4. $10 \div 1\dfrac{1}{5}$ 5. $4\dfrac{1}{2} \div 1\dfrac{3}{4}$

6. $\dfrac{2}{3} \div 6$ 7. $\dfrac{2\dfrac{1}{3}}{\dfrac{3}{4}}$ 8. $6\dfrac{5}{6} \div \dfrac{1}{2}$ 9. $5 \div 1\dfrac{1}{4}$ 10. $\dfrac{\dfrac{3}{4}}{6}$

Exercise D No answers are given.

Divide.

1. $\dfrac{5}{6} \div \dfrac{1}{2}$ 2. $8 \div \dfrac{1}{3}$ 3. $\dfrac{3}{4} \div \dfrac{5}{16}$ 4. $6 \div 1\dfrac{1}{2}$ 5. $6\dfrac{2}{5} \div 5\dfrac{1}{3}$

6. $2\dfrac{1}{2} \div 2$ 7. $\dfrac{\dfrac{3}{4}}{1\dfrac{1}{2}}$

REVIEW EXERCISES

Refer to the section listed if you have difficulty. Answers to these exercises can be found on page 280.

Section 10.1 Multiply.

1. $\dfrac{3}{8} \times \dfrac{6}{21}$ 2. $8 \times 7\dfrac{3}{4}$ 3. $2\dfrac{1}{6} \times 1\dfrac{1}{2}$

Section 10.2 Divide.

4. $\dfrac{8}{9} \div \dfrac{2}{3}$ 5. $2\dfrac{2}{5} \div 3$ 6. $4\dfrac{1}{6} \div 3\dfrac{1}{3}$ 7. $\dfrac{15}{\dfrac{3}{4}}$ 8. $\dfrac{\dfrac{5}{8}}{12}$

CHAPTER POST-TEST

This test should determine if you have mastered the topics in Chapter 10. Answers to this test can be found on page 280.

Multiply.

1. $\dfrac{9}{25} \times 1\dfrac{2}{3}$ 2. $5 \times \dfrac{3}{5}$ 3. $\dfrac{24}{35} \times 2\dfrac{2}{3} \times 5$

Divide.

4. $\dfrac{2}{5} \div \dfrac{3}{7}$ 5. $3\dfrac{3}{5} \div 9$ 6. $8 \div \dfrac{3}{4}$ 7. $\dfrac{1\frac{3}{16}}{4\frac{3}{4}}$ 8. $\dfrac{\frac{5}{9}}{3}$

11

applications

SECTION 11.1
APPLICATIONS OF FRACTIONS

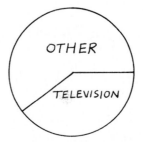

Approximately ⅖ of the total cost of advertising was spent on television advertising.

This section deals with applications of fractions.

Fractions   **EXAMPLE 1:**   How can 7 people equally share 3 oranges?
Divide 3 by 7.

$$3 \div 7 = \frac{3}{7}$$

To interpret the answer,

$$\dfrac{3}{7} \quad \text{means} \quad \dfrac{3 \text{ of}}{7 \text{ equal parts of a whole orange}}$$

Therefore, each person gets $\dfrac{3}{7}$ of an orange.

EXAMPLE 2: A book contains 136 pages. What fractional part of the book do 34 pages represent?

The 34 pages represent part of the whole book.

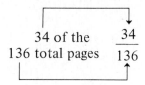

$$\dfrac{34 \text{ of the}}{136 \text{ total pages}} \quad \dfrac{34}{136}$$

Reducing,

$$\dfrac{34}{136} = \dfrac{\overset{1}{\cancel{2}} \times \overset{1}{\cancel{17}}}{\underset{1}{\cancel{2}} \times 2 \times 2 \times \underset{1}{\cancel{17}}}$$

$$= \dfrac{1}{4}$$

Therefore, 34 pages represent $\dfrac{1}{4}$ of the book.

EXAMPLE 3: A book of stamps contains 20 stamps. How many stamps represent $\dfrac{1}{5}$ of the book?

$$\dfrac{1}{5} = \dfrac{?}{20} \qquad \dfrac{1}{5} \text{ of the book equals how many of the 20 stamps?}$$

Build $\dfrac{1}{5}$ to the denominator 20. Because $20 \div 5 = 4$,

$$\dfrac{1}{5} = \dfrac{1 \times 4}{5 \times 4}$$

$$= \dfrac{4}{20}$$

Therefore, 4 stamps represent $\dfrac{1}{5}$ of the book.

Problems like Example 3 can be worked using the operation of multiplication, but for this section I want you to concentrate on the concept of fractions.

These three examples illustrate two special applications of fractions—(1) division of whole numbers and (2) parts of a whole. In Chapter 17, I discuss a third application of fractions—ratio and proportion.

The following is an application of ordering fractions.

EXAMPLE 4:   Two runners finish a race with times of $27\frac{1}{5}$ minutes and $27\frac{7}{30}$ minutes. Who won the race?

Build the fractional parts to a common denominator in order to compare the mixed numbers.

$$5 = 5^1$$
$$\underline{30 = 2^1 \times 3^1 \times 5^1}$$
$$\text{LCD} = 2 \times 3 \times 5$$
$$= 30$$

$$27\frac{1}{5} = 27\frac{1 \times 6}{5 \times 6} \qquad 27\frac{7}{30} = 27\frac{7 \times 1}{30 \times 1}$$
$$= 27\frac{6}{30} \qquad\qquad = 27\frac{7}{30}$$

Therefore, $27\frac{1}{5} < 27\frac{7}{30}$, and the person with the time $27\frac{1}{5}$ minutes came across the finish line first!

Now try to solve these problems.

Examples	Solutions
How can 3 people equally share 2 candy bars?	Dividing 2 by 3, $$2 \div 3 = \frac{2}{3}$$ Each person should get $\frac{2}{3}$ of a whole bar.
A book contains 8 sections. What fractional part of the book remains to be read if you have read 6 sections?	2 of the 8 sections of the book remain. Reduce by dividing by 2: $$\frac{2}{8} = \frac{1}{4}$$ Therefore, $\frac{1}{4}$ of the book remains to be read.
An exercise section contains 36 problems. How many problems make up $\frac{2}{3}$ of the exercises?	$\frac{2}{3}$ equals what part of the 36? $$\frac{2}{3} = \frac{?}{36}$$ Because $36 \div 3 = 12$, $$\frac{2}{3} = \frac{2 \times 12}{3 \times 12}$$ $$= \frac{24}{36}$$

Examples	Solutions
	Therefore, 24 problems make up $\frac{2}{3}$ of the exercises.
Which piece of wood is thicker, one $\frac{7}{16}$ inch thick or one $\frac{3}{8}$ inch thick?	Build both fractions to a common denominator. $$16 = 2^4$$ $$\frac{8 = 2^3}{LCD = 2^4}$$ $$= 16$$ $$\frac{7}{16} = \frac{7 \times 1}{16 \times 1} \qquad \frac{3}{8} = \frac{3 \times 2}{8 \times 2}$$ $$= \frac{7}{16} \qquad\qquad = \frac{6}{16}$$ Because 6 is less than 7, the piece $\frac{7}{16}$ inch thick is the thicker piece.

Now try these—no hints given.

Which is larger, $\frac{7}{30}$ or $\frac{6}{24}$?	$\frac{6}{24}$
How can 3 oranges be divided equally among 4 children?	Each child gets $\frac{3}{4}$ of an orange.
A book contains 124 pages. How many pages make up $\frac{3}{4}$ of the book?	93 pages.
There are 100 multiplication facts to memorize. If you have memorized 30 facts, what fractional part remains to be memorized?	$\frac{7}{10}$ of the facts.

If you had difficulty with the last three problems, please talk with an instructor or read this section again.

Exercises Part I

Answers to these exercises can be found on page 281.

1. A pizza is cut into 12 slices. What fractional part remains if 4 slices are eaten?

2. In a recent survey of 100 people, 40 people said they would vote for the referendum. What fractional part of the people surveyed were for the referendum?

3. How can 5 people equally share 3 pizzas?

4. The fall term at school is 10 weeks long. After 6 weeks, what fractional part of the term remains?

5. How can 4 people equally share 10 candy bars?

6. A pizza is cut into 12 slices. How many slices make up $\frac{2}{3}$ of the pizza?

7. The winter term at school is 10 weeks long. How many weeks make up $\frac{2}{5}$ of the term?

8. Which weighs more, $5\frac{1}{4}$ pounds of meat or $5\frac{5}{16}$ pounds of meat?

9. If a job is going to take 4 hours, how can 3 people equally share the work?

10. If an exercise section has 50 problems, how many problems make up $\frac{4}{5}$ of the exercise section?

Exercises Part II

1. A pizza is cut into 8 slices. What fractional part remains if 2 slices are eaten?
2. If a job is going to take 5 hours, how can 4 people equally share the work?
3. A pizza is cut into 8 slices. How many slices make up $\frac{1}{4}$ of the pizza?
4. If an exercise section has 15 problems, how many problems make up $\frac{4}{5}$ of the exercise section?

SECTION 11.2
APPLICATIONS OF THE ARITHMETIC OF FRACTIONS AND MIXED NUMBERS

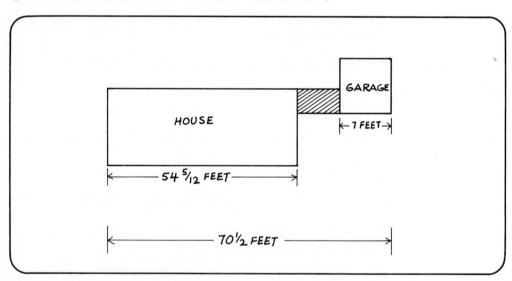

This section deals with applications of the arithmetic of fractions and mixed numbers.

Addition and Subtraction The operation of addition is used to find the total of values. The operation of subtraction is used to find the difference between two values, the change from one value to another, or what remains after some value is removed.

EXAMPLE 1:   What is the total distance around this figure?

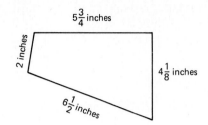

Adding up the distances,

$$5\frac{3}{4} = 5\frac{6}{8}$$

$$4\frac{1}{8} = 4\frac{1}{8}$$

$$6\frac{1}{2} = 6\frac{4}{8}$$

$$+2 = 2$$

$$17\frac{11}{8} = 17 + 1\frac{3}{8}$$

$$= 18\frac{3}{8}$$

The distance around the figure is $18\frac{3}{8}$ inches.

EXAMPLE 2:   A piece of cloth was $5\frac{1}{4}$ feet long before it was washed. If it measures $4\frac{5}{8}$ feet after washing, how much did the cloth shrink?

Subtracting to find the difference in the two measurements,

$$5\frac{1}{4} = 5\frac{2}{8} = 4\frac{10}{8} \qquad \text{Building and borrowing.}$$

$$-4\frac{5}{8} = 4\frac{5}{8} = 4\frac{5}{8}$$

$$\frac{5}{8}$$

The cloth shrank $\frac{5}{8}$ foot in length.

EXAMPLE 3:   On a trip to New York I bought $9\frac{1}{2}$ gallons of gas, and on the return trip I bought $10\frac{4}{5}$ gallons.

(a) How much gas did I buy on the trip?
(b) How much more gas did I buy on the return portion of the trip?

(a) Total the values.

$$9\frac{1}{2} = 9\frac{5}{10}$$

$$+\ 10\frac{4}{5} = 10\frac{8}{10}$$

$$19\frac{13}{10} = 20\frac{3}{10}$$

I purchased a total of $20\frac{3}{10}$ gallons of gas.

(b) Subtract to find the difference.

$$10\frac{4}{5} = 10\frac{8}{10}$$

$$-\ 9\frac{1}{2} = 9\frac{5}{10}$$

$$1\frac{3}{10}$$

I purchased $1\frac{3}{10}$ gallons more on the return trip.

EXAMPLE 4:   I cut a board $38\frac{3}{4}$ inches long from a board $47\frac{1}{2}$ inches long. If the cut wasted $\frac{1}{16}$ inch, how much of the $47\frac{1}{2}$-inch board remains?

Remove $38\frac{3}{4}$ inches by subtraction.

$$47\frac{1}{2} = 47\frac{2}{4} = 46\frac{6}{4} \qquad \text{Building and borrowing.}$$

$$-\ 38\frac{3}{4} = 38\frac{3}{4} = 38\frac{3}{4}$$

$$8\frac{3}{4}$$

But the cut wasted $\frac{1}{16}$ of an inch. Subtracting gives

$$8\frac{3}{4} = 8\frac{12}{16}$$

$$-\ \frac{1}{16} = \frac{1}{16}$$

$$8\frac{11}{16}$$

Therefore, $8\frac{11}{16}$ inches of the board remain.

Multiplication   Multiplication with fractions calculates fractional parts *of* another number. Multiplication with mixed numbers can also calculate repeated additions while accounting for the fraction parts.

EXAMPLE 5:   If cabbage is on sale for 24¢ per pound, how much do $3\frac{3}{4}$ pounds cost?

Because each pound costs 24¢, I need to add 24¢ for each pound and some additional amount for $\frac{3}{4}$ pound. Multiplication by $3\frac{3}{4}$ accounts for both parts.

$$\text{Three and three-fourths 24's} = 3\frac{3}{4} \times 24$$

$$= \frac{15}{4} \times \frac{24}{1}$$

$$= 90 \qquad \text{In lowest terms.}$$

Therefore, $3\frac{3}{4}$ pounds cost 90¢.

EXAMPLE 6:   A car traveled 45 miles per hour for $5\frac{1}{2}$ hours. How far did it go by the end of the trip?

The 45 miles must be added for each hour and an additional amount for $\frac{1}{2}$ hour. By multiplication

$$\text{Five and one-half 45's} = 5\frac{1}{2} \times 45$$

$$= \frac{11}{2} \times \frac{45}{1}$$

$$= 247\frac{1}{2} \qquad \text{As a mixed number.}$$

Therefore, the car traveled $247\frac{1}{2}$ miles in $5\frac{1}{2}$ hours.

EXAMPLE 7:   What is $\frac{1}{3}$ of a book that has 132 pages?

A fractional part *of* a number is easily found by *multiplication*.

$$\frac{1}{3} \text{ of } 132 = \frac{1}{3} \times 132$$

$$= \frac{1}{3} \times \frac{132}{1}$$

$$= 44 \qquad \text{In lowest terms.}$$

One-third of a book with 132 pages is 44 pages.

EXAMPLE 8:   Dan ate $\frac{1}{2}$ of a pie. Larry ate $\frac{1}{3}$ of what was left. What fractional part of the pie did Larry eat?

Larry ate $\frac{1}{3}$ of $\frac{1}{2}$ of the pie.

$$\frac{1}{3} \text{ of } \frac{1}{2} = \frac{1}{3} \times \frac{1}{2}$$

$$= \frac{1}{6}$$

Larry ate $\frac{1}{6}$ of a whole pie.

EXAMPLE 9: A recipe calls for $\frac{1}{2}$ cup of water and $1\frac{3}{4}$ cups of flour. How much of each of these ingredients should you use if you want to make $\frac{2}{3}$ of the recipe?

Take $\frac{2}{3}$ of each measurement.

$$\frac{2}{3} \text{ of } \frac{1}{2} = \frac{2}{3} \times \frac{1}{2}$$

$$= \frac{1}{3} \qquad \text{In lowest terms.}$$

Use $\frac{1}{3}$ cup of water.

$$\frac{2}{3} \text{ of } 1\frac{3}{4} = \frac{2}{3} \times \frac{7}{4}$$

$$= \frac{14}{12}$$

$$= 1\frac{1}{6}$$

Use $1\frac{1}{6}$ cups of flour (good luck measuring that!).

Division   Division is used to find the size of each equal part or to find the number of equal parts that make up a given quantity. Division is also used to reduce information to a per unit basis and to average a group of numbers.

EXAMPLE 10: A piece of fabric $6\frac{7}{8}$ yards long is divided into 11 pieces. How long is each piece?

In other words, eleven pieces of what size equal $6\frac{7}{8}$?

$$11 \times \boxed{} = 6\frac{7}{8}?$$

Eleven times what value equals $6\frac{7}{8}$ is a division question. Dividing the product by the known factor 11 gives the missing factor.

$$6\frac{7}{8} \div 11 = \frac{55}{8} \div \frac{11}{1}$$

$$= \frac{55}{8} \times \frac{1}{11}$$

$$= \frac{5}{8} \qquad \text{In lowest terms.}$$

Each of the 11 pieces is $\frac{5}{8}$ yard long.

EXAMPLE 11: How many $2\frac{1}{4}$-foot pieces can be cut from a board that is $15\frac{3}{4}$ feet long?

In other words, how many of the equal parts of size $2\frac{1}{4}$ equal $15\frac{3}{4}$?

$$\square \times 2\frac{1}{4} = 15\frac{3}{4}?$$

What value times $2\frac{1}{4}$ equals $15\frac{3}{4}$ is a division question. Dividing by the known factor,

$$15\frac{3}{4} \div 2\frac{1}{4} = \frac{63}{4} \div \frac{9}{4}$$

$$= \frac{63}{4} \times \frac{4}{9}$$

$$= 7 \qquad \text{In lowest terms.}$$

Therefore, 7 pieces $2\frac{1}{4}$ feet long can be cut from a board $15\frac{3}{4}$ feet long.

EXAMPLE 12: If I drove 252 miles in $4\frac{1}{2}$ hours, what was my average rate in miles per hour?

In other words, $4\frac{1}{2}$ times what rate equals 252 miles?

$$4\frac{1}{2} \times \square = 252?$$

Dividing,

$$252 \div 4\frac{1}{2} = \frac{252}{1} \div \frac{9}{2}$$

$$= \frac{252}{1} \times \frac{2}{9}$$

$$= 56 \qquad \text{In lowest terms.}$$

My average rate of speed was 56 miles per hour.

Clue: With per unit problems, the unit after the word *per* indicates the divisor. In the previous example,

$$\text{miles per hour}$$
$$\uparrow$$

hour follows the word per.

Therefore, you divide the number of miles by the number of hours.

$$252 \div 4\frac{1}{2} = ?$$
$$\uparrow$$

per hour

EXAMPLE 13:   If I drive 264 miles on $12\frac{4}{5}$ gallons of gas, how many miles do I get per gallon?

I am looking for miles per gallon; therefore, I divide by the number of gallons.

$$264 \div 12\frac{4}{5} = \frac{264}{1} \div \frac{64}{5}$$
$$\uparrow$$

per gallon

$$= \frac{264}{1} \times \frac{5}{64}$$

$$= \frac{165}{8} \qquad \text{In lowest terms.}$$

$$= 20\frac{5}{8} \qquad \text{As a mixed number.}$$

Therefore, I get $20\frac{5}{8}$ miles per gallon of gas.

EXAMPLE 14:   Over a 3-week period James bought $10\frac{3}{5}$ gallons, $9\frac{1}{10}$ gallons, and 12 gallons of gas. What is the average amount of gas James bought each week?

Average by adding the values and dividing the sum by the number of values.

$$\text{Average} = \left(10\frac{3}{5} + 9\frac{1}{10} + 12\right) \div 3$$

Adding,

$$10\frac{3}{5} = 10\frac{6}{10}$$

$$9\frac{1}{10} = 9\frac{1}{10}$$

$$+ 12 = 12$$

$$\overline{\qquad\qquad 31\frac{7}{10}}$$

Dividing by 3,

$$31\frac{7}{10} \div 3 = \frac{317}{10} \div \frac{3}{1}$$

$$= \frac{317}{10} \times \frac{1}{3}$$

$$= \frac{317}{30}$$

$$= 10\frac{17}{30}$$

James bought an average of $10\frac{17}{30}$ gallons each week.

Now it is your turn! Practice these problems using addition, subtraction, multiplication, or division.

Examples	Solutions
A large board is $5\frac{1}{2}$ feet long. A smaller board is $2\frac{3}{4}$ feet long.	
(a) What is the total length of the two boards laid end to end?	(a) Adding, $$5\frac{1}{2} = 5\frac{2}{4}$$ $$+ 2\frac{3}{4} = 2\frac{3}{4}$$ $$\overline{7\frac{5}{4} = 8\frac{1}{4}}$$ The total length is $8\frac{1}{4}$ feet.
(b) How much longer is the first board?	(b) Subtracting to find the difference, $$5\frac{1}{2} = 5\frac{2}{4} = 4\frac{6}{4}$$ $$-2\frac{3}{4} = 2\frac{3}{4} = 2\frac{3}{4}$$ $$\overline{\phantom{-2\frac{3}{4} = 2\frac{3}{4} = }2\frac{3}{4}}$$ It is $2\frac{3}{4}$ feet longer.
(c) How many boards $2\frac{3}{4}$ feet in length can be cut from the board $5\frac{1}{2}$ feet long?	(c) How many $2\frac{3}{4}$'s equal $5\frac{1}{2}$? $$\square \times 2\frac{3}{4} = 5\frac{1}{2}?$$

Examples	Solutions
	Dividing by the known factor $2\frac{3}{4}$,
	$5\frac{1}{2} \div 2\frac{3}{4} = \frac{11}{2} \div \frac{11}{4}$
	$= \frac{11}{2} \times \frac{4}{11}$
	$= \frac{2}{1}$ In lowest terms.
	$= 2$
	Two boards $2\frac{3}{4}$ feet in length can be cut from the board $5\frac{1}{2}$ feet long.
(d) What is $\frac{2}{3}$ of the length of the board that is $2\frac{3}{4}$ feet long?	(d) A fractional part of a number is found by multiplication.
	$\frac{2}{3}$ of $2\frac{3}{4} = \frac{2}{3} \times 2\frac{3}{4}$
	$= \frac{2}{3} \times \frac{11}{4}$
	$= \frac{11}{6}$ In lowest terms.
	$= 1\frac{5}{6}$
	$\frac{2}{3}$ of the length $2\frac{3}{4}$ is $1\frac{5}{6}$.
(e) If the board $5\frac{1}{2}$ feet long is cut into 3 pieces, how long is each piece?	(e) Three times what number equals $5\frac{1}{2}$?
	$3 \times \square = 5\frac{1}{2}?$
	Dividing by 3,
	$5\frac{1}{2} \div 3 = \frac{11}{2} \div \frac{3}{1}$
	$= \frac{11}{2} \times \frac{1}{3}$
	$= \frac{11}{6}$
	$= 1\frac{5}{6}$
	Each piece is $1\frac{5}{6}$ feet long.

Examples	Solutions
On a trip I used $5\frac{3}{10}$ gallons of gas to drive 106 miles, and $6\frac{4}{5}$ gallons to drive 136 miles.	
(a) How much gas did I use on the trip of 242 miles?	(a) Adding, $$5\frac{3}{10} = 5\frac{3}{10}$$ $$+6\frac{4}{5} = 6\frac{8}{10}$$ $$11\frac{11}{10} = 12\frac{1}{10}$$ I used a total of $12\frac{1}{10}$ gallons of gas.
(b) What was my average miles per gallon for the entire trip?	(b) The total length of the trip was 242 miles. The total gas used was $12\frac{1}{10}$ gallons. $$242 \div 12\frac{1}{10} = \frac{242}{1} \div \frac{121}{10}$$ $$\uparrow$$ per gallon $$= \frac{242}{1} \times \frac{10}{121}$$ $$= \frac{20}{1} \quad \text{In lowest terms.}$$ $$= 20$$ I got 20 miles per gallon.
(c) How many miles could I travel on $15\frac{1}{2}$ gallons of gas at the rate found in part b?	(c) At 20 miles for each of the $15\frac{1}{2}$ gallons, fifteen and one-half 20's $$= 15\frac{1}{2} \times 20$$ $$= \frac{31}{2} \times \frac{20}{1} = 310$$ I could travel 310 miles.

Answers to these exercises can be found on page 281.

1. What is the total distance around this figure?

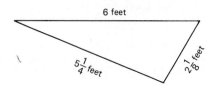

6 feet

$5\frac{1}{4}$ feet

$2\frac{1}{8}$ feet

2. A pound of meat weighed $\frac{7}{8}$ pound after being cooked. How much of the meat's weight was lost in cooking?

3. On a trip I stopped for gas three times. I bought $8\frac{3}{10}$ gallons, $7\frac{4}{5}$ gallons, and $9\frac{1}{2}$ gallons.

    (a) How much gas did I buy on the trip?
    (b) How much more gas did I buy at the last stop than at the first stop?

4. A board $27\frac{3}{8}$ inches long is cut from a board $46\frac{1}{2}$ inches long. If the cut wastes $\frac{1}{16}$ inch, how much of the $46\frac{1}{2}$-inch board remains?

5. If nails cost 20¢ per pound, how much do $3\frac{1}{2}$ pounds cost?

6. A car traveled 35 miles per hour for $6\frac{3}{4}$ hours. How far did it go?

7. How much is $\frac{1}{2}$ of $3\frac{1}{4}$ cups of water?

8. How long is each piece if a ribbon $43\frac{1}{3}$ inches long is cut into 13 pieces?

9. How many $3\frac{1}{2}$-foot pieces can be cut from a board that is $24\frac{1}{2}$ feet long?

10. If I drove 210 miles in $3\frac{3}{4}$ hours, what was my average rate in miles per hour?

1. Mr. Wratten placed an order for twenty $6\frac{1}{2}$-foot lengths of shelving.

    (a) If each foot of shelving costs $28\frac{1}{2}$¢, how much does one $6\frac{1}{2}$-foot section cost?
    (b) What was the cost of the entire order?
    (c) Only $5\frac{3}{4}$ feet of one board were used. How much of the board remains?
    (d) How many $\frac{1}{4}$-foot sections can be cut from one $6\frac{1}{2}$-foot board?

2. How much is $\frac{1}{2}$ of $2\frac{1}{4}$ cups of water?

$$A = P + PRT$$

Letters in **formulas** are used to represent, or to stand for, numbers that can change from one application to another.

$$R = \frac{D}{T}$$

This is the formula for determining the average rate of speed of a car (or other object). The particular letters chosen for a formula are not important, but those that remind you of the quantities they represent are often used.

Written verbally this formula states that the average rate of speed (R) equals the distance traveled (D) divided by the amount of time (T) it took to travel that distance.

$$\underline{R}ate = \frac{\underline{D}istance}{\underline{T}ime} \qquad R = \frac{D}{T}$$

Notice how this lengthy verbal expression translates into a simple formula. This formula can be used in many applications to calculate the rate by substituting values of distance and time for the letters.

EXAMPLE 1: If a bicyclist travels 60 miles in 4 hours, what is the average rate of speed?

The distance traveled is 60 miles.

$$D = 60$$

The time it took to travel that distance is 4 hours.

$$T = 4$$

To use the formula, replace each letter in the formula with its actual value.

$$R = \frac{D}{T}$$

Replace the letter D with 60.

$$R = \frac{60}{T}$$

Replace the letter T with 4.

$$R = \frac{60}{4}$$

Now calculate the answer for R using the arithmetic operation of division indicated by the formula.

$$R = \frac{60}{4}$$

$$= 15$$

Therefore, the bicyclist travels at the average rate of 15 miles per hour.

You can use the same formula with different values for the letters.

EXAMPLE 2:   If $D = 203$ and $T = 4$, what is the value of R if

$$R = \frac{D}{T}.$$

Replace the letters with their actual values.

$$R = \frac{D}{T}$$

$$= \frac{203}{4}$$

$$= 50\frac{3}{4}$$

The formula for selling price (S) of an article states that the price is equal to the sum of the markup (M) and the cost of the article (C). That is,

$$S = M + C$$

EXAMPLE 3:   If a winter coat cost the business owner $75 and the markup is $16, what is the selling price of that coat?
Here,

$$M = 16 \quad \text{and} \quad C = 75$$

Replace the letters in the formula with their actual values, then add.

$$S = M + C$$

$$S = 16 + 75$$

$$= 91$$

The selling price of the coat is $91.

The area (A) of a rectangle is calculated by multiplying the length (L) and the width (W) of the rectangle.

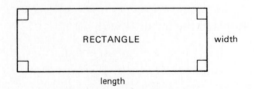

 In formulas, multiplications are represented by letters (or letters and a number) written close to each other in the absence of any other sign of operation.

$$A = LW$$

Multiplication is indicated by the formula because there is no operation sign between the L and W.

EXAMPLE 4: Find the area of a rectangle with length 5 inches and width $2\frac{1}{2}$ inches.

The values of L and W are

$$L = 5 \quad \text{and} \quad W = 2\frac{1}{2}$$

When replacing the values in a formula that has multiplication, you also have to supply the multiplication symbol.

$$A = LW$$

$$= 5 \times 2\frac{1}{2} \qquad \text{Length times width.}$$

$$= \frac{5}{1} \times \frac{5}{2}$$

$$= 12\frac{1}{2}$$

The area is $12\frac{1}{2}$ square inches.

EXAMPLE 5: In the formula

$$H = 2K$$

what is the value of H if K equals $7\frac{3}{8}$?

Because no operation sign appears between 2 and K, multiplication is understood. Therefore,

$$H = 2K$$

indicates that 2 is multiplied times the value of K to calculate the value of H.

Because $K = 7\frac{3}{8}$

$$H = 2K$$

$$= 2 \times 7\frac{3}{8} \qquad \text{2 times } K.$$

$$= \frac{2}{1} \times \frac{59}{8}$$

$$= 14\frac{3}{4}$$

In banking, the amount of money earned depends upon the amount of money deposited, or *principal* (P), the *rate* of interest (R), and the length of *time* (T) it is on deposit. The following formula can be used to compute the amount of money (A) you have after a period of time.

$$A = P + PRT$$

This formula is a more complicated one. The letters PRT indicate, by the absence of any operation sign, multiplication of P, R, and T. Their product is then added to P to obtain A.

$$A = P + PRT$$

A equals P added to the product of P, R, and T.

EXAMPLE 6:   Calculate A when $P = 80$, $R = \dfrac{3}{10}$, and $T = \dfrac{1}{4}$.

Replace the letters with their corresponding values and write multiplications where indicated by the absence of operation signs.

$A = P + PRT$

$\quad = 80 + 80 \times \dfrac{3}{10} \times \dfrac{1}{4}$ Product of P, R, and T.

$\quad = 80 + \dfrac{80}{1} \times \dfrac{3}{10} \times \dfrac{1}{4}$

$\quad = 80 + 6$

$\quad = 86$

Order of Operations

Example 6 also illustrates an important rule about the order in which operations are performed.

Grouping symbols, such as parentheses, brackets, or braces, can be used to indicate which operation needs to be evaluated first. For example,

$$2 \times 3 + 6$$

could possibly be calculated two different ways. But

$$2 \times (3 + 6)$$

clearly indicates a particular order in which to perform the calculation.

Operations within grouping symbols should be performed first. Therefore,

$2 \times (3 + 6)$ Perform operation within parentheses first.

$= 2 \times \quad 9$

$= \quad 18$

Order of Operations

Perform operations within grouping symbols first, starting with the innermost grouping.

Therefore, the expression $2 + [5 \times (7 + 2)]$ is evaluated using the following steps.

$2 + [5 \times (7 + 2)]$

$= 2 + [5 \times \quad 9]$ Innermost grouping first.

$= 2 + \quad 45$ Next grouping.

$= \quad 47$

Although grouping symbols can be used to clarify order of operations, they can be confusing when too many are used. Therefore, in arithmetic (and particularly with formulas) some additional rules are followed.

Order of Operations

In the absence of grouping symbols to control the order of operations, follow these steps in the order presented.

1. Do all multiplications and divisions in the order in which they appear from left to right.
2. Do all additions and subtractions in the order in which they appear from left to right.

These rules explain the order of operations used in Example 6.

$$A = P + PRT$$

The product PRT was evaluated before the value of P was added because multiplications are done before additions unless grouping symbols say otherwise. Under these rules, multiplication and division are considered higher priority operations than addition and subtraction.

EXAMPLE 7: Calculate T if $T = B + 2A$, $A = 7$, and $B = 5$.

Replace the letters with their corresponding values, remembering that $2A$ indicates multiplication.

$$T = B + 2A$$
$$= 5 + 2 \times 7 \qquad \text{Multiply first,}$$
$$= 5 + 14 \qquad \text{then add.}$$
$$= 19$$

 In formulas, multiplication is also indicated by writing a letter or number next to a grouping symbol with no operation sign separating the two.

$$5(T - 3)$$

indicates 5 times the expression $T - 3$.

The formula

$$A = P + PRT$$

is sometimes written as

$$A = P(1 + RT)$$

which indicates P times the expression $1 + RT$.

EXAMPLE 8: Calculate H if $H = \dfrac{5(T - 3)}{7}$ and $T = 9$.

$$H = \frac{5(T - 3)}{7}$$
$$= \frac{5(9 - 3)}{7}$$

Do the operation in the grouping first.

$$H = \frac{5(9-3)}{7}$$

$$= \frac{5(6)}{7}$$

The expression $5(6)$ has no sign of operation, so multiplication is understood.

$$H = \frac{5(6)}{7}$$

$$= \frac{30}{7}$$

$$= 4\frac{2}{7}$$

Therefore, $H = 4\frac{2}{7}$, when $T = 9$.

Now you try these.

Examples	Solutions
The distance traveled equals the rate (R) of speed times the time (T); that is, $$D = RT$$ How far will a car travel in $4\frac{1}{2}$ hours at a rate of 55 miles per hour?	Here, $$R = 55$$ $$T = 4\frac{1}{2}$$ Therefore, $$D = RT$$ $$= 55 \times 4\frac{1}{2}$$ $$= \frac{55}{1} \times \frac{9}{2}$$ $$= 247\frac{1}{2} \quad \text{miles}$$
Markup (M) equals selling price (S) minus the cost of the article (C), or $$M = S - C$$ If the cost is \$85 and the selling price is \$100, what is the markup?	$$S = 100$$ $$C = 85$$ Therefore, $$M = S - C$$ $$= 100 - 85$$ $$= 15 \quad \text{dollars}$$

Examples	Solutions
A temperature measured in Celsius (C) can be changed to Fahrenheit (F) by multiplying the Celsius temperature by $\frac{9}{5}$ and adding 32 degrees. More simply, $$F = \frac{9}{5}C + 32$$ A Celsius temperature of 20° equals how many degrees Fahrenheit?	Remember that $\frac{9}{5}C$ indicates multiplication because no operation sign appears between the $\frac{9}{5}$ and C. Replace C with its actual value of 20°. $$F = \frac{9}{5}C + 32$$ $$= \frac{9}{5} \times 20 + 32$$ $$= \frac{9}{5} \times \frac{20}{1} + 32$$ $$= 36 + 32 \quad \text{Multiply first, then add.}$$ $$= 68 \quad \text{degrees}$$
To find a Celsius temperature when a Fahrenheit one is given, use the formula $$C = \frac{5}{9}(F - 32)$$ A Fahrenheit temperature of 68° equals how many degrees Celsius?	$$C = \frac{5}{9}(F - 32)$$ $$= \frac{5}{9}(68 - 32) \quad \text{Grouping symbol—do subtraction first.}$$ $$= \frac{5}{9}(36)$$ $$= \frac{5}{9} \times \frac{36}{1} \quad \text{Absence of operation sign indicates multiplication.}$$ $$= 20 \quad \text{degrees}$$

Now try these—no hints given.

Calculate C, where $$C = S - M$$ and $S = 2\frac{1}{4}$ and $M = 1\frac{1}{2}$.	$\frac{3}{4}$
Find C, where $$C = \frac{5}{9}F - 17\frac{7}{9}$$ and $F = 68$.	20

If you had difficulty with these problems, talk with an instructor or read this section again.

**Exercises
Part I** Answers to these exercises can be found on page 281.

Calculate the value of H in the following formulas.

1. $H = a - b$ when $a = 3\frac{1}{2}$ and $b = 2\frac{3}{4}$

2. $H = \frac{1}{2}F$ when $F = 25$

3. $H = c + 2d$ when $c = 32$ and $d = 12$

4. $H = \frac{w}{x}$ when $w = 7$ and $x = 2$

5. $H = ts$ when $t = 8$ and $s = 9\frac{3}{4}$

6. $H = 2(c + d)$ when $c = 5$ and $d = 3$

7. $H = \frac{5(r - 9)}{t}$ when $r = 12$ and $t = 3$

**Exercises
Part II** Evaluate T in the following formulas.

1. $T = xy$ when $x = 2$ and $y = 1\frac{1}{12}$

2. $T = 3a + b$ when $a = 7$ and $b = 10$

3. $T = 3(a + b)$ when $a = 7$ and $b = 10$

REVIEW EXERCISES

Refer to the section listed if you have difficulty. Answers to these exercises can be found on page 281.

Section 11.1
1. How can 8 people equally share 3 pickles?
2. An exercise section contains 20 problems. What fractional part of the exercises remains if you have worked 15 problems?
3. Which weighs more, $6\frac{2}{3}$ pounds or $6\frac{5}{6}$ pounds?
4. A book contains 420 pages. What fractional part of the book do 105 pages represent?
5. How can 6 people share 2 pizzas?

Section 11.2
6. A book of stamps contains 20 stamps. How many stamps represent $\frac{1}{4}$ of the book?
7. The winter term at school is 10 weeks long. How many weeks represent $\frac{3}{5}$ of the term?
8. If a car travels 55 miles per hour for $6\frac{3}{4}$ hours, how far does it go?

9. A recipe calls for $\frac{1}{2}$ cup of flour and 3 cups of water. How much of each of these ingredients should you use if you want to make $\frac{1}{3}$ of the recipe?

10. A piece of fabric $6\frac{2}{3}$ yards long is divided into 5 equal pieces. How long is each piece?

11. Dan ran $6\frac{1}{10}$ miles the first day and $5\frac{3}{5}$ miles the second. What is the average number of miles Dan ran per day?

12. At the beginning of winter there was a stack of wood $8\frac{1}{4}$ feet high. At the end of winter, the stack was $2\frac{3}{4}$ feet high. How many feet were used during the winter?

13. How many $2\frac{1}{3}$-foot pieces can be cut from a board that is $16\frac{1}{3}$ feet long?

14. If I drive $78\frac{2}{5}$ miles using $5\frac{3}{5}$ gallons of gas, how many miles do I get per gallon?

15. If a printer types 585 words in $6\frac{1}{2}$ minutes, how many words does it type per minute?

Section 11.3 16. Find A when $x = 5$, $y = 2\frac{1}{3}$, and $A = xy$.

17. Find H when $a = 7$, $b = 2$, and $H = 3a + b$.

CHAPTER POST-TEST

This test should determine if you have mastered the topics in Chapter 11. Answers to this test can be found on page 281.

1. How can 9 people equally share 3 oranges?
2. A book contains 500 pages. What fractional part remains to be read if I have read 250 pages?
3. What is $\frac{3}{4}$ of a trip of 200 miles?

4. A recipe calls for $2\frac{1}{2}$ cups of water and $3\frac{3}{4}$ cups of milk. What is the total amount of liquid called for in the recipe?

5. If bananas are on sale for 45¢ per pound, how much do $2\frac{1}{5}$ pounds cost?

6. If a piece of fabric was originally $3\frac{1}{4}$ inches long, and after being washed it was $2\frac{7}{8}$ inches long, how much did it shrink?

7. If a board $9\frac{1}{3}$ feet long is divided into 4 equal pieces, how long is each piece?

8. If Johnnie can run a mile in $9\frac{3}{4}$ minutes, how long will it take her to run 3 miles?

9. I have three stakes in my garage, $3\frac{5}{8}$ feet, $2\frac{3}{4}$ feet, and $5\frac{1}{2}$ feet long. What is the average length of each of the three stakes?

10. Vera types 253 words in $5\frac{3}{4}$ minutes. How many words can she type per minute?

11. Find H when $a = 5$, $b = 7$, and $H = a + 6b$.

12. Find T when $x = 4\frac{1}{2}$, $y = 7\frac{5}{12}$, and $T = xy$.

UNIT TWO
FRACTIONS AND MIXED NUMBERS
REVIEW AND POST-TEST

Fractions are used to represent (1) parts of wholes and (2) division of whole numbers.

1. $\frac{2}{3}$ means   2 of
                     3 equal parts
                     of a whole.

2. $\frac{2}{3}$ also means $2 \div 3$

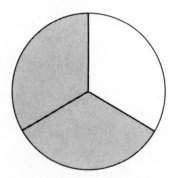

Improper fractions and mixed numbers are related by the following conversions.

1. $\frac{52}{3} = 17\frac{1}{3}$ because

$$
\begin{array}{r}
17\frac{1}{3} \\
3\overline{)52} \\
3 \\
\hline
22 \\
21 \\
\hline
1
\end{array}
$$

2. $7\dfrac{1}{8} = \dfrac{57}{8}$ because $7\dfrac{1}{8} = \dfrac{(8 \times 7) + 1}{8}$

$$= \dfrac{56 + 1}{8}$$

$$= \dfrac{57}{8}$$

The arithmetic of fractions will be reviewed by examples.

EXAMPLE 1:   Add: $2\dfrac{5}{6} + 1\dfrac{3}{8}$.

When adding and subtracting fractions you must build each fraction to an equivalent fraction with the LCD as the denominator.

$$6 = 2^1 \times 3^1$$

$$8 = 2^3$$

$$\text{LCM of 6 and 8} = 2^3 \times 3^1$$

$$= 8 \times 3$$

$$= 24$$

Therefore,

$$2\dfrac{5}{6} = 2\dfrac{5 \times 4}{6 \times 4} = 2\dfrac{20}{24}$$

$$+ 1\dfrac{3}{8} = 1\dfrac{3 \times 3}{8 \times 3} = 1\dfrac{9}{24}$$

$$3\dfrac{29}{24}$$

$$= 3 + 1\dfrac{5}{24}$$

$$= 4\dfrac{5}{24}$$

EXAMPLE 2:   Subtract: $4\dfrac{1}{8} - 2\dfrac{3}{4}$.

Again, build each fraction to an equivalent fraction with the LCD as the denominator.

$$8 = 2^3$$

$$4 = 2^2$$

$$\text{LCM of 8 and 4} = 2^3$$

$$= 8$$

build↘ borrow↘

$$4\dfrac{1}{8} = 4\dfrac{1}{8} = \overset{3}{4} + 1\dfrac{1}{8} = 3\dfrac{9}{8}$$

$$-2\dfrac{3}{4} = 2\dfrac{6}{8} = \qquad 2\dfrac{6}{8} = 2\dfrac{6}{8}$$

$$1\dfrac{3}{8}$$

EXAMPLE 3:   Multiply: $\frac{2}{3} \times 2\frac{2}{5}$.

$\frac{2}{3} \times 2\frac{2}{5} = \frac{2}{3} \times \frac{12}{5}$ Change mixed number to an improper fraction.

$= \frac{2}{\underset{1}{3}} \times \frac{\overset{4}{12}}{5}$ Reduce.

$= \frac{8}{5}$ Multiply numerators and multiply denominators.

$= 1\frac{3}{5}$

EXAMPLE 4:   Divide: $28 \div \frac{4}{5}$.

To divide by a fraction, invert the divisor and multiply.

$28 \div \frac{4}{5} = \frac{28}{1} \div \frac{4}{5}$

$= \frac{28}{1} \times \frac{5}{4}$ Invert.

$= \frac{\overset{7}{28}}{1} \times \frac{5}{\underset{1}{4}}$ Reduce.

$= \frac{35}{1}$

$= 35$

Before working the Unit Two Post-Test, work (or rework)

Chapter 6 Review Exercises, page 174
Chapter 7 Review Exercises, page 201
Chapter 8 Review Exercises, page 217
Chapter 9 Review Exercises, page 229
Chapter 10 Review Exercises, page 248
Chapter 11 Review Exercises, page 272

Study the sections for which you cannot work the sample problems.

UNIT TWO POST-TEST

Answers to this test can be found on page 281.

1. Write $\frac{27}{4}$ as a mixed number.

2. Write $12\frac{1}{2}$ as a mixed number.

3. Determine the larger fraction of $\frac{3}{4}$ and $\frac{7}{12}$.

4. Find the LCM of 8, 10, and 15.

5. Perform the indicated operation.

   (a) $\dfrac{5}{6} + \dfrac{4}{15}$ (b) $2\dfrac{8}{9} + 1\dfrac{11}{12}$ (c) $\dfrac{11}{18} - \dfrac{11}{30}$ (d) $6\dfrac{2}{3} - 4\dfrac{3}{4}$

   (e) $7 - 5\dfrac{2}{5}$ (f) $1\dfrac{1}{4} + 1\dfrac{7}{8}$ (g) $\dfrac{\dfrac{15}{5}}{9}$ (h) $\dfrac{6}{8} \times \dfrac{4}{9}$

   (i) $9 \times 2\dfrac{2}{3}$ (j) $\dfrac{24}{35} \times 2\dfrac{2}{3} \times 5$

6. If nails cost 22¢ per pound, how much do $3\dfrac{1}{2}$ pounds cost?

7. On a trip I stopped for gas twice. I bought $8\dfrac{2}{5}$ gallons and $11\dfrac{3}{10}$ gallons. (a) How much gas did I buy on the trip? (b) How much more gas did I buy the second time?

8. If I drove 210 miles in $3\dfrac{3}{4}$ hours, what was my average rate in miles per hour?

9. What is $\dfrac{1}{4}$ of 136 pages?

10. Find the value of A if $A = 2a + b$, $a = 3$, and $b = 7$.

ANSWERS

Unit Two Pretest 1. $\dfrac{1}{3}$ 2. $\dfrac{5}{6}$ 3. $7\dfrac{1}{4}$ 4. $\dfrac{38}{3}$ 5. $\dfrac{24}{103}$ 6. $\dfrac{2}{15}$ 7. no   8. $>$

9. $3^2 \times 5 \times 7$ 10. $\dfrac{8}{9}$ 11. 126   12. $\dfrac{7}{9}$ 13. (a) $1\dfrac{1}{20}$ (b) $4\dfrac{5}{24}$

14. $11\dfrac{1}{4}$ hours   15. (a) $\dfrac{35}{48}$ (b) $4\dfrac{5}{6}$ (c) $6\dfrac{5}{8}$ 16. (a) $\dfrac{1}{4}$ (b) $2\dfrac{11}{12}$

17. (a) 62   (b) $\dfrac{1}{4}$ 18. 34 pages   19. $178\dfrac{3}{4}$ miles   20. 60   21. 12

Chapter 6

Section 6.1   1. $\dfrac{4}{9}$ 2. $\dfrac{5}{6}$ 3. $\dfrac{5}{7}$ 4. $\dfrac{5}{8}$ 5. $\dfrac{3}{20}$ 6. Each gets $\dfrac{1}{3}$.   7. Each gets $\dfrac{2}{7}$.

Section 6.2   1. $5\dfrac{3}{4}$ 2. 1   3. 9   4. $3\dfrac{5}{6}$ 5. $\dfrac{47}{5}$ 6. $\dfrac{99}{8}$ 7. $\dfrac{58}{5}$ 8. $\dfrac{93}{7}$ 9. $86\dfrac{1}{4}$

10. $8389\dfrac{4}{15}$

Section 6.3   1. $\dfrac{32}{56}$ 2. $\dfrac{20}{30}$ 3. $\dfrac{10}{65}$ 4. $\dfrac{14}{147}$ 5. $\dfrac{23}{209}$ 6. $\dfrac{72}{90}$ 7. $\dfrac{105}{330}$ 8. $\dfrac{156}{195}$

9. 42   10. 350

Section 6.4   **1.** $\dfrac{4}{7}$ **2.** $\dfrac{2}{3}$ **3.** $\dfrac{2}{13}$ **4.** $\dfrac{4}{5}$ **5.** $\dfrac{7}{22}$ **6.** $\dfrac{5}{16}$ **7.** $\dfrac{12}{35}$ is already reduced.
8. $\dfrac{16}{35}$

Section 6.5   **1.** No.   **2.** Yes.   **3.** No.   **4.** No.   **5.** Yes.   **6.** Yes.   **7.** No.

Exercise A   **1.** $\dfrac{7}{49}$ **2.** $\dfrac{54}{72}$ **3.** $\dfrac{104}{143}$ **4.** $\dfrac{12}{32}$ **5.** $\dfrac{84}{144}$ **6.** $\dfrac{9}{63}$ **7.** $\dfrac{12}{16}$ **8.** $\dfrac{16}{22}$
9. $\dfrac{42}{112}$ **10.** $\dfrac{105}{180}$

Exercise C   **1.** $\dfrac{1}{7}$ **2.** $\dfrac{3}{4}$ **3.** $\dfrac{8}{11}$ **4.** $\dfrac{3}{8}$ **5.** $\dfrac{7}{12}$ **6.** $\dfrac{1}{7}$ **7.** $\dfrac{3}{4}$ **8.** $\dfrac{8}{11}$ **9.** $\dfrac{3}{8}$
10. $\dfrac{7}{12}$

Exercise E   **1.** $5\dfrac{4}{7}$ **2.** $3\dfrac{7}{15}$ **3.** $15\dfrac{11}{18}$ **4.** $\dfrac{15}{4}$ **5.** $\dfrac{59}{8}$ **6.** $\dfrac{134}{9}$

Review Exercises   **1.** $\dfrac{5}{9}$ **2.** $\dfrac{3}{8}$ **3.** $6\dfrac{1}{4}$ **4.** 8   **5.** $5\dfrac{4}{5}$ **6.** 6   **7.** $\dfrac{48}{5}$ **8.** $\dfrac{47}{3}$ **9.** $\dfrac{12}{1}$
10. $\dfrac{31}{12}$ **11.** $\dfrac{27}{45}$ **12.** $\dfrac{42}{90}$ **13.** $\dfrac{46}{69}$ **14.** $\dfrac{45}{126}$ **15.** $\dfrac{2}{3}$ **16.** $\dfrac{2}{5}$ **17.** $\dfrac{2}{3}$
18. $\dfrac{7}{8}$ **19.** No.   **20.** Yes.

Post-Test   **1.** (a) $\dfrac{34}{5}$ (b) $\dfrac{103}{4}$ (c) $\dfrac{27}{1}$ **2.** (a) $2\dfrac{4}{5}$ (b) $7\dfrac{2}{9}$ **3.** (a) $\dfrac{8}{12}$ (b) $\dfrac{117}{135}$
4. (a) $\dfrac{3}{4}$ (b) $\dfrac{2}{3}$ (c) $\dfrac{3}{7}$ (d) $\dfrac{13}{17}$ **5.** (a) No.   (b) No.

Chapter 7

Section 7.2   **1.** $2 \times 3 \times 7$ **2.** $2^2 \times 3 \times 7$ **3.** 2^8 **4.** $2 \times 3^3 \times 7$ **5.** $2 \times 7^2 \times 11$
6. $\dfrac{4}{7}$ **7.** $\dfrac{2}{3}$ **8.** $\dfrac{2}{13}$ **9.** $\dfrac{4}{5}$ **10.** $\dfrac{7}{22}$ **11.** $\dfrac{5}{16}$ **12.** $\dfrac{16}{35}$

Section 7.3   **1.** 36   **2.** 30   **3.** 28   **4.** 60   **5.** 252   **6.** 84   **7.** 120   **8.** 135
9. $\dfrac{1}{3}$ **10.** $\dfrac{3}{4}$ **11.** $\dfrac{2}{3}$ **12.** $\dfrac{7}{30}$

Exercise A   **1.** 24   **2.** 45   **3.** 96   **4.** 546   **5.** 48   **6.** 2,016   **7.** 36   **8.** 24

Review Exercises   **1.** <   **2.** >   **3.** <   **4.** <   **5.** 3^2 **6.** 2×5 **7.** 5×7 **8.** $2^3 \times 7$
9. $2^3 \times 3^2$ **10.** $2^5 \times 3 \times 5$ **11.** $3 \times 5 \times 7$ **12.** $2 \times 3 \times 5 \times 11$
13. $\dfrac{9}{26}$ **14.** $\dfrac{7}{9}$ **15.** $\dfrac{9}{16}$ **16.** 18   **17.** 10   **18.** 105   **19.** 294
20. 504   **21.** 72   **22.** 126   **23.** 30   **24.** 360   **25.** $\dfrac{15}{21}$ **26.** $\dfrac{17}{18}$

Post-Test **1.** $2^4 \times 3$ **2.** $2^3 \times 3^2 \times 5$ **3.** 5×61 **4.** $2^2 \times 3^3 \times 5$ **5.** 3×5^3
6. 42   **7.** 210   **8.** 300   **9.** 135   **10.** $\dfrac{57}{64}$

Chapter 8

Section 8.1   **1.** $\dfrac{3}{8}$ **2.** $\dfrac{2}{3}$ **3.** $1\dfrac{1}{3}$ **4.** $1\dfrac{4}{9}$ **5.** $1\dfrac{1}{2}$ **6.** $1\dfrac{1}{9}$ **7.** 1   **8.** 1

Section 8.2   **1.** $1\dfrac{5}{12}$ **2.** $1\dfrac{1}{10}$ **3.** $\dfrac{5}{8}$ **4.** $\dfrac{71}{120}$ **5.** $1\dfrac{1}{36}$ **6.** $\dfrac{27}{28}$ **7.** $\dfrac{151}{252}$ **8.** $\dfrac{34}{45}$
9. $1\dfrac{5}{24}$ **10.** $\dfrac{106}{135}$ **11.** $\dfrac{7}{8}$ pound

Section 8.3   **1.** $1\dfrac{2}{5}$ **2.** $\dfrac{5}{6}$ **3.** $1\dfrac{1}{15}$ **4.** $1\dfrac{5}{8}$ **5.** $\dfrac{29}{48}$ **6.** $\dfrac{53}{192}$ **7.** 10   **8.** $4\dfrac{1}{36}$
9. $16\dfrac{1}{18}$ **10.** $12\dfrac{5}{12}$ **11.** $4\dfrac{29}{36}$ **12.** $5\dfrac{5}{24}$ **13.** $20\dfrac{1}{6}$ **14.** $8\dfrac{7}{8}$
15. $11\dfrac{106}{135}$ **16.** $14\dfrac{1}{4}$ hours   **17.** $23\dfrac{4}{5}$ gallons

Exercise A   **1.** $1\dfrac{5}{8}$ **2.** $1\dfrac{7}{15}$ **3.** $1\dfrac{13}{16}$ **4.** $1\dfrac{29}{38}$ **5.** $1\dfrac{23}{48}$ **6.** $1\dfrac{29}{30}$ **7.** $1\dfrac{17}{48}$
8. $\dfrac{227}{420}$ **9.** $\dfrac{73}{220}$ **10.** $\dfrac{38}{63}$

Exercise C   **1.** $16\dfrac{5}{6}$ **2.** $9\dfrac{7}{12}$ **3.** $26\dfrac{6}{7}$ **4.** $2\dfrac{13}{15}$ **5.** $7\dfrac{1}{24}$ **6.** $12\dfrac{5}{8}$ **7.** $15\dfrac{2}{3}$
8. $8\dfrac{1}{2}$ **9.** $11\dfrac{1}{12}$ **10.** $17\dfrac{17}{45}$

Review Exercises   **1.** $\dfrac{4}{5}$ **2.** $1\dfrac{1}{2}$ **3.** $\dfrac{1}{3}$ **4.** $\dfrac{31}{35}$ **5.** $\dfrac{53}{60}$ **6.** $\dfrac{39}{48}$ **7.** $8\dfrac{5}{12}$ **8.** $9\dfrac{5}{24}$
9. $6\dfrac{45}{112}$ **10.** $3\dfrac{3}{8}$ yards

Post-Test   **1.** $1\dfrac{1}{3}$ **2.** 6   **3.** $14\dfrac{8}{15}$ **4.** $\dfrac{13}{24}$ **5.** $11\dfrac{27}{35}$ **6.** $8\dfrac{17}{36}$ **7.** $5\dfrac{71}{135}$
8. $30\dfrac{1}{5}$ gallons

Chapter 9

Section 9.1   **1.** $\dfrac{1}{12}$ **2.** $\dfrac{17}{30}$ **3.** $\dfrac{5}{24}$ **4.** $\dfrac{27}{40}$ **5.** $\dfrac{29}{36}$ **6.** $\dfrac{15}{28}$ **7.** $\dfrac{7}{15}$ **8.** $\dfrac{11}{45}$
9. $\dfrac{53}{135}$ **10.** $\dfrac{1}{8}$

Section 9.2   1. $\frac{1}{5}$ 2. $\frac{1}{3}$ 3. $\frac{1}{8}$ 4. $1\frac{1}{3}$ 5. $1\frac{17}{36}$ 6. $3\frac{17}{30}$ 7. $4\frac{2}{3}$ 8. $4\frac{1}{4}$
9. $1\frac{11}{12}$ 10. $1\frac{35}{36}$ 11. $1\frac{3}{5}$ 12. $\frac{5}{9}$ 13. $2\frac{101}{120}$ 14. $5\frac{23}{27}$
15. $2\frac{3}{4}$ hours   16. $4\frac{1}{8}$ yards

Exercise A   1. $\frac{1}{8}$ 2. $\frac{2}{15}$ 3. $\frac{1}{16}$ 4. $\frac{13}{48}$ 5. $\frac{37}{48}$ 6. $\frac{11}{420}$ 7. $\frac{5}{12}$ 8. $\frac{83}{180}$
9. $\frac{19}{54}$ 10. $\frac{1}{42}$

Exercise C   1. $2\frac{1}{6}$ 2. $1\frac{11}{12}$ 3. $3\frac{13}{14}$ 4. $1\frac{1}{8}$ 5. $\frac{10}{21}$

Review Exercises   1. $\frac{1}{9}$ 2. $\frac{1}{2}$ 3. $\frac{11}{35}$ 4. $\frac{2}{5}$ 5. $\frac{19}{60}$ 6. $\frac{29}{48}$ 7. $3\frac{1}{12}$ 8. $3\frac{13}{24}$
9. $5\frac{1}{3}$ 10. $6\frac{3}{8}$

Post-Test   1. $\frac{4}{13}$ 2. $1\frac{1}{6}$ 3. $5\frac{2}{15}$ 4. $7\frac{13}{35}$ 5. $2\frac{25}{36}$ 6. $4\frac{106}{135}$ 7. $2\frac{2}{3}$ 8. $\frac{8}{9}$
9. $5\frac{1}{4}$ feet   10. $\frac{3}{4}$ hour

Chapter 10

Section 10.1   1. $\frac{1}{4}$ 2. $\frac{1}{3}$ 3. $\frac{3}{8}$ 4. $\frac{1}{5}$ 5. $\frac{1}{17}$ 6. $1\frac{1}{2}$ 7. 24   8. $3\frac{3}{4}$ 9. $\frac{8}{9}$
10. $17\frac{1}{2}$

Section 10.2   1. $\frac{5}{6}$ 2. $\frac{4}{5}$ 3. 2   4. 4   5. $\frac{2}{15}$ 6. 12   7. $1\frac{3}{7}$ 8. 10   9. $1\frac{1}{2}$
10. $\frac{2}{3}$ 11. $\frac{5}{24}$ 12. 27

Exercise A   1. 1   2. $\frac{1}{10}$ 3. $\frac{1}{11}$ 4. $\frac{2}{7}$ 5. 10   6. $\frac{3}{5}$ 7. 6   8. $\frac{4}{15}$ 9. $10\frac{1}{2}$
10. 1

Exercise C   1. $1\frac{3}{4}$ 2. 9   3. $\frac{5}{28}$ 4. $8\frac{1}{3}$ 5. $2\frac{4}{7}$ 6. $\frac{1}{9}$ 7. $3\frac{1}{9}$ 8. $13\frac{2}{3}$
9. 4   10. $\frac{1}{8}$

Review Exercises   1. $\frac{3}{28}$ 2. 62   3. $3\frac{1}{4}$ 4. $1\frac{1}{3}$ 5. $\frac{4}{5}$ 6. $1\frac{1}{4}$ 7. 20   8. $\frac{5}{96}$

Post-Test   1. $\frac{3}{5}$ 2. 3   3. $9\frac{1}{7}$ 4. $\frac{14}{15}$ 5. $\frac{2}{5}$ 6. $10\frac{2}{3}$ 7. $\frac{1}{4}$ 8. $\frac{5}{27}$

Chapter 11

Section 11.1   1. $\frac{2}{3}$ 2. $\frac{2}{5}$ 3. Each gets $\frac{3}{5}$ of a pizza.   4. $\frac{2}{5}$ 5. Each gets $2\frac{1}{2}$ bars.

6. 8   7. 4   8. $5\frac{5}{16}$ pounds   9. $1\frac{1}{2}$ hours   10. 40

Section 11.2   1. $13\frac{3}{8}$ feet   2. $\frac{1}{8}$ pound   3. (a) $25\frac{3}{5}$ gallons (b) $1\frac{1}{5}$ gallons

4. $9\frac{1}{16}$ inches   5. 70¢   6. $236\frac{1}{4}$ miles   7. $1\frac{5}{8}$ cups   8. $3\frac{1}{3}$ inches
9. 7 pieces   10. 56 miles per hour

Section 11.3   1. $\frac{3}{4}$ 2. $12\frac{1}{2}$ 3. 56   4. $3\frac{1}{2}$ 5. 78   6. 16   7. 5

Review Exercises   1. Each gets $2\frac{2}{3}$ pickles.   2. $\frac{1}{4}$ 3. $6\frac{5}{6}$ 4. $\frac{1}{4}$ 5. Each gets $\frac{1}{3}$ of a pizza.

6. 5   7. 6   8. $371\frac{1}{4}$ miles   9. $\frac{1}{6}$ cup flour and 1 cup water

10. $1\frac{1}{3}$ yard   11. $5\frac{17}{20}$ miles   12. $5\frac{1}{2}$ 13. 7   14. 14   15. 90

16. $11\frac{2}{3}$ 17. 23

Post-Test   1. Each gets $\frac{1}{3}$ of an orange.   2. $\frac{1}{2}$ 3. 150 miles   4. $6\frac{1}{4}$ cups   5. 99¢

6. $\frac{3}{8}$ inch   7. $2\frac{1}{3}$ feet   8. $29\frac{1}{4}$ minutes   9. $3\frac{23}{24}$

10. 44 words per minute   11. 47   12. $33\frac{3}{8}$

Unit Two Post-Test

1. $6\frac{3}{4}$ 2. $\frac{25}{2}$ 3. $\frac{3}{4}$ 4. 120

5. (a) $1\frac{1}{10}$ (b) $4\frac{29}{36}$ (c) $\frac{11}{45}$ (d) $1\frac{11}{12}$ (e) $1\frac{3}{5}$ (f) $3\frac{1}{8}$ (g) 27 (h) $\frac{1}{3}$
(i) 24 (j) $9\frac{1}{7}$

6. 77¢   7. (a) $19\frac{7}{10}$ gallons (b) $2\frac{9}{10}$ gallons   8. 56 miles per hour
9. 34 pages   10. 13

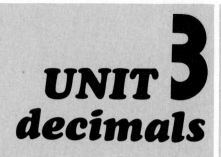

UNIT 3
decimals

This test is given *before* you read the material to see what you already know. If you cannot work each sample problem, you need to read the indicated chapters carefully. Answers to this test can be found on page 363.

CHAPTER 12

1. Write each number as you would read it.
   (a) .073   (b) 752.12   (c) 103.004

2. Write the numerals for each of the following.
   (a) twenty-four and fifteen thousandths
   (b) forty-seven hundredths

3. Write each decimal number as a fraction or mixed number.
   (a) .082   (b) 17.59

4. Round each number to the nearest hundredth.
   (a) 5.129   (b) 72.132   (c) 17.295

CHAPTER 13

5. Add.
   (a) 1.25 + 9.99   (b) 17.5 + 19.631   (c) 15.1 + 27 + 19.38

6. Subtract.
   (a) .97 - .89   (b) 15.2 - 10.98   (c) 12 - 7.89

CHAPTER 14

7. Multiply.
   (a) 17.25 × 1.35   (b) 12 × 7.2   (c) .705 × 100

CHAPTER 15

8. Divide (round, if necessary, to the nearest thousandth).
   (a) 18.75 ÷ 15   (b) .036176 ÷ 1.03   (c) 1.73 ÷ 10

CHAPTER 16

9. If you have $728.19 in your bank account, how much money do you have left after writing a check for $157.98 and depositing $69.72?

10. I pay my babysitter, Carol, $1.50 per hour. How much do I owe her for 3 hours?

11. A can of beans contains 16 ounces. How much does it cost per ounce if the can sells for 59¢?

12

decimal numbers

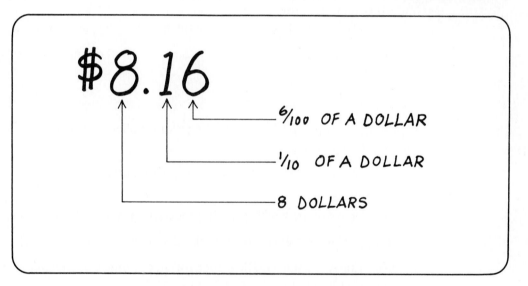

After reading this section you will know the place values of decimal numbers, as illustrated in this example.

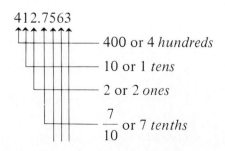

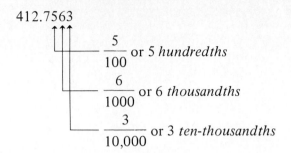

412.7563

$\dfrac{5}{100}$ or 5 *hundredths*

$\dfrac{6}{1000}$ or 6 *thousandths*

$\dfrac{3}{10,000}$ or 3 *ten-thousandths*

If you think you already know this information, turn to the exercises and work each problem. If you have difficulty, turn back and read this section completely.

Place Value   In Chapter 1 you learned that the place value of whole numbers is based on powers of 10.

Powers of Ten	Names of Place Values
1	ones
$10^1 = 10$	tens
$10^2 = 10 \times 10 = 100$	hundreds
$10^3 = 10 \times 10 \times 10 = 1,000$	thousands
$10^4 = 10 \times 10 \times 10 \times 10 = 10,000$	ten thousands
and so forth.	

The values of digits after the decimal point are based on fractions with powers of 10 as denominators.

You should already be familiar with two such positions that are used in writing money amounts. A dime is $\dfrac{1}{10}$ of a dollar, and a penny is $\dfrac{1}{100}$ of a dollar.

$35.16

$\dfrac{1}{10}$ of a dollar, or one dime

$\dfrac{6}{100}$ of a dollar, or 6 pennies

The *position* in which a digit appears in a numeral determines the *value* of that digit. The first position after the decimal point is tenths with value equal to $\dfrac{1}{10}$.

The value of each position that follows is one more power of 10 (in the denominator) compared to the position to its left.

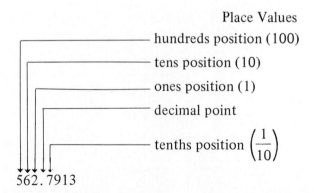

Place Values

hundreds position (100)

tens position (10)

ones position (1)

decimal point

tenths position $\left(\dfrac{1}{10}\right)$

562.7913

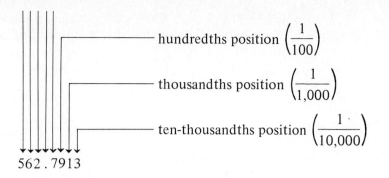

562.7913

Because the place values after the decimal point are fractional (and proper), any decimal number can be separated into two parts: (1) the whole number and (2) the fractional part.

$$562 . 7913$$

whole number   fractional part

In fact, it is correct to think of this four-place decimal number (four digits after the decimal point) as a mixed number.

> In determining the place value for any digit in a decimal number, start at the ones position. For a digit in the whole number part, count to that position by powers of 10: 10, 100, 1,000, and so forth. For a digit in the fractional part, count to that position by fractional powers of 10: $\dfrac{1}{10}, \dfrac{1}{100}, \dfrac{1}{1,000}$, and so on.

EXAMPLE 1:   What is the position of the digit 5 in the number 62.357? Starting at the ones position and counting,

ones

$$6 \; 2 \; . \; 3 \; 5 \; 7$$
$$\quad\;\; \tfrac{1}{10} \;\; \tfrac{1}{100}$$

The digit 5 is in the hundredths position.

EXAMPLE 2:   What is the position of the digit 2 in the number 1,234.56? Starting at the ones position and counting,

ones

$$1 \; , \; 2 \; 3 \; 4 \; . \; 5 \; 6$$
$$\quad 100 \; 10$$

The digit 2 is in the hundreds position.

EXAMPLE 3:   What is the position of the digit 7 in the number 15.3675?

ones

$$1 \; 5 \; . \; 3 \; 6 \; 7 \; 5$$
$$\quad\;\; \tfrac{1}{10} \;\; \tfrac{1}{100} \;\; \tfrac{1}{1000}$$

The digit 7 is in the thousandths position.

EXAMPLE 4:   What is the position of the digit 4 in the number .05634?

There is no digit in the ones position, but consider that position as your starting point.

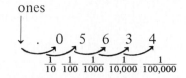

The digit 4 is in the hundred-thousandths position.

Because of the possibility of overlooking the decimal point, many people place a 0 in the ones position in a number like .05634. That is,

.05634   can be written as   0.05634

I will not use this zero.

Now you try these.

Examples	Solutions
What is the whole number part of 59.2731?	The decimal point separates the number into two parts. 59 . 2731 whole number fractional part 59 is the whole number part of 59.2731.
What is the fractional part of the number 159.285?	.285
Write the names of place values of each digit in 59.2731	5 9 . 2 7 3 1 tens, ones, tenths, hundredths, thousandths, ten-thousandths
What is the position of the 2 in each of the following numbers? (a) 53.521 (b) 2135.06 (c) 31.602 (d) 956.27	(a) Starting at the ones position and counting, ones 5 3 . 5 2 1 $\frac{1}{10}$ $\frac{1}{100}$ The 2 is in the hundredths position.

Examples	Solutions
	(b) Counting this time in the whole part, ones ↓ 2 1 3 5 . 0 6 1000 100 10 The 2 is in the thousand<u>s</u> position.
	(c) ones 3 1 . 6 0 2 $\frac{1}{10}$ $\frac{1}{100}$ $\frac{1}{1000}$ The 2 is in the thousand<u>ths</u> position.
	(d) 9 5 6 . 2 7 $\frac{1}{10}$ The digit is in the tenths position.

What digit of 217.5638 is in each of the indicated positions. (a) hundredths (b) hundreds (c) ones (d) tenths (e) thousandths	2 1 7 . 5 6 3 8 hundreds / tens / ones / tenths / hundredths / thousandths / ten-thousandths / decimal point (a) The digit 6 is in the hundredths position. (b) 2 (c) 7 (d) 5 (e) 3

Now try these—no hints given.

What is the position of the 5 in each of the following numbers? (a) 79.053 (b) 127.5 (c) 9.735	(a) hundredths (b) tenths (c) thousandths

Exercises Part I   Answers to these exercises can be found on page 363.

1. What is the position of the 7 in each of the following numbers?

   (a) 1.37   (b) 70.25   (c) 65.957   (d) 7,152.06
   (e) .0357   (f) 7.25   (g) 756.1   (h) 29.0597
   (i) .7   (j) .07   (k) .007

2. What digit of 1,253.6749 is in the following positions?
   (a) tens   **(b)** thousandths   **(c)** tenths   **(d)** thousands
   (e) ones   **(f)** hundredths   **(g)** ten-thousandths   **(h)** hundreds

Exercises
Part II

1. What is the position of the 7 in each of the following numbers?
   (a) 1.07   **(b)** 70.5   **(c)** 39.76

2. What digit of 1,234.5678 is in the indicated positions?
   (a) thousandths   **(b)** ones

SECTION 12.2
READING AND WRITING
DECIMAL NUMBERS

A quarter-mile race (440 yards) can be compared with a 400-meter race by the following:

400 meters ≅ 437.4 yards
quarter mile = 440 yards

That is, four hundred meters is approximately equal to four hundred thirty-seven and four-tenths yards.

After studying this section you will know how to read and write decimal numbers, as illustrated by these two examples.

1. 27.567 is read twenty-seven and five hundred sixty-seven thousandths.
2. Three and seventy-two thousandths is written 3.072.

If you think you know this information, turn to the exercises and work *each* problem. If you have difficulty, turn back and read this section completely.

Reading Decimal Numbers

Reading Decimal Numbers

1. Read the whole number as in Chapter 1.
2. Say *and* for the decimal point.
3. Read the fractional part *as if* it were a whole number, and say the name of place value of the rightmost digit.

EXAMPLE 1:   The number 16.253 is read as follows:

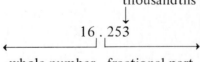

thousandths place for rightmost digit

16 . 253

whole number   fractional part

By reading the whole number,

Sixteen . . .

saying *and* for the decimal,

Sixteen and . . .

reading the fractional part as if it were a whole number,

Sixteen and two hundred fifty-three . . .

and saying the name of place value of the rightmost digit; in this case, the digit 3 is in the thousandths position.

Sixteen and two hundred fifty-three thousandths

EXAMPLE 2:   Write the number 703.0521 as you would read it.

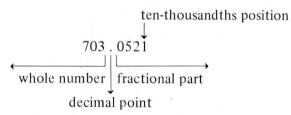

Because of the position of the digit 1, the fractional part is read as ten-thousandths; therefore, 703.0521 is read

seven hundred three and five hundred twenty-one ten-thousandths

Notice that the zero in the fractional part is ignored in the reading, but counted in determining the position of the digit 1.

EXAMPLE 3:   Write the number 20.07 as you would read it.

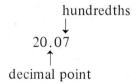

The fractional part is read as hundredths. The number 20.07 is read

twenty and seven hundredths

EXAMPLE 4:   Write 12,513.8 as you would read it.

tenths
↓
12,513 . 8
←————————| |————————→
whole number   fractional part

The number 12,513.8 is read

twelve thousand, five hundred thirteen and eight tenths

Writing Decimal Numbers   EXAMPLE 5:   Write the numeral for four hundred and twenty-seven hundredths.

Underline the word *and*, which signals the decimal point.

four hundred <u>and</u> twenty-seven hundredths

Write the numeral for the whole number as in Chapter 1, followed by the decimal point.

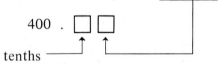

four hundred and twenty-seven hundredths

400 .

Next, determine the number of places that will be used after the decimal point; hundredths means two decimal places.

four hundred and twenty-seven <u>hundredths</u>

400 . ☐ ☐

tenths

Fill in the boxes with the number of hundredths—in this case twenty-seven.

four hundred and twenty-seven hundredths

400 . ② ⑦

Therefore, four hundred and twenty-seven hundredths is written

400.27

EXAMPLE 6:   Write the numeral for thirty-five and eleven thousandths.
Write the numeral for the whole number and write the decimal point for the word *and*.

thirty-five and eleven thousandths

35 .

Determine the number of places that will be used after the decimal point.

thirty-five and eleven <u>thousandths</u>

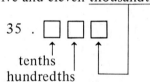

35 . ☐ ☐ ☐

tenths
hundredths

Fill in the boxes with the number of thousandths—11. Because the digits must end in the thousandths position, the 11 must be positioned as follows:

35.☐① ①

To show the space between the decimal point and the 11, insert a zero.

35.⓪① ①

Therefore, thirty-five and eleven thousandths is written

35.011

Writing Decimal Numbers

1. Write the whole number.
2. Replace the word *and* with a decimal point.
3. Determine the number of places that will be used in the fractional part.
4. Fill in the places with the digits in the fractional part so that the digits end in the named position, inserting zeros where necessary.

EXAMPLE 7:   Write the numeral for one thousand, five hundred four and eight thousandths.

Write the whole number part and the decimal point.

$$1,504.$$

Determine the number of places after the decimal point.

1,504. ☐☐☐

    ↑ ↑ ↑

tenths | |

hundredths |

thousandths

Position the 8 in the thousandths position, inserting zeros where necessary.

1,504. |0||0||8|

Therefore, one thousand, five hundred four and eight thousandths is written

$$1,504.008$$

EXAMPLE 8:   Write the numerals for one hundred fifty-two thousandths.

There is no word *and* in the statement because this number does not have a whole number part.

Determine the number of places after the decimal point.

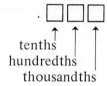

. ☐☐☐

   ↑ ↑ ↑

tenths | |

hundredths |

thousandths

Position the 152 so that it ends in the thousandths position.

. |1||5||2|

Therefore, one hundred fifty-two thousandths is written

$$.152$$

Now you try these.

Examples	Solutions
Write the number 52.19 as you would read it.	<u>52</u> . <u>19</u> ↓ fifty-two and nineteen hundredths The rightmost digit determines the name of the fractional part. In this example the 9 is in the hundredths position.
Write 205.013 as you would read it.	<u>205</u> . <u>013</u> ↓ two hundred five and thirteen thousandths
Write 0.0715 as you would read it.	seven hundred fifteen ten-thousandths This number has no whole number part, therefore the *and* is not needed for a separator.

Examples	Solutions
Write the numeral for the number two and fifty-nine thousandths.	2 . Whole number and decimal. 2 . ☐☐☐ Fractional part will have 3 places (thousandths). 2 . 0 5 9 Position the 59. Answer: 2.059
Write the numeral for one thousand, five hundred sixty-three ten-thousandths.	Be careful! There is no whole number part because the separator *and* is missing. .1563

Ready to continue? No hints!

Write the number 103.05 as you would read it.	one hundred three and five hundredths
Write the numeral for the number two hundred fifty-four thousandths.	.254

If you had difficulty with these two examples, please talk with an instructor or read this section again *before* working the exercises.

Exercises Part I

Answers to these exercises can be found on page 363.

1. Write each number as you would read it.
    - (a) 53.15
    - (b) 172.5
    - (c) 7.153
    - (d) 6.04
    - (e) 605.009
    - (f) .713
    - (g) .9
    - (h) .003

2. Write the numeral for each of the following numbers.
    - (a) thirty-five and eighty-three hundredths
    - (b) nine and one hundred sixteen thousandths
    - (c) five and three thousandths
    - (d) two hundred five and six hundredths
    - (e) one hundred fifty-nine ten-thousandths
    - (f) three hundredths

Exercises Part II

1. Write each number as you would read it.
    - (a) 72.123
    - (b) .96
    - (c) .004

2. Write the numeral for each of the following numbers.
    - (a) twenty-two and eight tenths
    - (b) three hundred and four hundredths
    - (c) fifteen thousandths

341

_____ 19 _____

PAY TO THE
ORDER OF _____ $ | 27.39 |

TWENTY-SEVEN AND $^{39}/_{100}$ _____ DOLLARS

BANK OF TNCC

MEMO _____ _____

⑆0 2ι 2ι4 2ι 0 56 2⑈ 40 2 ι00 5 3 8 2⑈ 2 0 0 0 9 6⑆

In this section you will learn how to change a decimal number to a fraction or mixed number, as illustrated by these examples.

(a) $.027 = \dfrac{27}{1,000}$

(b) $24.5 = 24\dfrac{5}{10}$

$= 24\dfrac{1}{2}$ In lowest terms.

If you think you know this information, turn to the exercises and work *each* problem. If you have difficulty turn back and read this section completely.

Changing Decimals to Fractions If you can correctly read a decimal number, you can write it as a fraction.

EXAMPLE 1:   Change .13 to a fraction.
The decimal number .13 is read thirteen hundredths. The last word describes the denominator.

$$\text{hundredths} \longrightarrow \overline{100}$$

The number of hundredths describes the numerator.

$$\text{thirteen} \longrightarrow \frac{13}{100}$$
$$\text{hundredths} \longrightarrow$$

Therefore, $.13 = \dfrac{13}{100}$

Changing Decimals to Fractions
1. Read the number.
2. Write the denominator of the fraction as described by the place value.
3. Write the numerator as described by the number of that place value.

EXAMPLE 2:   Change .027 to a fraction.
The decimal number .027 is read twenty-seven thousandths.

The denominator is described by thousandths.

$$\text{thousandths} \quad \overline{}1000$$

The number of thousandths is twenty-seven.

$$\text{twenty-seven} \quad \dfrac{27}{1000}$$
$$\text{thousandths}$$

 Always reduce the fractional form to lowest terms.

EXAMPLE 3:   Write .5 as a fraction.
The decimal number .5 is read five tenths.

$$\text{five} \quad \dfrac{5}{10}$$
$$\text{tenths}$$

However, $\dfrac{5}{10}$ can be reduced to $\dfrac{1}{2}$ by dividing numerator and denominator by 5.
Therefore,

$$.5 = \frac{5}{10}$$
$$= \frac{1}{2} \qquad \text{In lowest terms.}$$

If a decimal number has a whole number part, write it as the whole number part of a mixed number.

EXAMPLE 4:   Write 27.45 as a mixed number.
The decimal number 27.45 is read twenty-seven and forty-five hundredths.

The whole number is twenty-seven.

$$\text{twenty-seven} \qquad 27$$

The fractional part is forty-five hundredths.

$$27\frac{45}{100} \qquad \begin{array}{l}\text{forty-five}\\ \text{hundredths}\end{array}$$

Therefore,

$$27.45 = 27\frac{45}{100}$$

$$= 27\frac{9}{20} \quad \text{In lowest terms.}$$

EXAMPLE 5: Change 173.217 to a mixed number.

one hundred
seventy-three $\quad 173\frac{217}{1,000}\quad$ two hundred seventeen
thousandths

EXAMPLE 6: How to Write the Numbers on a Check

1. Write the amount using digits close to the dollar sign.
2. Write the amount using words starting as far left as possible. Express cents as a fraction. Insert a wavy line between the amount and the word *dollars* to prevent tampering.

Here is the correct way to fill in the check shown with the amount $17.33. Because 17.33 means seventeen and thirty-three hundredths:

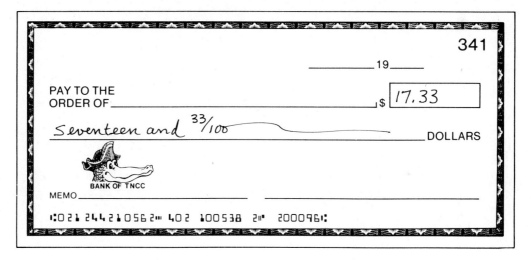

Now you try these.

Examples	Solutions
Change .9 to a fraction.	.9 is read nine tenths. $.9 = \dfrac{9}{10}$ $\quad$ nine tenths
Change .15 to a fraction.	.15 is read fifteen hundredths. $.15 = \dfrac{15}{100}$ $\quad$ fifteen hundredths $= \dfrac{3}{20}$ $\quad$ In lowest terms.

Examples	Solutions
Write .052 as a fraction.	.052 is read fifty-two thousandths. $$.052 = \frac{52}{1000}$$ $$= \frac{26}{500}$$ $$= \frac{13}{250} \quad \text{In lowest terms.}$$
Change 93.05 to a mixed number.	93.05 is read ninety-three and five hundredths. ninety-three $\quad 93\frac{5}{100} \quad$ five hundredths $$93.05 = 93\frac{1}{20} \quad \text{In lowest terms.}$$

Now try these—no hints given.

Write each decimal number as a fraction or mixed number.	
(a) .193	(a) $\dfrac{193}{1,000}$
(b) .12	(b) $\dfrac{3}{25}$
(c) 51.25	(c) $51\dfrac{1}{4}$

If you had difficulty with these examples, please talk with an instructor or read this section again.

Exercises Part I

Answers to these exercises can be found on page 363.

Write each decimal number as a fraction or mixed number, reducing fractions to lowest terms whenever possible.

1. .123	2. .56	3. .109	4. .08	5. 59.3
6. 67.8	7. 27.105	8. 6.333	9. 19.75	10. 125.005

Exercises Part II

Write each decimal number as a fraction or mixed number, reducing fractions to lowest terms whenever possible.

1. .17	·2. .46	3. 3.91	4. .008	5. 124.02

```
┌──────────────────────────────────────┐
│   ┌────────────────────────────┐      │
│   │     SAVE  $1.99             │      │
│   │                            │      │
│   │  LARGE-COUNT TRASH BAGS    │      │
│   │        $6.00               │      │
│   │                            │      │
│   │   REG. $7.99               │      │
│   └────────────────────────────┘      │
└──────────────────────────────────────┘
```

I have already noted that a number like .27 is often written as 0.27 for clarity; that is,

$$.27 = 0.27$$

Here is another use of zeros.

> It is possible to **annex** zeros to a decimal number without changing the value of the number. That is, writing zeros to the right of the last digit following the decimal point does not change the value of a number.

EXAMPLE 1:   Are 17.5 and 17.50 equal?
The decimal number 17.5 is a one-place decimal because there is only one digit after the decimal point. The decimal number 17.50 is a two-place decimal because there are two digits after the decimal point. However, adding zero hundredths does not change the value of 17.5; that is,

$$17.5 = 17.50$$
$$\uparrow$$
$$0 \text{ hundredths, or } \frac{0}{100}$$

Look at the numbers in fractional form.

$$17.5 = 17\frac{5}{10} \qquad 17.50 = 17\frac{50}{100}$$
$$= 17\frac{1}{2} \qquad\qquad = 17\frac{1}{2}$$

In fractional form you can clearly see they are equal. If you annex one zero to 17.5 you get an equal two-place form for 17.5.

EXAMPLE 2: Annex two zeros to the decimal number 153.72.

$$153.72 = 153.7200$$

0 ten-thousandths, or $\dfrac{0}{10,000}$

0 thousandths, or $\dfrac{0}{1,000}$

Adding zero thousandths and zero ten-thousandths does not change the value of the number.

EXAMPLE 3: Are 157.2 and 157.02 equal?

No! In one case the digit 2 is in the tenths position; in the other, the 2 is in the hundredths position.

In fractional form

$$157.2 = 157\frac{2}{10}$$

$$= 157\frac{1}{5}$$

but,

$$157.02 = 157\frac{2}{100}$$

$$= 157\frac{1}{50}$$

The two numbers are not equal.

You need to know how to annex zeros properly to be able to add, subtract, and divide decimal numbers. Additionally, you need to know how to annex zeros to a whole number.

EXAMPLE 4: Where is the decimal point in the number 63 understood to be?

To answer this, read 63 as sixty-three and no fractional part. Therefore, you would write,

$$63 = 63.$$

with the decimal point after the ones position.

EXAMPLE 5: Annex three zeros to the number 175.

Because the decimal point is written after the ones position,

$$175 = 175.$$

Annexing three zeros,

$$175 = 175.000$$

EXAMPLE 6: Write 210.3 as a three-place decimal number.

In order not to change the value of the number, annex zeros.

$$210.3 = 210.300$$

as a three-place decimal number.

EXAMPLE 7:   Write 625 as a two-place decimal number.

Place the decimal after the ones position of the whole number and annex two zeros.

$$625 = 625.00$$

Now try these.

Examples	Solutions
Are 7.03 and 7.030 equal?	Yes! $7.03 = 7.030$ ↑ annexing 0 thousandths
Are 7.30 and 7.03 equal?	No! $7.30 \neq 7.03$ ↑ ↑ 3 3 tenths hundredths
Annex two zeros to the following numbers. (a) 89.57 (b) .201 (c) 27	(a) 89.5700 (b) .20100 (c) With a whole number, place the decimal point after the ones position before annexing zeros. $27 = 27.00$
Write each number as a three-place decimal. (a) 2.3 (b) 57.12 (c) 93	(a) 2.300 (b) 57.120 (c) 93.000

**SECTION 12.5
ROUNDING**

One kilometer is approximately equal to .6 mile, to the nearest tenth of a mile.

This section explains the rounding procedure for decimals, as illustrated by these examples.

(a) 12.573 rounded to the tenths position becomes 12.6.

(b) 12.573 rounded to the hundredths position becomes 12.57.

If you think you know this information, turn to the exercises and work each problem. If you have difficulty with the exercises, turn back and read this section completely.

Rounding The procedure for rounding decimal numbers is the same as for whole numbers.

Rounding

When rounding to a certain position, drop and replace all digits to the immediate right of this position with zeros. Then consider the first digit dropped to the immediate right of the rounding position.

1. If this digit is less than 5, leave the digit in the rounding position unchanged.
2. If this digit is 5 or greater, add 1 to the rounding position.

EXAMPLE 1: Round 17.536 to the nearest tenth.
Replace all digits after the tenths position with zeros.

$$\overset{00}{17.8\cancel{36}}$$

Because the digit to the right of the rounding position is less than 5, leave the digit in the tenths position unchanged.

$$\overset{00}{17.8\cancel{36}} \text{ becomes } 17.800 \quad \text{to the nearest tenth}$$

Because 17.800 = 17.8, drop the zeros in the final answer.

EXAMPLE 2: Round 17.836 to the nearest hundredth.
Replace the digit after the hundredths position with a zero.

$$\overset{0}{17.83\cancel{6}}$$

Because the digit dropped was 5 or greater, add 1 to the rounding position.

$$\overset{0}{17.83\cancel{6}} \cong 17.840$$

$$\cong 17.84 \quad \text{Drop the zero.}$$

EXAMPLE 3: Round 725.396 to the tenths place.

$$\overset{00}{725.3\cancel{96}} \cong 725.400 \quad \text{Adding 1 to the rounding position.}$$

$$\cong 725.4$$

EXAMPLE 4: Round 725.396 to the nearest hundredth.

$$\overset{0}{725.39\cancel{6}} \cong 725.400$$

└ add 1 ─┘

When 1 is added to 9, the fractional part .39 becomes .40.

In final form, $725.396 \cong 725.40$ to the nearest hundredth. Notice that the zero in the hundredths position is *not* dropped because the answer *requires* hundredths.

EXAMPLE 5: On federal income tax forms, you are allowed to round all figures to the nearest dollar. Round each of these numbers to the nearest dollar.

(a) $15.27

(b) $28.50

(c) $139.63

In each problem the digits after the decimal point are dropped.

(a) $15.~~27~~ \cong \$15.00$ because the digit 2 is less than 5.

(b) $28.~~50~~ \cong \$29.00$ because the digit 5 is 5 or more.

(c) $139.~~63~~ \cong \$140.00$ because the digit 6 is more than 5.

EXAMPLE 6: Round 18.335 to a two-place decimal number.
Rounding to a two-place decimal number means to round to hundredths.

$18.3~~35~~ \cong 18.340$ Adding 1 to the rounding position.

$\cong 18.34$

EXAMPLE 7: Round 18.335 to a one-place decimal number.
Rounding to a one-place number means to round to tenths.

$18.3~~35~~ \cong 18.300$ Position unchanged.

$\cong 18.3$

Now you try these.

Examples	Solutions
Round 519.3497 to the nearest tenth.	$519.3~~497~~ \cong 519.3$ ↑ tenths Because the first digit dropped is less than 5, the digit 3 in the rounding position remains unchanged.
Round 519.3497 to the hundredths place.	$519.34~~97~~ \cong 519.35$ ↑ hundredths Because the digit 9 is 5 or more, add 1 to the rounding position.

Examples	Solutions
Round 519.3497 to the nearest thousandth.	$$\overset{0}{519.3497} \cong 519.350$$ ↑ thousandths Add 1 to the rounding position; .349 becomes .350.
Round 519.3497 to a two-place decimal number.	519.35 A two-place decimal number ends in the hundredths position.

Now try these—no hints given.

Round 6.1951 to the indicated position. (a) tenths (b) hundredths (c) a three-place decimal	(a) 6.2 (b) 6.20 (c) 6.195

If you had difficulty with the last examples, please read this section again or see an instructor.

Exercises Part I   Answers to these exercises can be found on page 363.

1. Round each number to the nearest hundredth.
   (a) 27.5921   (b) 395.125   (c) 29.695   (d) 14.54321
2. Round each number to the nearest thousandth.
   (a) 27.5921   (b) 395.1259   (c) 29.6898   (d) 14.54321
3. Round each number to a one-place decimal number.
   (a) 27.5921   (b) 395.125   (c) 29.696   (d) 14.54321

Exercises Part II   Round each number to the nearest hundredth.

1. 15.172   2. 732.1581   3. 82.1963

Round each number to a two-place decimal number.

4. 15.132   5. 4.1582   6. .0863

REVIEW MATERIAL

Review Skill Reading and Writing Decimal Numbers   To read a decimal number, read the whole number, say *and* for the decimal point, and read the fractional part as if it were a whole number, but say the name of place value of the rightmost digit.

25.103 is read twenty-five and one hundred three thousandths
Four and twenty-seven ten-thousandths is written 4.0027

Answers to this exercise can be found on page 364.

Write the following numbers as you would read them.

    1. 193.27   **2.** 6.004   **3.** .8   **4.** 105.507

Write numerals for the following numbers.

    5. two and nineteen hundredths
    6. eighty-five thousandths
    7. one hundred four and one hundred fifty-two thousandths

Exercise B No answers are given.

Write the following numbers as you would read them.

    1. 204.38   **2.** 7.005   **3.** .9   **4.** 206.608

Write the numerals for the following numbers.

    5. three and eighteen hundredths
    6. eighty-four thousandths
    7. one hundred five and one hundred fifty-three thousandths

Review Skill
Decimals
to Fractions and
Mixed Numbers

To change a decimal to a fraction or mixed number, *write* it the way you *read* it. The decimal number 24.53 is read twenty-four *and* fifty-three hundredths.

$$24.53 = 24\frac{53 \longleftarrow \text{fifty-three}}{100 \longleftarrow \text{hundredths}}$$

twenty-four

Exercise C Answers to this exercise can be found on page 364.

Change each decimal number to a fraction or mixed number.

    1. .23   **2.** .8   **3.** .107   **4.** 62.5   **5.** 135.071   **6.** 27.25

Exercise D No answers are given.

Change each decimal number to a fraction or mixed number.

    1. .13   **2.** .6   **3.** .203   **4.** 59.8   **5.** 153.091   **6.** 28.75

Review Skill
Rounding
Decimal Numbers

Decimal numbers are rounded in much the same way as are whole numbers. Consider 35.7259.

$$\overset{000}{35.7259} \cong 35.7 \qquad \text{To the nearest tenth.}$$

$$35.7259 \cong 35.73 \qquad \text{To the nearest hundredth.}$$
$$\overset{0}{35.7259} \cong 35.726 \qquad \text{To the nearest thousandth.}$$

Exercise E Answers to this exercise can be found on page 364.

Round to the nearest tenth.

1. 85.425   2. 74.356   3. 2.895

Round to the nearest thousandth.

4. .0535   5. 1.0622   6. 1.0598
7. 103.0582

Exercise F No answers are given.

Round to the nearest tenth.

1. 74.315   2. 85.456   3. 3.782

Round to the nearest thousandth.

4. .0425   5. 2.0532   6. 7.1297
7. 207.0694

REVIEW EXERCISES

Refer to the section listed if you have difficulty. Answers to these exercises can be found on page 364.

Section 12.1
1. What is the place value of the 3 in each of the following numbers?
   (a) 1.38   (b) 16.053   (c) 359.2   (d) .03
2. In the number 135.7246, name the digit in the following position.
   (a) hundredths   (b) tens   (c) thousandths
   (d) tenths   (e) hundreds

Section 12.2
3. Write each number as you would read it.
   (a) 57.54   (b) 1,652.059   (c) .9   (d) .05
4. Write the numerals for each of the numbers.
   (a) twenty-three and fifty-two hundredths
   (b) five and twenty-seven thousandths
   (c) one hundred forty-three thousandths

Section 12.3
5. Write each decimal number as a fraction or mixed number.
   (a) .73   (b) .2   (c) 203.123   (d) 59.095

6.   Round each number to the nearest hundredth.
    (a) .537   (b) .424   (c) .6985   (d) .0352

CHAPTER POST-TEST

This test should determine if you have mastered the topics in Chapter 12. Answers to this test can be found on page 364.

1.   What is the place value of the 4 in each number?
    (a) 27.054   (b) 3.495
2.   In the number 275.369, name the digit in the following positions.
    (a) tens   (b) tenths   (c) hundreds   (d) thousandths
3.   Write each number as you would read it.
    (a) 289.057   (b) .275
4.   Write the numerals for each of the numbers.
    (a) six and fifty-two thousandths
    (b) forty-seven ten-thousandths
5.   Write each decimal number as a fraction or mixed number.
    (a) .105   (b) 6.23
6.   Round each number to the nearest tenth.
    (a) 27.59   (b) 8.35   (c) 28.98

13

addition and subtraction of decimal numbers

SECTION 13.1
ADDITION OF DECIMAL NUMBERS

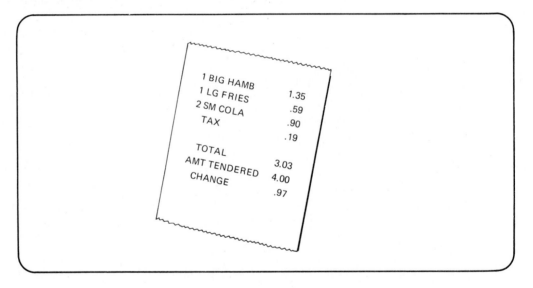

```
1 BIG HAMB
1 LG FRIES        1.35
2 SM COLA          .59
TAX                .90
                   .19

TOTAL
AMT TENDERED      3.03
CHANGE            4.00
                   .97
```

This section explains the procedure for adding decimal numbers, as illustrated by this example.

$$5 + 6.72 + 25.135: \qquad
\begin{array}{r}
5.000 \\
6.720 \\
+25.135 \\
\hline
36.855
\end{array}$$

If you think you know this information, turn to the exercises and work each problem. If you have difficulty with the exercises, turn back and read this section completely.

**Addition of
Decimal Numbers**

> The procedure for adding decimal numbers is similar to that for adding whole numbers. Arrange the digits in columns. Line up digits with the same place value. Because the place values of both the whole and fractional parts are based on powers of ten, the carrying procedure is the same as before.

EXAMPLE 1:   Add 12.13 and 69.29.

First, arrange the digits in columns. Line up digits with the same place value.

$$\begin{array}{r} 12.13 \\ +69.29 \\ \hline \end{array}$$ Notice that this lines up the decimal points.

Now add the digits in each position, starting from the right, carrying to the next position whenever necessary.

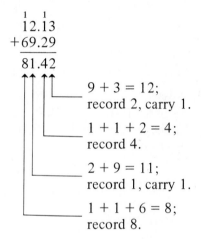

$$\begin{array}{r} \overset{1}{1}\overset{1}{2}.13 \\ +69.29 \\ \hline 81.42 \end{array}$$

9 + 3 = 12;
record 2, carry 1.

1 + 1 + 2 = 4;
record 4.

2 + 9 = 11;
record 1, carry 1.

1 + 1 + 6 = 8;
record 8.

Notice that the decimal point in the answer is lined up with the decimal points in the addends. Therefore, 12.13 + 69.29 = 81.42.

As with all arithmetic, check that the sum is roughly equal to an estimation. With decimals, round each number to a one-digit number; that is,

12.13 + 69.29 is approximately 10 + 70, which equals 80

which agrees well with the actual answer 81.42.

Do not expect your estimation to equal the actual answer. With an estimation of 80, I would *not* expect an answer of 8 or 800, but some number reasonably close to 80 in value.

EXAMPLE 2:   Add 59.3 and 167.99.

Line up the decimal points and digits with the same place value.

Actual	Estimation
59.3	60
+167.99	+200
	260

Some people prefer to give each number the same number of places after the decimal by annexing zeros. I can make both numbers two-place decimals by annexing a zero to 59.3.

$$\begin{array}{r} 59.30 \\ +167.99 \\ \hline \end{array}$$

Now add.

$$\begin{array}{r} \overset{1\ 1\ 1}{59.30} \\ +167.99 \\ \hline 227.29 \end{array}$$ Which agrees well with the estimation.

EXAMPLE 3:   Add 17.4, 95.98, and 127.569.
Line up the decimal points.

$$\begin{array}{r} 17.4 \\ 95.98 \\ +127.569 \end{array}$$

Annex zeros to make each number a three-place decimal number and add.

$$\begin{array}{r} \overset{1\ 2\ 1\ \ 1}{17.400} \\ 95.980 \\ +127.569 \\ \hline 240.949 \end{array}$$

EXAMPLE 4:   Add 27 and 32.56.
In the whole number 27, the decimal point is understood to be after the ones position.

$$\begin{array}{r} 27. \\ +32.56 \end{array}$$

Annex zeros to make each number a two-place decimal number and add.

$$\begin{array}{r} 27.00 \\ +32.56 \\ \hline 59.56 \end{array}$$

EXAMPLE 5:   Add 15.381, 59.98, and 24.

$$\begin{array}{r} \overset{1\ 1\ \ 1}{15.381} \\ 59.980 \\ +24.000 \\ \hline 99.361 \end{array}$$

Now you try these.

Examples	Solutions
.981 + .456 = ?	Line up the decimal points. $$\begin{array}{r} \overset{1}{.981} \\ +\ .456 \\ \hline 1.437 \end{array}$$
56.2 + 1.593 = ?	Line up the decimal points. $$\begin{array}{r} 56.2 \\ +\ 1.593 \end{array}$$

Examples	Solutions
	Annex zeros to make each number a three-place decimal number. $$\begin{array}{r} 56.200 \\ +\ 1.593 \\ \hline 57.793 \end{array}$$

$59.87 + 2.9 + 62 = ?$	Line up the decimal points. Remember that the decimal point is after the ones position in the whole number.
	    Annex zeros to make each number a two-place decimal number.
	
	              Estimation
	
	$\begin{array}{r} \overset{1\ 1}{59.87} \\ 2.90 \\ +\ 62.00 \\ \hline 124.77 \end{array}$ $\begin{array}{r} 60 \\ 3 \\ +\ 60 \\ \hline 123 \end{array}$ Rounding to one digit.

Now try these—no hints given.

$58.89 + 1.999 = ?$	60.889
$27 + 5.2 + 1.32 = ?$	33.52

If you had difficulty with the last two examples, read this section again or talk with an instructor.

Exercises
Part I

Answers to these exercises can be found on page 364.

Add.

1. $.93 + .85$
2. $.107 + .989$
3. $12.3 + 15.9$
4. $66.6 + 77.77$
5. $59.21 + 68.589$
6. $7.2 + 5.983$
7. $27 + 5.93$
8. $6.02 + 5$
9. $1.596 + .03 + 59$
10. $6 + 93.03 + .275$
11. Find the total restaurant bill (without tax and tip).

The POT		
1	HOT DOG	1.99
1	SPECIAL	.74
2	MED. COLA	1.20

12. Find the total price for the merchandise.

DIRECT MAIL ORDER BLANK

Catalog Number	How Many	Color Size	Name of Item	Price Each	Total Price
3262	1		Vac	84.95	84.95
9352	1		Sewing Head	199.95	199.95
5021	1	12 in	B&W TV	99.99	99.99

TOTAL FOR GOODS. . . . .

**Exercises
Part II**

Add.

1. .15 + 8.15
3. 69.123 + 17.9
5. 90.02 + 57 + .079

2. 17.2 + 29.53
4. 5 + 6.02

SECTION 13.2
SUBTRACTION OF DECIMAL NUMBERS

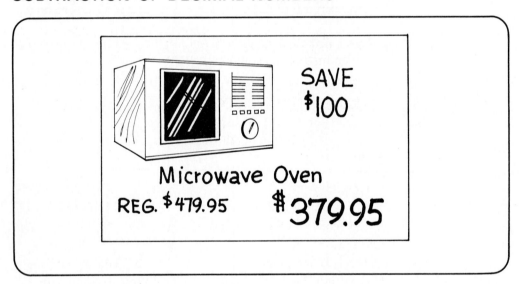

SAVE $100

Microwave Oven
REG. $479.95 $379.95

This section explains the procedure for subtraction of decimal numbers, as illustrated by this example.

Subtract 9.75 from 27.5.

$$\begin{array}{r} 27.50 \\ -\ \ 9.75 \\ \hline 17.75 \end{array}$$

If you think you know this information, turn to the exercises and work *each* problem. If you have difficulty, turn back and read this section completely.

The procedure for subtraction of decimal numbers is very similar to that for subtraction of whole numbers. Arrange the digits in columns. Line up digits with the same place value. Annex zeros to give each number the same number of decimal positions, then subtract, borrowing whenever necessary.

EXAMPLE 1:   Subtract 12.5 from 19.2.
Line up the decimal points.

$$\begin{array}{r} 19.2 \\ -12.5 \\ \hline \end{array}$$

Now subtract the digits in each position, starting from the right, borrowing from the next position whenever necessary.

$$\begin{array}{r} \overset{8\ 1}{1\cancel{9}.2} \\ -12.5 \\ \hline 6.7 \end{array}$$

12 − 5 = 7
8 − 2 = 6
1 − 1 = 0   (not shown)

Again, you can estimate to check if your answer is reasonable.

$$19.2 - 12.5 \cong 20 - 10 = 10$$

which agrees well with the actual answer of 6.7.

However, the best check for subtraction is addition. Draw a line under the answer and add as follows.

$$\begin{array}{rl} & 19.2 \\ & -12.5 \\ \text{Answer:} & \underline{6.7} \\ \text{Check:} & 19.2 \end{array} \quad \text{add}$$

The addition gives the number at the top—the minuend—when the answer is correct.

EXAMPLE 2:   Subtract 8.98 from 15.2.
Line up the decimal points.

$$\begin{array}{r} 15.2 \\ -\ 8.98 \\ \hline \end{array}$$

Annex zeros to make each number a two-place decimal number, and subtract.

$$\begin{array}{r} \overset{1}{}\overset{4\ \ 1_1}{1\cancel{5}.\cancel{2}0} \\ -\ 8.98 \\ \hline 6.22 \end{array}$$

EXAMPLE 3:   Subtract 5 from 24.3.

$$\begin{array}{r} 24.3 \\ -\ 5.0 \\ \hline 19.3 \end{array}$$

EXAMPLE 4:   Subtract 1.03 from 27.

$$
\begin{array}{r}
2\overset{6}{\cancel{7}}.\overset{9}{\cancel{0}}\overset{1}{0} \\
-\ 1.03 \\
\hline
25.97
\end{array}
$$

EXAMPLE 5:
How to Balance a Checkbook

1. Sort canceled checks by date or number.
2. Start with your current checkbook balance.
3. *Subtract* any service charges.
4. Check for and make a list of all outstanding checks—those not yet processed by the bank and returned to you.
5. *Add* the amounts of any outstanding checks.
6. Check for deposits that have not been processed by the bank and *subtract* these deposits.
7. Compare your final result with your bank statement.

Suppose that your bank statement shows a final balance of $1,247.77, whereas your checkbook shows a balance of $1,289.13. There are two outstanding checks, for $69.89 and $5.19. A deposit for $115.00 has not been processed. Inspection of the bank statement shows a service charge of $1.44.

Here is the work needed to check your balance. As you notice steps 1 and 4 have already been done.

2. Current balance in checkbook $1,289.13
3. Subtract service charge . . . . −$ 1.44
                                                                         $1,287.69

5. Add outstanding checks.

$$
\begin{array}{r}
69.89 \\
+\ 5.19 \\
\hline
75.08
\end{array}
$$

                75.08 . . . . +$ 75.08
                                                                  $1,362.77

6. Subtract unprocessed deposits. . . . . −$ 115.00

7. Revised balance. $1,247.77
Balance as shown on bank statement. $1,247.77

If these two balances do not agree, there is a mistake somewhere.

That is easy enough! Now you try these.

Examples	Solutions
.981 − .456 = ?	Line up the decimal points. $\begin{array}{r}.981\\-.456\\\hline.525\end{array}$
56.2 − 1.593 = ?	Line up the decimal points. $\begin{array}{r}56.2\ \ \ \\-\ 1.593\end{array}$

Examples	Solutions
	Annex zeros.
	$$\begin{array}{r} {\scriptstyle 1\,9\,1} \\ 56.200 \\ -\ 1.593 \\ \hline 54.607 \end{array}$$ $Check:$ $$\begin{array}{r} 56.200 \\ -\ 1.593 \\ \hline 54.607 \\ \hline 56.200 \end{array}$$ add
$62 - 2.9 = ?$	Line up the decimal points. Remember that the decimal point is after the ones position in the whole number. $$\begin{array}{r} 62. \\ -\ 2.9 \\ \hline \end{array}$$ Annex a zero. $$\begin{array}{r} {\scriptstyle 5\ 1\ 1} \\ \not{6}\not{2}.0 \\ -\ 2.9 \\ \hline 59.1 \end{array}$$

Now try these—no hints given.

$58.89 - 1.999 = ?$	56.891
$27 - 1.32 = ?$	25.68

If you had difficulty with the last two examples, read this section again or talk with an instructor.

Exercises Part I   Answers to these exercises can be found on page 364.

Subtract.

1. $.93 - .85$
2. $.989 - .107$
3. $15.9 - 12.3$
4. $77.77 - 66.6$
5. $68.589 - 59.21$
6. $7.2 - 5.983$
7. $27 - 5.93$
8. $6.02 - 5$

```
SAVE $30 - $60
    on Dryers

3-cycle model
Reg. $259.95        Now $230.99

Automatic Model
Reg. $299.95        Now $256

Large-Capacity Model
Reg. $339.95        Now $280.99
```

9. What is the savings on the 3-cycle model?

10. What is the savings on the large-capacity model?

11. Complete the check stub for a check for $18.19 with a balance forward of $689.03.

BAL. FOR'D.	$689	03
DEPOSIT		
TOTAL		
THIS PAYMENT		
OTHER DED.		
BAL. FOR'D.		

12. Complete the check stub for a check for $29.89 with a balance forward of $1,019.08.

BAL. FOR'D.	$1,019	08
DEPOSIT		
TOTAL		
THIS PAYMENT		
OTHER DED.		
BAL. FOR'D.		

Exercises Part II

Subtract.

1. $.91 - .83$
2. $17.9 - 1.03$
3. $12 - 7.9$
4. $89.5 - 13$
5. $1003.2 - 2.37$
6. $5.004 - .54$

REVIEW MATERIAL

Review Skill Addition and Subtraction of Decimal Numbers

To add or subtract decimal numbers, line up the decimal points. Next, annex zeros to give each number the same number of places after the decimal. Finally, add or subtract as you would whole numbers. The decimal point in the answer is lined up with the decimal points in the problem.

$$13.09 + 1.72: \quad \begin{array}{r} \overset{1}{13.09} \\ + \ 1.72 \\ \hline 14.81 \end{array} \qquad 28 - 4.32: \quad \begin{array}{r} \overset{7}{2}\overset{9}{8}.\overset{1}{0}0 \\ - \ 4.32 \\ \hline 23.68 \end{array}$$

Exercise A

Answers to this exercise can be found on page 364.

Add or subtract as indicated.

1. $.15 + .07$
2. $12.3 + .179$
3. $.76 - .28$
4. $1.12 - .395$
5. $12 + 1.56$
6. $15 - 3.25$
7. $9.15 - 7$
8. $12.5 + 7 + 9.32$
9. $17.1 - .03$
10. $143.72 + 8.1 + 32$

Exercise B

No answers are given.

Add or subtract as indicated.

1. .15 − .07
2. 12.3 − 1.79
3. .76 + .28
4. 1.12 + .395
5. 12 − 1.56
6. 15 + 3.25
7. 5.62 − 3

REVIEW EXERCISES

Refer to the section listed if you have difficulty. Answers to these exercises can be found on page 364.

Section 13.1 Add.

1. .99 + .27
2. 15.5 + 7.98
3. 17.596 + 8.58
4. 9.7 + 15 + 16.98

Section 13.2 Subtract.

5. .97 − .89
6. 15.5 − 7.98
7. 17.586 − 8.59
8. 15 − 9.7

CHAPTER POST-TEST

This test should determine if you have mastered the topics in Chapter 13. Answers to this test can be found on page 364.

Add or subtract as indicated.

1. 27.593 + 59.98
2. 59.98 − 27.593
3. 145.2 + 15.79 + 19
4. 105.003 + 9.07
5. 105.003 − 9.07
6. 99.99 + 9.9
7. 597 − 12.53
8. 175.02 − 19

14
multiplication of decimal numbers

SECTION 14.1
MULTIPLICATION OF DECIMAL NUMBERS

Catalog Number	How Many	Color Size	Name of Item	Price Each	Total Price
5702	2	32 gal.	Trash Container	9.99	19.98
8515	3	33 gal.	Trash Bags	2.00	6.00

Total for Goods 25.98

After reading this section you will know the procedure for multiplying decimal numbers, as illustrated by this example.

$$
\begin{array}{r}
12.7 \\
\times\ \ .705 \\
\hline
635 \\
000\ \ \\
8\ 89\ \ \ \ \\
\hline
8.9535
\end{array}
$$

If you think you know this material, work all the exercises at the end of the section. If you have difficulty, turn back and read this section completely.

The rule for placing the decimal point in multiplication is based on the fractional form of the decimal numbers. For example,

$$.3 \times .71 = \frac{3}{10} \times \frac{71}{100}$$

tenths

hundredths

$$= \frac{213}{1,000}$$

$$= .213$$

thousandths

Because tenths × hundredths = thousandths, a one-place decimal times a two-place decimal number equals a three-place decimal number (The sum of 1 and 2 is 3).

The preceding problem illustrates the rule for multiplication of decimal numbers.

Multiplication of Decimal Numbers

To find the product of two decimal numbers, multiply the numbers as if they were whole numbers. To find the number of decimal places in the product, add the number of decimal places in the numbers being multiplied.

EXAMPLE 1: Multiply:

$$1.23 \times 1.75$$

```
      1.23  ←————— 2-place decimal
  ×   1.75  ←————— 2-place decimal
  ————————
      6 15
     86 1
    1 23
  ————————
    2.15 25        4-place decimal (2 + 2 = 4)
```

This makes sense because

hundredths × hundredths = ten-thousands

2 places   2 places   4 places

EXAMPLE 2: Multiply:

$$1.257 \times 0.51$$

You do not need to line up the decimal points when multiplying.

```
      1.257  ←————— 3-place decimal
  ×    .51   ←————— 2-place decimal
  —————————
      1257
     6285
  —————————
     .64107        5-place decimal (3 + 2 = 5)
```

To check if your answer is reasonable, round each number to one digit and multiply.

Estimation

$$1 \times .5 = .5$$

This is a good estimate of the actual product, .64107.

EXAMPLE 3:   Multiply:

.123 × .8

.123 ←——— 3-place decimal
X   .8 ←——— 1-place decimal
.0984   4-place decimal (3 + 1 = 4)

A zero is used to show an additional
position so that 984 ends in the
ten-thousandths position.

You may have to add zeros to the left of the whole number product in order to have the answer end in the correct position.

EXAMPLE 4:   Multiply:

.251 × .0021

.251   3-place decimal
X   .0021   4-place decimal
251
502
000
000
.0005271   7-place decimal (3 + 4 = 7)

Add zeros so that the 5271
ends in the seventh position.

EXAMPLE 5:   Multiply:

27 × .053

27   0-place decimal (whole number)
X   .053   3-place decimal
81
1 35
00 0
01.431   3-place decimal (0 + 3 = 3)

This zero is not needed.

Therefore, 27 × .053 = 1.431

EXAMPLE 6:   Multiply:

.073 × 125

$$\begin{array}{r} .073 \\ \times \underline{125} \\ 365 \\ 1\ 46 \\ \underline{07\ 3} \\ 09.125 \quad \text{or} \quad 9.125 \\ \uparrow \end{array}$$

This zero is not needed.

Your turn!

Examples	Solutions
$1.23 \times .5 = ?$	The number of decimal places in the product is the sum of the number of decimal places in the numbers being multiplied. 1.23 —— 2-place decimal $\times \underline{.5}$ —— 1-place decimal $.615$ ___ 3-place decimal $(2 + 1 = 3)$ $\uparrow\underline{\vert}$
$27.32 \times .005 = ?$	$$\begin{array}{r} 27.32 \\ \times \underline{.005} \\ 13660 \\ 0000 \\ \underline{0\ 000} \\ 0.13660 \quad \text{or} \quad .1366 \end{array}$$ $(2 + 3 = 5 \text{ places})$
$.03 \times .1 = ?$	You may have to add zeros to the left of the whole number product to correctly position the decimal point. $$\begin{array}{r} .03 \\ \times \underline{.1} \\ .003 \\ \uparrow \end{array}$$ Added to give a 3-place decimal.
$52 \times 1.03 = ?$	$$\begin{array}{r} 52 \quad \text{0-place decimal number} \\ \times \underline{1.03} \\ 1\ 56 \\ 0\ 0 \\ \underline{52} \\ 53.56 \end{array}$$

Remember: Do not line up the decimal points to multiply, just add the number of places after the decimal points to determine the position of the decimal point in the product.

Answers to these exercises can be found on page 364.

Exercises Part I

Multiply.

1. .15 × .17
2. 57.93 × 1.59
3. .27 × .3
4. 19.53 × 1.072
5. .053 × .06
6. 15 × 7.32
7. .03 × 18
8. 59.3 × .752
9. .007 × .03
10. .013 × .0005

Exercises Part II

Multiply.

1. 27.3 × .09
2. .89 × .7
3. 12 × 17.5
4. .007 × .005
5. .013 × 19

SECTION 14.2
MULTIPLICATION BY POWERS OF TEN

BLANK CHECKS
$3.98 per 100

Consider these examples.

```
      .135            .135             .135              .135
  ×    10        ×    100         ×   1,000         ×  10,000
     000             000              000               000
   1 35            0 00             0 00              0 00
   1.350          13 5             00 0              00 0
                  13.500           135               000
                                   135.000           135
                                                    1350.000
```

There is a quick way to do these multiplications mentally.

Multiplication by Powers of 10

To multiply by a power of 10, move the decimal point to the *right* as many places as there are zeros in the power of 10.

EXAMPLE 1:   Here are the same examples.

(a) $\overbrace{\text{one zero}}$

.135 $\times$ 10 = 1.35 = 1.35

moved one place

(b) $\overbrace{\text{two zeros}}$

.135 $\times$ 100 = 13.5 = 13.5

moved two places

(c) $\overbrace{\text{three zeros}}$

.135 $\times$ 1,000 = 135. = 135

moved three places

(d) $\overbrace{\text{four zeros}}$

.135 $\times$ 10,000 = 1350. = 1,350

moved four places,
zero added

EXAMPLE 2: Multiply .0351 by 100.

$\overbrace{\text{two zeros}}$

.0351 $\times$ 100 = 03.51 = 3.51

moved two places

EXAMPLE 3: Multiply .0351 by 10,000.

$\overbrace{\text{four zeros}}$

.0351 $\times$ 10,000 = 0351. = 351

moved four places

EXAMPLE 4: Multiply 37.5 by 10.

$\overbrace{\text{one zero}}$

37.5 $\times$ 10 = 375. = 375

moved one place

EXAMPLE 5: Multiply 37.5 by 1,000.
If you forget the rule (or do not trust your answer), multiply the long way.

$$
\begin{array}{r}
37.5 \\
\times \quad 1,000 \\
\hline
00\ 0 \\
000 \\
000 \\
375 \\
\hline
37,500.0 = 37,500
\end{array}
$$

The rule allows you to do the multiplication mentally!

$$\overbrace{\text{three zeros}}$$
$$37.5 \times 1{,}000 = 37500. = 37{,}500$$

moved three places,
zeros added

EXAMPLE 6:   Multiply 3.8 by 100.
$$3.8 \times 100 = 380. = 380$$

EXAMPLE 7:   Multiply 5 by 1,000.
Remember that with a whole number the decimal is understood to be after the ones position.
$$5 \times 1{,}000 = 5. \times 1{,}000$$
$$= 5000.$$
$$= 5{,}000$$

Now you try these.

Examples	Solutions
.052 × 100 = ?	To multiply by a power of 10, move the decimal point to the right as many places as there are number of zeros in the power of 10. $\overbrace{\text{two zeros}}$ $.052 \times 100 = 05.2 = 5.2$ moved two places
.3 × 10,000 = ?	You may need to add zeros to show the correct position. $.3 \times 10{,}000 = 3000. = 3{,}000$
15 × 100 = ?	$15 \times 100 = 15. \times 100$ $= 1500.$ $= 1{,}500$

Now try these—no hints given.

1.15 × .732 = ?	.84180
.03 × .07 = ?	.0021
1.15 × 10 = ?	11.5
.732 × 100 = ?	73.2
.03 × 10,000 = ?	300
52 × 100 = ?	5,200

If you had difficulty, please read the last two sections again.

Answers to these exercises can be found on page 364.

Multiply.

1. 15.9 × 1.73   2. 27 × 3.2
3. 8.15 × 5   4. .003 × .071
5. 1.73 × 10   6. 3.2 × 1000
7. 15.9 × 100   8. 15 × 100
9. .003 × 100   10. .003 × 1000

Multiply.

1. 17.5 × 1.3   2. 27 × 100
3. .007 × 10   4. .019 × 1000
5. .27 × 100

REVIEW MATERIAL

**Review Skill
Multiplication of
Decimal Numbers**

To find the product of two decimal numbers multiply the numbers as if they were whole numbers. To find the number of decimal places in the product, add the number of decimal places in the numbers being multiplied.

To multiply by a power of 10, move the decimal point to the right as many places as there are zeros in the power of 10.

$$\begin{array}{r} 12.1 \\ \times\ 3.72 \\ \hline 242 \\ 8\ 47 \\ 36\ 3 \\ \hline 45.012 \end{array}$$

12.1 — 1-place decimal
3.72 — 2-place decimal
45.012 — 3-place decimal

$.07 \times 10{,}000 = 0700. = 700$

moved four places

Exercise A

Answers to this exercise can be found on page 364.

Multiply.

1. .13 × .15   2. 1.72 × .3
3. 17 × .003   4. 17.5 × 100
5. .03 × 1,000   6. 27 × 1.3
7. .038 × .07   8. .003 × 100
9. .07 × 10,000   10. 15 × 100

Exercise B

No answers are given.

Multiply.

1. .14 × .15   2. 2.81 × .4
3. 19 × .007   4. 18.6 × 100
5. .07 × 1000

REVIEW EXERCISES

Refer to the section listed if you have difficulty. (Answers to these exercises can be found on page 365.

Section 14.1   Multiply.

1.   17.593 X 1.42   2.   .057 X .013
3.   1.17 X .007   4.   12 X 11.72
5.   .0125 X .03

Section 14.2   Multiply.

6.   17.561 X 100   7.   .705 X 10
8.   3.72 X 10,000

CHAPTER POST-TEST

This test should determine if you have mastered the topics in Chapter 14. Answers to this test can be found on page 365.

Multiply.

1.   15.56 X .139   2.   .021 X 11.3
3.   62.5 X .005   4.   9 X 11.3
5.   .0072 X .005   6.   15.9 X 10
7.   317.59 X 1,000   8.   27.526 X 100

division of decimal numbers

Tomato Sauce
3 for 89¢
Net Wt. 8 oz

Compare at 3.71¢ per ounce

In this section you will learn how to position the decimal point when dividing a decimal number by a whole number, as illustrated by this example.

$$
\begin{array}{r}
1.35 \\
27\overline{)36.45} \\
\underline{27} \\
9\,4 \\
\underline{8\,1} \\
1\,35 \\
\underline{1\,35} \\
0
\end{array}
$$

If you think you know this information, turn to the exercises and work *each* problem. If you have difficulty, turn back and read this section completely.

Division by Whole Numbers

EXAMPLE 1:   Divide 379.2 by 4.
Divide as if the dividend were a whole number.

```
          94 8
     4 ) 379.2
         36
         ‾‾
         ,19
          16
          ‾‾
           3 2
           3 2
           ‾‾‾
             0
```

But where does the decimal go in the quotient? To find out, let me check the division by multiplying the divisor times the quotient.

```
      948        ?-place decimal.
   X    4        0-place decimal.
   ‾‾‾‾‾‾
     3792        Should be a 1-place decimal.
```

Because the answer should be 379.2, a one-place decimal, and the divisor is a zero-place decimal, 948 will have to be a one-place decimal. Therefore,

$$379.2 \div 4 = 94.8$$

or

```
          94.8
     4 ) 379.2
         36
         ‾‾
          19
          16
          ‾‾
           3 2
           3 2
           ‾‾‾
             0
```

The previous example illustrates the rule for dividing a decimal number by a whole number.

> To divide a decimal number by a whole number, place the decimal point in the quotient directly above the decimal point in the dividend.

EXAMPLE 2:   Divide 82.16 by 52.
First, place the decimal point in the quotient directly above the decimal point in the dividend.

```
              .
     52 ) 82.16
              ↑
```

Now divide as if both numbers were whole numbers. Be careful to place the partial quotients in the correct positions.

$$
\begin{array}{r}
1.58 \\
52\overline{\smash{\big)}\,82.16} \\
\underline{52} \\
30\ 1 \\
\underline{26\ 0} \\
4\ 16 \\
\underline{4\ 16} \\
0
\end{array}
$$

Therefore, $82.16 \div 52 = 1.58$.

$$
\begin{array}{rl}
\textit{Check:} \quad 1.58 & \text{Quotient} \\
\times \quad 52 & \text{times divisor} \\
\hline
3\ 16 & \\
79\ 0 & \\
\hline
82.16 & \\
0 & \text{plus remainder} \\
\hline
82.16 & \text{equals dividend.}
\end{array}
$$

EXAMPLE 3: Divide .675 by 25.

Place the decimal point in the quotient above the decimal point in the dividend.

$$
25\overline{\smash{\big)}\,\overset{.}{\wedge}675}
$$

Notice that the division begins over the digit 7.

$$
\begin{array}{r}
.\ 27 \\
25\overline{\smash{\big)}\,.675} \\
\underline{50} \\
175 \\
\underline{175} \\
0
\end{array}
$$

To indicate the correct place value, place a zero over the 6 to show that the division started 2 places after the decimal point. That is,

$$
\begin{array}{r}
.027 \\
25\overline{\smash{\big)}\,.675} \\
\underline{50} \\
175 \\
\underline{175} \\
0
\end{array}
$$

Therefore, $.675 \div 25 = .027$

EXAMPLE 4: Divide .087 by 29.

Again, position the decimal point above the decimal point in the dividend.

$$
29\overline{\smash{\big)}\,\overset{.}{\wedge}087}
$$

Notice that the division begins over the digit 7. Use zeros to show this.

```
       .003
29 ) .087
       87
        0
```

Therefore, .087 ÷ 29 = .003

Division does not always result in a zero remainder.

EXAMPLE 5:   Divide 4.5 by 18.

```
       .2
18 ) 4.5
     3 6
       9
```

This time there is a remainder. However, it is possible to continue the division by annexing zeros to the dividend, one at a time.

```
       .2
18 ) 4.50
     3 6           Annex a zero and
      90           bring it down.
```

Continue the division.

```
       .25
18 ) 4.50
     3 6
      90
      90
       0
```

Therefore, 4.5 ÷ 18 = .25

It is sometimes necessary to annex more than one zero.

EXAMPLE 6:   Divide 2.6 by 64.

```
       .0
64 ) 2.6
```

I need to annex a zero before the division can begin.

```
       .04
64 ) 2.60
     2 56
        4
```

Because there is a remainder, I annex another zero.

```
       .040
64 ) 2.600
     2 56
        40        64 divides 40 zero times.
        00
        40
```

It is again necessary to annex a zero.

```
        .0406
  64 | 2.6000
        2 56
        ────
          40
          00
        ────
         400
         384
        ────
          16
```

The remainder is still not zero, so I annex another zero.

```
        .04062
  64 | 2.60000
        2 56
        ────
          40
          00
        ────
         400
         384
        ────
         160
         128
        ────
          32
```

I annex another zero.

```
        .040625
  64 | 2.600000
        2 56
        ────
          40
          00
        ────
         400
         384
        ────
         160
         128
        ────
         320
         320
        ────
           0
```

By annexing 5 zeros, I finally obtained a zero remainder!

This will not always happen but for now try these problems.

Examples	Solutions
Divide 379.2 by 2.	When dividing a decimal number by a whole number, place the decimal point directly above the decimal point in the dividend.

Examples	Solutions
	$$\begin{array}{r} 189.6 \\ 2\overline{)379\,{}_\wedge2} \\ \underline{2} \\ 17 \\ \underline{16} \\ 19 \\ \underline{18} \\ 1\,2 \\ \underline{1\,2} \end{array}$$
Divide 0.126 by 3.	$3\overline{)0.126}$

Use a zero to show that the division begins two places after the decimal point.

$$\begin{array}{r} .042 \\ 3\overline{)0.126} \\ \underline{12} \\ 06 \\ \underline{6} \\ 0 \end{array}$$ |
| Divide 7.8 by 52. | $$\begin{array}{r} .1 \\ 52\overline{)7.8} \\ \underline{5\,2} \\ 2\,6 \end{array}$$

Annex a zero to continue the division.

$$\begin{array}{r} .15 \\ 52\overline{)7.80} \\ \underline{5\,2} \\ 2\,60 \\ \underline{2\,60} \end{array}$$ |

Ready to continue?

Rounding in Division It is not always possible to get a zero remainder no matter how many zeros you annex to the dividend. Therefore, it is the usual practice to set the number of places the division will be carried out and round the quotient to that place value. It still may be necessary to annex zeros to do this.

EXAMPLE 7: Divide 27.3 by 11. Round the answer to the nearest tenth.

Place the decimal point in the quotient.

$$11\overline{)27.3}$$

Now divide. It is necessary to carry out the division two places past the decimal in order to round to tenths, so annex one zero.

$$
\begin{array}{r}
2.48 \\
11\overline{)27.30} \\
22 \\
\hline
5\,3 \\
4\,4 \\
\hline
90 \\
88 \\
\hline
2
\end{array}
$$

Round the number 2.48 to one place by dropping the 8 and rounding up.

$$2.48 \cong 2.5 \qquad \text{Add 1 to the tenths position.}$$

Therefore, $27.3 \div 11 = 2.5$, to the nearest tenth.

To round to a particular place value, carry out the division one extra place.

EXAMPLE 8: Divide .0736 by 12. Round the answer to the nearest thousandth.

Place the decimal point in the quotient and carry out the division to four places in order to round to three places.

$$
\begin{array}{r}
.0061 \\
12\overline{).0736} \\
72 \\
\hline
16 \\
12 \\
\hline
4
\end{array}
$$

Rounding to thousandths,

$$.0061 \cong .006$$

Therefore, $.0736 \div 12 = .006$, to the nearest thousandth.

EXAMPLE 9: Divide 7 by 9. Round the answer to the nearest hundredth.

Both of these numbers are whole numbers. Remember that the decimal point is understood to be after the ones position; therefore,

$$7 = 7.$$

↑
decimal point

Position the decimal in the quotient and divide three places in order to round to hundredths. Annex three zeros after the decimal point.

$$
\begin{array}{r}
.777 \\
9\overline{)7.000} \\
6\,3 \\
\hline
70 \\
63 \\
\hline
70 \\
63 \\
\hline
7
\end{array}
$$

Rounding to hundredths,

$$.777 \cong .78$$

Therefore, $7 \div 9 = .78$, to the nearest hundredth.

Now it is your turn!

Examples	Solutions
Divide 5.973 by 17. Round the answer to the nearest hundredth.	$\begin{array}{r} .351 \\ 17\overline{)5.973} \\ 5\,1 \\ \hline 87 \\ 85 \\ \hline 23 \\ 17 \\ \hline 6 \end{array}$ .351 = .35 to the nearest hundredth. Therefore, 5.973 ÷ 17 = .35, to the nearest hundredth.
Divide 5.973 by 17. Round the answer to the nearest thousandth.	Annex a zero to carry out the division four places in order to round to three places. $\begin{array}{r} .3513 = .351 \\ 17\overline{)5.9730} \\ 5\,1 \\ \hline 87 \\ 85 \\ \hline 23 \\ 17 \\ \hline 60 \\ 51 \\ \hline 9 \end{array}$ To the nearest thousandth.
Divide 6 by 9. Round the answer to the nearest tenth.	Divide two places after the decimal point to round to one place. $\begin{array}{r} .66 = .7 \\ 9\overline{)6.00} \\ 5\,4 \\ \hline 60 \\ 54 \\ \hline 6 \end{array}$ To the nearest tenth.

Now try these—no hints given.

Divide and round the answers as indicated.	
(a) 17.54 ÷ 27, to the nearest hundredth.	(a) .65

Examples	Solutions
(b) .435 ÷ 15, to the nearest hundredth.	(b) .03
(c) 15 ÷ 23, to the nearest thousandth.	(c) .652

Ready for the exercises?

Exercises Part I Answers to these exercises can be found on page 365.

Divide (do not round).

1. $67.5 \div 25$ 2. $38.228 \div 19$ 3. $.225 \div 9$
4. $.06 \div 15$ 5. $3 \div 5$

Divide and round the answers to the nearest hundredth.

6. $68.2 \div 25$ 7. $38.228 \div 20$ 8. $.275 \div 9$
9. $.16 \div 14$ 10. $8 \div 9$

Exercises Part II Divide (do not round).

1. $187.5 \div 15$ 2. $9.315 \div 9$ 3. $100.624 \div 8$

Divide and round the answers to the nearest tenth.

4. $217.8 \div 15$ 5. $282.192 \div 8$

SECTION 15.2
DIVISION BY POWERS OF TEN

Consider these examples.

$$
\begin{array}{r}
2.35 \\
10\overline{)23.50} \\
\underline{20} \\
3\,5 \\
\underline{3\,0} \\
50 \\
\underline{50}
\end{array}
\qquad
\begin{array}{r}
.235 \\
100\overline{)23.500} \\
\underline{20\,0} \\
3\,50 \\
\underline{3\,00} \\
500 \\
\underline{500}
\end{array}
$$

There is a quick way to do these divisions mentally.

Division by Powers of 10

To divide by a power of 10, move the decimal point to the *left* as many places as there are zeros in the power of 10.

EXAMPLE 1:   Here are the same examples.

one zero

(a) $23.5 \div 10 = 2.35 = 2.35$

moved
one place

two zeros

(b) $23.5 \div 100 = .235 = .235$

moved
two places

EXAMPLE 2:   Divide 27.593 by 100.

two zeros

$27.593 \div 100 = .27593 = .27593$

moved
two places

EXAMPLE 3:   Divide 27.95 by 10.

one zero

$27.95 \div 10 = 2.795 = 2.795$

moved
one place

EXAMPLE 4:   Divide 27.95 by 1,000.

In this problem a zero is needed to show the movement of the decimal point three places.

$27.95 \div 1,000 = .02795 = .02795$

EXAMPLE 5:   Divide 59 by 1,000.
Remember,

$$59 = 59.$$

↑
decimal point

Therefore,

$$59 \div 1,000 = .059 = .059$$

EXAMPLE 6:   (a) Multiply 27.59 by 100.
(b) Divide 27.59 by 100.

Which way does the decimal point move?

(a) Multiplication by a power of 10 produces a *larger* number—move the decimal point to the *right*.

$$27.59 \times 100 = 2759. = 2759 \qquad \text{Changed from 27.59 to 2759 (larger).}$$

(b) Division by a power of 10 makes the number *smaller*—move the decimal point to the *left*.

$$27.59 \div 100 = .2759 = .2759 \qquad \text{Changed from 27.59 to .2759 (smaller).}$$

Now you try these.

Examples	Solutions
$151.23 \div 10 = ?$	To divided by a power of 10, move the decimal point to the left as many places as there are zeros in the power of ten. $$151.23 \div 10 = 15.123 = 15.123$$ moved one place
$17.5 \div 100 = ?$	$$17.5 \div 100 = .175 = .175$$ moved two places
$18.93 \div 1,000 = ?$	$$18.93 \div 1,000 = .01893 = .01893$$ Zero had to be added.
$28 \div 100 = ?$	$$28 \div 100 = .28 = .28$$
$28 \div 10,000 = ?$	$$28 \div 10,000 = .0028 = .0028$$

Ready for the exercises?

Answers to these exercises can be found on page 365.

Exercises Part I

Divide.

1. $19.2 \div 10$ 2. $125.35 \div 100$ 3. $89.2 \div 1{,}000$ 4. $135.2 \div 10{,}000$
5. $59 \div 10$ 6. $63 \div 100$ 7. $4 \div 10$ 8. $49 \div 1{,}000$

Exercises Part II

1. $17.3 \div 10$ 2. $1.03 \div 100$ 3. $275.32 \div 1{,}000$ 4. $15.95 \div 1{,}000$

SECTION 15.3
DIVISION OF DECIMAL NUMBERS BY DECIMAL NUMBERS

Computing Fuel Usage in Miles per Gallon

1. Take odometer reading when tank is full.

2. Take a second reading when tank is refilled.

3. Divide the difference in the two odometer readings by the number of gallons of fuel needed to refill the tank.

In this section the rule for division by a decimal number is given. Here is an example.

$$
\begin{array}{r}
1\,.\,53 \\
.027_\wedge \overline{\smash{)}.041_\wedge 31} \\
27 \\
\hline
14\ 3 \\
13\ 5 \\
\hline
81 \\
81 \\
\hline
\end{array}
$$

If you think you know this rule, turn to the exercises and work *each* problem. If you have difficulty, turn back and read this section completely.

Division by Decimal Numbers

The rule for division by decimal numbers is based on the fractional form of the decimal numbers.

EXAMPLE 1: Consider the division

$$.027\overline{\smash{)}.04131}$$

How do I divide by .027? To answer this, I shall change the problem so that the divisor is a whole number. First, I change the decimal numbers to fractions.

$$.04131 \div .027 = \frac{4131}{100,000} \div \frac{27}{1,000}$$

thousandths

hundred-thousandths

$$= \frac{4131}{\overset{}{\underset{100}{\cancel{100,000}}}} \times \frac{\overset{1}{\cancel{1,000}}}{27} \qquad \text{Inverting and reducing the powers of 10.}$$

$$= \frac{4131}{100} \times \frac{1}{27}$$

$$= 41\frac{31}{100} \times \frac{1}{27} \qquad \text{Changing } \frac{4131}{100} \text{ to a mixed number.}$$

$$= 41.31 \times \frac{1}{27} \qquad \text{Changing } 41\frac{31}{100} \text{ to a decimal number.}$$

$$= 41.31 \div 27 \qquad \text{Changing the multiplication back to division by inverting.}$$

Therefore,

$$.027\overline{)\,.04131\,} \quad \text{is the same as} \quad 27\overline{)\,41.31\,}$$

Notice that .027 has been changed to the whole number 27, and .04131 has become 41.31. Both decimal points have been moved three places to the right.

$$.027\overline{)\,.041\;31\,}$$

moved   moved

three places   three places

Now place the decimal point in the quotient above the decimal in the dividend (because the divisor is now a whole number) and divide as before.

```
              1 . 53
  .027 ) .041 . 31
         27
         14  3
         13  5
            81
            81
```

 Caution: Never try to repeat this procedure. Use, instead, the following simple rule, which is based on the preceding example.

Division by Decimal Numbers

Move the decimal point in the *divisor* the number of places necessary to make it a whole number. Then move the decimal point in the *dividend* the same number of places to the right, annexing zeros where necessary.

I use a caret to show the new position of the decimal points.

<table>
<tr><td>Move the decimal point three places to make .027 a whole number.</td><td>$.027_\wedge \overline{).041_\wedge 31}$</td><td>Move the decimal point three places in the dividend.</td></tr>
</table>

After you have carefully positioned the decimal point in the quotient, line up your division exactly.

From above, .04131 ÷ .027 = 1.53. Check your answer as you would any other division.

$$
\begin{array}{rl}
1.53 & \text{Quotient} \\
\times \quad .027 & \text{times divisor} \\
\hline
1071 & \\
306 \quad\; & \\
000 \quad\quad\; & \\
\hline
.04131 & \\
.00000 & \text{plus remainder} \\
\hline
.04131 & \text{equals dividend.}
\end{array}
$$

EXAMPLE 2:   Divide 2.7 by .03.
To make the divisor a whole number, move the decimal point two places to the right.

$$.03_\wedge \overline{)2.7}$$

To move the decimal point two places in the dividend, annex a zero.

$$.03 \overline{)2.70_\wedge}$$

Divide.

$$
\begin{array}{r}
90. \\
.03_\wedge \overline{)2.70_\wedge} \\
2\,7 \quad\;\; \\
\hline
0 \\
\underline{0}
\end{array}
$$

Therefore, 2.7 ÷ .03 = 90.

EXAMPLE 3:   Divide and round to the nearest hundredth: $.9\overline{)1.5.}$
Move the decimal points one place.

$$.9_\wedge \overline{)1.5_\wedge}$$

Annex three zeros so that the division can be carried out to three places past the decimal point *in the quotient*, then divide and round to two places.

$$
\begin{array}{r}
1.666 \cong 1.67 \quad \text{To the nearest hundredth.} \\
.9_\wedge \overline{)1.5_\wedge 000} \\
9 \quad\quad\quad\; \\
\hline
6\;0 \quad\quad \\
5\;4 \quad\quad \\
\hline
60 \quad \\
54 \quad \\
\hline
60 \\
54 \\
\hline
6
\end{array}
$$

EXAMPLE 4:   Divide 6 by .063 and round to the nearest thousandth.
In a whole number the decimal point follows the ones position.

$$.063 \overline{\smash{)}6.}$$

Move each decimal point three places, annexing zeros.

$$.063_\wedge \overline{\smash{)}6.000_\wedge}$$

Annex four more zeros in order to carry out the division four places after the
decimal point in the quotient.

$$
\begin{array}{r}
95.2380 \cong 95.238 \qquad \text{To the nearest thousandth.} \\
.063_\wedge \overline{\smash{)}6.000_\wedge 0000} \\
\underline{5\ 67} \\
330 \\
\underline{315} \\
15\ \ 0 \\
\underline{12\ \ 6} \\
2\ \ 40 \\
\underline{1\ \ 89} \\
510 \\
\underline{504} \\
60 \\
\underline{0} \\
60
\end{array}
$$

Checking can be lengthy if you use multiplication. Estimate to obtain a rough
check of your quotient.

Estimation

$$
\begin{array}{r}
1\ 00. \\
.06_\wedge \overline{\smash{)}6.00_\wedge}
\end{array}
$$

which approximates the answer of 95.238.

Ready to try some?

Examples	Solutions
.5367 ÷ .03 = ?	$$\begin{array}{r} 17.89 \\ .03_\wedge \overline{\smash{)}.53_\wedge 67} \\ \underline{3} \\ 23 \\ \underline{21} \\ 2\ 6 \\ \underline{2\ 4} \\ 27 \\ \underline{27} \end{array}$$

Examples	Solutions
$69.68 \div 1.3 = ?$	$$\begin{array}{r} 53.6 \\ 1.3_\wedge\overline{)69.6_\wedge8} \\ 65 \\ \hline 46 \\ 39 \\ \hline 7\;8 \\ 7\;8 \\ \hline \end{array}$$
$6.8 \div .002 = ?$	$$\begin{array}{r} 3\,400. \\ .002_\wedge\overline{)6.800_\wedge} \\ 6 \\ \hline 8 \\ 8 \\ \hline 0 \\ 0 \\ \hline 0 \\ 0 \\ \hline \end{array}$$
$273 \div .03 = ?$	$$\begin{array}{r} 91\,00. \\ .03_\wedge\overline{)273.00_\wedge} \\ 27 \\ \hline 3 \\ 3 \\ \hline 0 \\ 0 \\ \hline 0 \\ 0 \\ \hline \end{array}$$
$.0828 \div 2.3 = ?$	$$\begin{array}{r} .036 \\ 2.3_\wedge\overline{).0_\wedge828} \\ 69 \\ \hline 138 \\ 138 \\ \hline \end{array}$$
$7.593 \div 1.7 = ?$	$$\begin{array}{r} 4.466 \cong 4.47 \\ 1.7_\wedge\overline{)7.5_\wedge930} \\ 6\;8 \\ \hline 7\;9 \\ 6\;8 \\ \hline 1\;13 \\ 1\;02 \\ \hline 110 \\ 102 \\ \hline 8 \\ \end{array}$$ I carried out the division three places to round to two. I annexed a zero to do this.

Examples	Solutions
No hints for these!	
$.3752 \div .07 = ?$	5.36
$.0048 \div 1.5 = ?$	.0032
$15.732 \div .73 = ?$ Round to the nearest tenth.	21.6

Please talk with an instructor or read this section again if you had difficulty with the last three problems.

Exercises Part I

Answers to these exercises can be found on page 365.

Divide.

1.	$5.475 \div 1.5$	2. $.00108 \div .09$	3. $12 \div .03$
4.	$685.848 \div 8.2$	5. $5.169976 \div .113$	

Divide and round to the nearest hundredth.

6.	$758 \div 1.5$	7. $.0395 \div .09$	8. $8 \div 0.3$
9.	$15.362 \div 8.2$	10. $715.9 \div .113$	11. $1.05887 \div .15$
12.	$3.4086 \div 1.7$		

Exercises Part II

Divide.

1.	$13.65 \div 10.5$	2. $8.1892 \div .08$	3. $2.1888 \div 0.171$

Divide and round to the nearest tenth.

4.	$8 \div 0.9$	5. $15.9 \div .27$

REVIEW MATERIAL

Review Skill Division with Decimal Numbers

1. To divide a decimal number by a whole number, place the decimal point in the quotient directly above the decimal point in the dividend.

$$
\begin{array}{r}
94.8 \\
4\,\overline{)\,379.2} \\
36 \\
\hline
19 \\
16 \\
\hline
3\,2 \\
3\,2 \\
\hline
\end{array}
$$

2. To divide by a power of 10, move the decimal point to the left as many places as there are zeros in the power of 10.

$$27.95 \div 100 = .2795 = .2795$$

3. To divide by a decimal number, move the decimal point in the divisor the number of places necessary to make it a whole number. Then move the decimal point in the dividend the same number of places to the right, annexing zeros where necessary.

$$
\begin{array}{r}
1.53 \\
.027_\wedge \overline{).041_\wedge 31} \\
\underline{27} \\
14\ 3 \\
\underline{13\ 5} \\
81 \\
\underline{81}
\end{array}
$$

Exercise A   Answers to this exercise can be found on page 365.

Divide.

1. .615 ÷ .05	2. .026 ÷ 1.3	3. 12.852 ÷ 1.02
4. 17.3 ÷ 100	5. 24 ÷ 10	6. 55.35 ÷ 45

Divide and round to the nearest tenth.

7. 1.03 ÷ .7	8. 62 ÷ .15	9. 3.6521 ÷ .002
10. 1.335 ÷ 1.3		

Exercise B   No answers are given.

Divide.

1. .0615 ÷ .05	2. .0143 ÷ 1.3	3. 2.3868 ÷ 1.02
4. 18.5 ÷ 100	5. 16 ÷ 10	6. .5715 ÷ 45

Divide and round to the nearest tenth.

7. 2.71 ÷ .7	8. 38 ÷ .15	9. 7.2953 ÷ .002

REVIEW EXERCISES

Refer to the section listed if you have difficulty. Answers to these exercises can be found on page 365.

Section 15.1   Divide (do not round).

1. 3.375 ÷ 27	2. .117 ÷ 9	3. 2 ÷ 8

Divide and round to the nearest hundredth.

4. 17.93 ÷ 27	5. 1.578 ÷ 9

Section 15.2 Divide.

  6. $18.5 \div 100$ **7.** $1.73 \div 1,000$ **8.** $5 \div 10$

Section 15.3 Divide (do not round).

  9. $2 \div .16$ **10.** $.01 \div .5$

Divide and round to the nearest tenth.

  11. $.123 \div .16$ **12.** $4.093 \div .5$ **13.** $8 \div .3$

CHAPTER POST-TEST

This test should determine if you have mastered the topics in Chapter 15. Answers to this test can be found on page 365.

Divide.

  1. $28.672 \div 28$ **2.** $2.35 \div 1,000$ **3.** $.03 \div 10$
  4. $17.6 \div 2.5$ **5.** $.4836 \div .39$

Divide and round to the nearest hundredth.

  6. $5 \div 7$ **7.** $27.15 \div 28$ **8.** $15 \div 2.6$
  9. $1.596 \div .39$ **10.** $1.835 \div 17.2$

16 applications

GROSS EARNINGS	$ 999.88
Deductions	
OASDI	66.99
Federal Tax	211.18
State Tax	43.19
Ret.	49.99
Gp. Ins.	7.20
Misc. Ded.	150.00
TAKE-HOME PAY	$ 471.33

The *arithmetic* of this section is not new, but recognition of *which operation* to select is sometimes difficult. As before, I shall use the following operations.

Addition to find *total* amounts.

Subtraction to find the amount of *change* or difference between two quantities, or to find what amount *remains* after removal of a quantity.

Multiplication to find *repeated additions* of the same quantity.

Division to find the *size of a number of equal parts* that total a number, or to find the *number of equal parts* of a certain value that equal a total, or to change information to a *per unit* basis.

Keep these in mind as I work the following examples.

EXAMPLE 1:   What is the total charge for materials and copying if correction liquid is $1.04, paper is $11.95, and copying is $51.96?
The *total:*

$$
\begin{array}{r}
1.04 \\
11.95 \\
+51.96 \\
\hline
64.95 \quad \text{dollars}
\end{array}
$$

EXAMPLE 2:   Which is thicker, .025 inch or .026 inch?
These numbers differ only in the thousandths position. The number .026 has the larger number of thousandths and is, therefore, the larger number.

EXAMPLE 3:   The total trip is 500.8 miles. If you have driven 275.9 miles, how many miles remain to be driven?
What *remains:*

$$
\begin{array}{r}
500.8 \\
-275.9 \\
\hline
224.9 \quad \text{miles}
\end{array}
$$

EXAMPLE 4:   If you have $159.24 in your bank account, how much money do you have after writing checks for $12.26, $1.97, $16.90, and $12.03?
Removing the amounts:

$$
\begin{array}{r}
159.24 \\
-\ 12.26 \\
\hline
146.98 \\
-\ \ 1.97 \\
\hline
145.01 \\
-\ 16.90 \\
\hline
128.11 \\
-\ 12.03 \\
\hline
116.08 \quad \text{dollars}
\end{array}
$$

EXAMPLE 5:   If the monthly payments were $10.83 for 24 months, what was the total payment over the 2 years?
Repeated addition:

24 ten and eighty-threes = ?

$$24 \quad \times \quad 10.83 \quad = \ ?$$

Multiplying,

$$
\begin{array}{r}
10.83 \\
\times \quad 24 \\
\hline
43\ 32 \\
216\ 6 \\
\hline
259.92 \quad \text{dollars}
\end{array}
$$

EXAMPLE 6:   Steaks were on sale for $1.84 per pound. If the total bill was $7.36, how many pounds were bought?
Number of equal parts of a certain value:

How many 1.84's equal 7.36?

$$\Box \quad \times \quad 1.84 \quad = \quad 7.36$$

What number times 1.84 equals 7.36 is a division question. Divide the product by the known factor 1.84 to find the missing factor.

$$1.84_\wedge \overline{\smash{\big)}\ 7.36_\wedge}$$
$$\phantom{1.84_\wedge \overline{\smash{\big)}\ }} \underline{7\ 36}$$

quotient $4.$

The answer is 4 pounds.

EXAMPLE 7:   If a board 5 feet long is cut into 4 equal pieces, how long is each piece?

Size of a number of fixed equal parts:

4 times what equals 5 feet?

$$4\ \times\ \square\ =\ 5$$

Four times what number is equal to 5 is a division question. The missing factor is found by division.

$$4\overline{\smash{\big)}\ 5.00}$$

quotient 1.25

$$\underline{4}$$
$$1\ 0$$
$$\underline{\ \ 8}$$
$$\ \ 20$$
$$\underline{\ \ 20}$$

Each piece is 1.25 feet long, or $1\frac{1}{4}$ feet long if the answer is expressed as a mixed number.

EXAMPLE 8:   A 15-pound turkey costs \$8.85. How much does it cost per pound?

Per unit:

cost per pound
↑
Indicates divisor.

$$8.85 \div 15 = ?$$

Dividing,

$$15\overline{\smash{\big)}\ 8.85}$$

quotient .59

$$\underline{7\ 5}$$
$$1\ 35$$
$$\underline{1\ 35}$$

It costs 59¢ per pound.

EXAMPLE 9:   Average 8.5, 9.2, 7.9, and 8.8

Average = *total* of quantities divided by the *number* of quantities

$$= (8.5 + 9.2 + 7.9 + 8.8) \div 4$$
$$= (34.4) \div 4$$
$$= 8.6$$

EXAMPLE 10:   Which is the better buy, 5 light bulbs for $5.03 or 7 bulbs for $7.14?

Change each to a per bulb basis to compare.

cost per bulb
↑
Indicates divisor.

$5.03 \div 5$ $7.14 \div 7$

```
        1.006                               1.02
   5 | 5.030                            7 | 7.14
       5                                    7
       ___                                  ___
       0                                    1
       0                                    0
       ___                                  ___
       3                                    14
       0                                    14
       ___                                  ___
       30
       30
       ___
```

Cost is about $1.01 per bulb. | Cost is $1.02 per bulb.

Five bulbs at $5.03 is the better buy because $1.01 per bulb is cheaper than $1.02 per bulb.

Now you try these, using the above examples as a guide.

**Exercises
Part I**

Answers to these exercises can be found on page 365.

1.   Which is thicker, .031 inch or .04 inch?
2.   Which is thicker, .023 inch or .0231 inch?
3.   What was the total gas bill for the trip if I bought gas for $6, $10.54, and $12.04?
4.   The math book cost $12.94, the English book cost $5.96, and the history book cost $10.04. What was the total cost of the books?
5.   Dan ran the race in 45.8 seconds. Larry ran the same race in 50.2 seconds. By how many seconds did Dan beat Larry in the race?
6.   A steak weighed 1.15 pounds. After cooking, it weighed .96 pounds. How much weight was lost in cooking the meat?
7.   If Terry receives $9.16 per hour, how much does she make in a 40-hour week?
8.   If hamburger is on sale for $1.89 per pound, how much do 3 pounds cost?
9.   If gas sells for $1.78 per gallon, and the bill was $16.02, how much was bought?
10. How much should be paid each month in order to pay bill of $1,992.62 in 9 months?
11. Which is the best buy?

ABC coffee, 1-pound can $2.69
ABC coffee, 2-pound can $5.34
ABC coffee, 3-pound can $8.04

12. Which is the better buy?

Trash bags, 10 in a box $2.00
Trash bags, 15 in a box $2.85

13. Suppose your checkbook balance shows a balance of $589.19. Looking at the canceled checks returned, you see that there are checks for $8.91, $21.50, and $5 outstanding. There is also a service charge of $1.14. What should the balance read on your bank statement (see page 347)?

14. Compute the fuel usage in miles per gallon if the first odometer reading was 1850.7, the second odometer reading was 2003.1, and it took 9.3 gallons to fill the tank (see page 347).

Exercises Part II

1. If hamburger is on sale for $1.99 per pound, how much do 5 pounds cost?
2. If gas sells for $1.50 per gallon, and the bill was $21.30, how much was bought?
3. The math book cost $19.70, and the English book cost $7.59.
   (a) What was the total cost for the books?
   (b) How much more did the math book cost?

SECTION 16.2
MIXED ARITHMETIC

This section gives the procedure for changing fractions to decimals.

$$\frac{2}{3} = .66666\ldots \qquad \frac{3}{8} = 0.375$$
$$= .\overline{6}$$

It also shows you how to work with numbers that are not all the same type.

Fractions to Decimals

Fractions to Decimals

To change a fraction to a decimal number, think of the fraction as division of whole numbers and divide the numerator by the denominator of the fraction.

EXAMPLE 1:   Change $\dfrac{3}{4}$ to a decimal number.

The fraction $\dfrac{3}{4}$ means $3 \div 4$. Use division with decimal numbers to get a decimal form.

$$
\begin{array}{r}
.75 \\
4\overline{\smash{)}3.00} \\
\underline{2\ 8} \\
20 \\
\underline{20} \\
\end{array}
$$

Therefore,

$$\dfrac{3}{4} = .75$$

EXAMPLE 2:   Change $\dfrac{3}{8}$ to a decimal number.

$$\dfrac{3}{8} = 3 \div 8 \qquad
\begin{array}{r}
.375 \\
8\overline{\smash{)}3.000} \\
\underline{2\ 4} \\
60 \\
\underline{56} \\
40 \\
\underline{40} \\
\end{array}$$

Therefore,

$$\dfrac{3}{8} = .375$$

 Not all decimal forms will **terminate** (end) as in the examples above. However, the decimal form of these fractions will have a repeating pattern.

EXAMPLE 3:   Change $\dfrac{5}{12}$ to a decimal number.

$$\dfrac{5}{12} = 5 \div 12 \qquad
\begin{array}{r}
.41666 \\
12\overline{\smash{)}5.00000} \\
\underline{4\ 8} \\
20 \\
\underline{12} \\
80 \\
\underline{72} \\
80 \\
\underline{72} \\
80 \\
\underline{72} \\
8 \\
\end{array}$$

Repeating pattern begins.

Therefore,

$$\frac{5}{12} = .41666\ldots$$

The division continues with 6 as the partial quotient without end.

EXAMPLE 4:   Change $\frac{3}{7}$ to a decimal number.

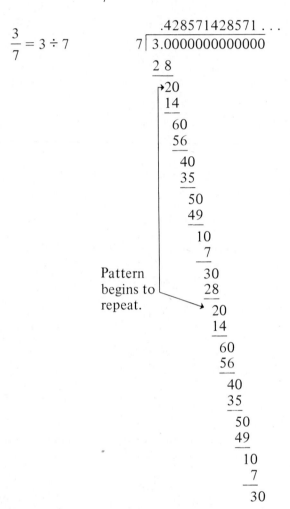

$$\frac{3}{7} = 3 \div 7$$

Pattern begins to repeat.

Therefore,

$$\frac{3}{7} = .428571428571428571\ldots$$

A **bar notation** is used to show the repeating pattern of nonterminating decimal numbers.

.416666 ... is written .41$\overline{6}$

.428571428571 ... is written $\overline{.428571}$

The bar is shown over the digit or digits that repeat infinitely.

Now you try these.

Examples	Solutions
Change $\frac{1}{8}$ to a decimal number.	Divide numerator by denominator. $$\begin{array}{r} .125 \\ 8\overline{)1.000} \\ \underline{8} \\ 20 \\ \underline{16} \\ 40 \\ \underline{40} \end{array}$$
Change $\frac{5}{9}$ to a decimal number.	$$\begin{array}{r} .55\ldots \\ 9\overline{)5.00} \\ \underline{4\,5} \\ 50 \\ \underline{45} \\ 5 \end{array}$$ Therefore, $\frac{5}{9} = .\overline{5}$

Ready to continue?

Mixed Arithmetic

EXAMPLE 5: Add 15.2 and $5\frac{3}{4}$.

Because one number is a decimal number and the other is a mixed number, one of them must be changed.

Method One Change the decimal number to a mixed number.

$$15.2 = 15\frac{2}{10}$$

$$= 15\frac{1}{5}$$

Therefore, $15.2 + 5\frac{3}{4} = 15\frac{1}{5} + 5\frac{3}{4}$.

$$\begin{array}{r} 15\frac{1}{5} = 15\frac{4}{20} \\ + \ 5\frac{3}{4} = \ 5\frac{15}{20} \\ \hline 20\frac{19}{20} \end{array}$$

Method Two Change the mixed number to a decimal number.

fraction
to decimal

$$5\frac{3}{4} = 5.75$$

whole number

Therefore, $15.2 + 5\frac{3}{4} = 15.2 + 5.75.$

$$\begin{array}{r} 15.20 \\ + \ \ 5.75 \\ \hline 20.95 \end{array}$$

Are these two answers the same? Yes!

$$20.95 = 20\frac{95}{100}$$

$$= 20\frac{19}{20} \qquad \text{In lowest terms.}$$

With calculations involving decimals, fractions, or mixed numbers—that is, mixed arithmetic—change all numbers either (1) to fractions (or mixed numbers) or (2) to decimals.

Unless the decimal form of the fraction is nonterminating, I change everything to decimal numbers, because working with decimals is more like working with whole numbers.

EXAMPLE 6:   Subtract $3\frac{1}{2}$ from 17.95.

First, change $3\frac{1}{2}$ to a decimal by changing $\frac{1}{2}$ to a decimal.

$$\frac{1}{2} = 1 \div 2 = .5$$

Therefore,

$$3\frac{1}{2} = 3.5$$

fraction to decimal

Now subtract.

$$\begin{array}{r} 17.95 \\ - \ \ 3.50 \\ \hline 14.45 \end{array}$$

EXAMPLE 7:   Multiply $5\frac{2}{3}$ by 17.1.

In this example,

$$5\frac{2}{3} = 5.666\ldots$$

$$= 5.\overline{6} \qquad \text{Nonterminating.}$$

Because it is impossible to work with this decimal form (how can you multiply by a decimal number that has no ending digit?), I shall change 17.1 to a mixed number and work with fractional forms.

$$17.1 = 17\frac{1}{10}$$

Therefore,

$$5\frac{2}{3} \times 17.1 = 5\frac{2}{3} \times 17\frac{1}{10}$$

$$= \frac{17}{\overset{3}{\cancel{3}}_{1}} \times \frac{\overset{57}{\cancel{171}}}{10}$$

$$= \frac{969}{10}$$

$$= 96\frac{9}{10} \qquad \text{Or 96.9 in decimal form.}$$

EXAMPLE 8:   Divide 15.3 by $\frac{2}{3}$.

Again, because $\frac{2}{3} = .666\ldots = .\overline{6}$, I shall change 15.3 to fractional form.

$$15.3 \div \frac{2}{3} = 15\frac{3}{10} \div \frac{2}{3}$$

$$= \frac{153}{10} \div \frac{2}{3}$$

$$= \frac{153}{10} \times \frac{3}{2}$$

$$= \frac{459}{20}$$

$$= 22\frac{19}{20} \qquad \text{Or 22.95 in decimal form.}$$

Your turn!

Examples	Solutions
$18.25 + \dfrac{1}{4} = ?$	Change to decimal numbers if the decimal form of the fraction terminates. $$\frac{1}{4} = .25$$ Therefore, $$18.25 + \frac{1}{4} = 18.25 + .25$$ $$= 18.50$$ $$= 18.5$$

Examples	Solutions
$3.2 \times \dfrac{9}{10} = ?$	$\dfrac{9}{10} = .9$ Therefore, $$3.2 \times \frac{9}{10} = 3.2 \times .9$$ $$= 2.88$$
$15.4 \div \dfrac{2}{9} = ?$	$\dfrac{2}{9} = .222\ldots$ $\quad = .\overline{2}$ Nonterminating. Therefore, work with fractions. $$15.4 = 15\frac{4}{10}$$ $$= 15\frac{2}{5}$$ Now the problem becomes $$15.4 \div \frac{2}{9} = 15\frac{2}{5} \div \frac{2}{9}$$ $$= \frac{77}{5} \div \frac{2}{9}$$ $$= \frac{77}{5} \times \frac{9}{2}$$ $$= \frac{693}{10}$$ $$= 69\frac{3}{10} \quad \text{Or 69.3 in decimal form.}$$

Please read this section again if you did not understand the last three problems. You also may want to review the chapters on fractions–Chapters 6 through 11.

**Exercises
Part I**
Answers to these exercises can be found on page 365.

Change each fraction or mixed number to a decimal.

1. $\dfrac{1}{4}$

2. $3\dfrac{4}{5}$

3. $\dfrac{7}{12}$

4. $5\dfrac{4}{9}$

5. $\dfrac{6}{7}$

Perform the indicated operations.

6. $1.12 + 4\dfrac{3}{8}$ 7. $2\dfrac{1}{3} - 1.5$ 8. $6.35 \times 2\dfrac{3}{4}$

9. $7.8 \div \dfrac{5}{12}$ 10. $5\dfrac{2}{3} + 6.75$

**Exercises
Part II**   Change each to a decimal.

1. $\dfrac{3}{4}$ 2. $\dfrac{1}{9}$ 3. $7\dfrac{4}{7}$

Perform the indicated operations.

4. $71.23 + 3\dfrac{3}{4}$ 5. $5\dfrac{1}{3} - 2.15$

SECTION 16.3
HAND CALCULATORS

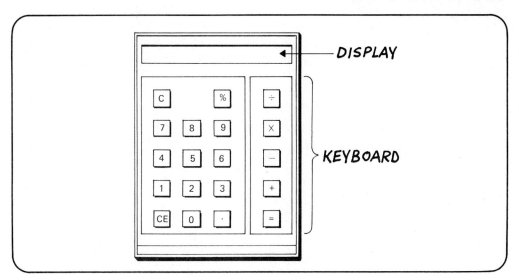

This section discusses the use of hand calculators in arithmetic.

Keyboard   I shall use the calculator pictured at the beginning of this section; consult your calculator manual for specific instructions for your machine.
There are 13 keys used for data entry.

$\boxed{0}$ through $\boxed{9}$ The **digit keys** are used to enter digits of a number into the display.

$\boxed{.}$ The **decimal point key** enters a decimal point.
$\boxed{CE}$ The **clear entry key** clears (erases) last entry in the display.
$\boxed{C}$ The **clear key** clears the information from the calculator and the display.

These 5 keys are used for the arithmetic operations.

$\boxed{+}$ The **add key** adds to the previous entry or result the next quantity entered.

$\boxed{-}$ The **subtract key** subtracts from the previous entry or result the next quantity entered.

$\boxed{\times}$ The **multiply key** multiplies the previous entry or result by the next quantity entered.

$\boxed{\div}$ The **divide key** divides the previous entry or result by the next quantity entered.

$\boxed{=}$ The **equals key** completes the calculation of all previously entered numbers and operations. It can be used for both intermediate and final results.

Arithmetic Calculations

Most of the tedious work of arithmetic can be done quickly and accurately with a calculator. I should not want to be without one because it allows me to spend more time on problem solving and less on the basic operations. However, a calculator can never completely replace a working knowledge of arithmetic. It does not tell you what operation to use.

There are many simple everyday tasks that do not warrant locating and using a calculator because basic arithmetic skills are adequate. In many situations, a calculator may not be handy. And practically speaking, what would happen if the battery died or if the calculator began to malfunction? A calculator can be used to great advantage, but I should not want to be completely dependent upon one.

The following examples illustrate the use of a calculator in arithmetic.

EXAMPLE 1:   Add 17.5 and 4.35.
Press the following keys to enter 17.5.

$$\boxed{1}, \boxed{7}, \boxed{.}, \boxed{5}$$

Press the addition key.

$$\boxed{+}$$

Press the following keys to enter 4.35.

$$\boxed{4}, \boxed{.}, \boxed{3}, \boxed{5}$$

Finally, press the equals key to complete the calculation.

$$\boxed{=}$$

The answer 21.85 appears in the display.

Notice that the sequence of keys pressed is the same sequence in which you write

$$17.5 + 4.35 =$$

Numbers and arithmetic operations are entered in the same sequence as the expression is written.

EXAMPLE 2:   Calculate $27.5 \div .37$.
Both of these mean the same thing.

$$27.5 \div .37 \qquad .37\overline{)27.5}$$

Enter a division problem in the calculator as it is written using the $\div$ sign.

$$27.5 \div .37$$

Press the following keys.

$$\boxed{2},\boxed{7},\boxed{.},\boxed{5}$$ Enters 27.5.

$\boxed{\div}$ Division key.

$\boxed{.},\boxed{3},\boxed{7}$ Enters .37.

$\boxed{=}$ Completes the calculation.

The display should read 74.32432432, which is the quotient rounded to 10 digits.

Of course, if you push the wrong key, your answer will not be correct, so it is a good idea to estimate the answer in order to check the accuracy of your entries.

In this example,

$$25.5 \div .37 \quad \text{approximately equals} \quad 30 \div .4$$

$$
\begin{array}{r}
75. \\
.4_\wedge \overline{\smash)30.0_\wedge} \\
28 \\
\hline
2\,0 \\
2\,0 \\
\hline
\end{array}
$$

which is close to the actual answer.

It is important for you to understand the rules of arithmetic in order to know how to estimate the answer because it is easy to enter the wrong number by entering a wrong digit or to misplace the decimal point, and because many calculators begin to give erratic answers when the battery begins to weaken. You are thus still dependent upon your arithmetic skills when using a calculator.

EXAMPLE 3:   Starting with a balance of $671.98, process the following checks and deposits to find the new balance.

Check for $19.72
Check for $5.06
Check for $12.15
Deposit of $75.00
Check for $195.03

With most calculators, you can process a sequence of additions and subtractions by using the equals key only once.

Press the following sequence of keys.

$\boxed{6},\boxed{7},\boxed{1},\boxed{.},\boxed{9},\boxed{8}$

$\boxed{-}$ Subtract check for $19.72.

$\boxed{1},\boxed{9},\boxed{.},\boxed{7},\boxed{2}$

$\boxed{-}$ Subtract check for $5.06.

$\boxed{5},\boxed{.},\boxed{0},\boxed{6}$

$\boxed{-}$ Subtract check for $12.15.

$\boxed{1},\boxed{2},\boxed{.},\boxed{1},\boxed{5}$

$\boxed{+}$ Add deposit of $75.00.

$\boxed{7},\boxed{5},\boxed{.},\boxed{0},\boxed{0}$

$\boxed{-}$ Subtract check for $195.03.

$$\boxed{1},\boxed{9},\boxed{5},\boxed{.},\boxed{0},\boxed{3}$$

$$\boxed{=}$$ Complete the processing of the entire sequence.

The display should read 515.02.

The percent key $\boxed{\%}$ will be discussed in Unit Four. More complex order of operations will be discussed in Unit Six.

Exercises Using a calculator, rework the exercises in Unit Three, starting on page 311, ending on page 344.

REVIEW MATERIAL

Review Skill Fractions to Decimal Numbers

To change a fraction or mixed number to a decimal, divide the numerator by the denominator of the fraction.

$$\frac{5}{6} = 5 \div 6$$

$$\begin{array}{r} .833\ldots \\ 6\overline{)5.000} \\ 4\ 8 \\ \hline 20 \\ 18 \\ \hline 20 \\ 18 \\ \hline 2 \end{array}$$

Therefore, $\frac{5}{6} = .8333\ldots = .8\overline{3}$

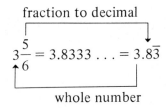

$$3\frac{5}{6} = 3.8333\ldots = 3.8\overline{3}$$

whole number

Exercise A Answers to this exercise can be found on page 365.

Change each fraction or mixed number to a decimal number.

1. $\dfrac{5}{8}$ 2. $\dfrac{3}{15}$ 3. $2\dfrac{3}{4}$ 4. $5\dfrac{1}{9}$

5. $3\dfrac{6}{7}$ 6. $5\dfrac{3}{8}$ 7. $12\dfrac{1}{15}$ 8. $\dfrac{17}{35}$

Exercise B No answers are given.

Change each fraction or mixed number to a decimal number.

1. $\dfrac{3}{8}$ 2. $\dfrac{2}{15}$ 3. $2\dfrac{1}{4}$ 4. $6\dfrac{4}{9}$

Refer to the section listed if you have difficulty. Answers to these exercises can be found on page 365.

Section 16.1

1. Which is thicker, .73 inch or .745 inch?
2. During the month of May, I made the following deposits: $53.49, $27, and $147.14. What was the total of my deposits for the month of May?
3. If you have $532.19 in your bank account, how much money do you have after writing a check for $127.89 and depositing $59.56?
4. For company use of her own car, Muriel receives 19¢ per mile. How much does she receive for driving her own car 27 miles?
5. We pay our babysitter, Rick, $1.50 per hour. How much do we owe him for 5 hours?
6. Jim receives $421.50 for working 6 days. How much does he average per day?

Section 16.2 Change each to a decimal.

7. $\dfrac{7}{9}$

8. $\dfrac{5}{16}$

9. $2\dfrac{3}{8}$

Perform the indicated operations.

10. $3\dfrac{3}{16} + 5.759$

11. $2\dfrac{1}{9} - 1.2$

12. $3\dfrac{1}{3} \times 5.3$

CHAPTER POST-TEST

This test should determine if you have mastered the topics in Chapter 16. Answers to this test can be found on page 365.

1. Which is thicker, .72 inch or .721 inch?
2. Which is thicker, .902 inch or .91 inch?
3. Joanne drove 253.5 miles the first day and 296.3 miles the second day. How far did she drive during the two days?
4. At the end of the business day, $1,563.23 should have been in the register. If only $1,527.96 was in the register, how much was missing?
5. How much change should you receive if you use a $20 bill to pay for a $16.27 dinner bill?
6. A pump can extract 3.2 gallons of water in a minute. How many gallons can be extracted in 55 minutes?
7. Tom paid $27.25 as a down payment and paid $30.27 per month for 12 months. How much did he pay all together?
8. A trip is 2,654 miles long. How far must you drive each day to make the trip in 10 days?
9. If you can drive 138.6 miles on 9 gallons of gas, how far can you drive per gallon?

Perform the indicated operations.

10. $6\dfrac{3}{16} + 9.562$ **11.** $8\dfrac{5}{6} - 2.3$ **12.** $2\dfrac{3}{8} \div 17.2$

UNIT THREE
DECIMAL NUMBERS REVIEW
AND POST-TEST

Use the following rules to perform the four basic operations for decimal numbers.

To add or subtract decimal numbers, line up the decimal points and add or subtract as if you were working with whole numbers.

$$
17.5 + 29.59: \qquad
\begin{array}{r}
17.50 \\
+\,29.59 \\
\hline
47.09
\end{array}
$$

$$
18 - 1.7: \qquad
\begin{array}{r}
18.0 \\
-1.7 \\
\hline
16.3
\end{array}
$$

To find the product of two decimal numbers, multiply the numbers as if they were whole numbers. To find the number of decimal places in the product, add the number of decimal places in the numbers being multiplied.

$$
\begin{array}{rl}
1.257 & \text{3-place decimal.} \\
\times.51 & \text{2-place decimal.} \\
\hline
1257 & \\
6285 & \\
\hline
.64107 & \text{5-place decimal.}
\end{array}
$$

To divide by a decimal number, move the decimal point in the divisor the number of places necessary to make it a whole number. Then move the decimal point in the dividend the same number of places to the right, annexing zeros where necessary.

$$
.04131 \div .027: \qquad
\begin{array}{r}
1.53 \\
.027\overline{)\,.041\ 31} \\
\underline{27} \\
14\ 3 \\
\underline{13\ 5} \\
81 \\
\underline{81}
\end{array}
$$

Before working the Unit Three Post-Test work (or rework)

Chapter 12 Review Exercises, page 306
Chapter 13 Review Exercises, page 317
Chapter 14 Review Exercises, page 326
Chapter 15 Review Exercises, page 344
Chapter 16 Review Exercises, page 361

Study the sections for which you cannot work the sample problems.

Answers to this test can be found on page 365.

1. Write each as a fraction or mixed number.
   (a) .013   (b) 7.8   (c) 107.049
2. Round each number to the nearest tenth.
   (a) 6.122   (b) 7.153   (c) 17.980
3. Perform the indicated operation.
   (a) 317.59 + 19.99   (b) 12 − 7.3   (c) 13.2 × .07
   (d) .003302 ÷ .013   (e) 17.983 + 45 + 2.09   (f) 14.5 × .9
   (g) .019 × 100   (h) 7.12 ÷ 1,000
4. If steak is on sale for $3.98 per pound, how much do 9 pounds cost?
5. If you have $392.16 in your bank account, how much money do you have after writing a check for $216.98 and depositing $65.00?
6. If gas sells for $1.78 per gallon, and the gas bill is $16.02, how much gas was bought?

ANSWERS

Unit Three Pretest

1. (a) seventy-three thousandths
   (b) seven hundred fifty-two and twelve hundredths
   (c) one hundred three and four thousandths   2. (a) 24.015 (b) .47
3. (a) $\frac{41}{50}$ (b) $17\frac{59}{100}$ 4. (a) 5.13   (b) 72.13   (c) 17.30
5. (a) 11.24   (b) 37.131   (c) 61.48   6. (a) .08   (b) 4.22   (c) 4.11
7. (a) 23.2875   (b) 86.4   (c) 70.5   8. (a) 1.25   (b) .035   (c) .173
9. $639.93   10. $4.50   11. 3.6875¢

Chapter 12

Section 12.1

1. (a) hundredths   (b) tens   (c) thousandths   (d) thousands
   (e) ten-thousandths   (f) ones   (g) hundreds   (h) ten-thousandths
   (i) tenths   (j) hundredths   (k) thousandths   2. (a) 5   (b) 4   (c) 6
   (d) 1   (e) 3   (f) 7   (g) 9   (h) 2

Section 12.2

1. (a) fifty-three and fifteen hundredths
   (b) one hundred seventy-two and five tenths
   (c) seven and one hundred fifty-three thousandths   (d) six and four hundredths
   (e) six hundred five and nine thousandths
   (f) seven hundred thirteen thousandths   (g) nine tenths   (h) three thousandths
   2. (a) 35.83   (b) 9.116   (c) 5.003   (d) 205.06   (e) .0159   (f) .03

Section 12.3

1. $\frac{123}{1,000}$ 2. $\frac{14}{25}$ 3. $\frac{109}{1,000}$ 4. $\frac{2}{25}$ 5. $59\frac{3}{10}$ 6. $67\frac{4}{5}$ 7. $27\frac{21}{200}$

8. $6\frac{333}{1,000}$ 9. $19\frac{3}{4}$ 10. $125\frac{1}{200}$

Section 12.5

1. (a) 27.59   (b) 395.13   (c) 29.70   (d) 14.54
2. (a) 27.592   (b) 395.126   (c) 29.690   (d) 14.543
3. (a) 27.6   (b) 395.1   (c) 29.7   (d) 14.5

Exercise A 1. one hundred ninety-three and twenty-seven hundredths
2. six and four thousandths   3. eight tenths
4. one hundred five and five hundred seven thousandths   5. 2.19
6. .085   7. 104.152

Exercise C 1. $\dfrac{23}{100}$ 2. $\dfrac{4}{5}$ 3. $\dfrac{107}{1,000}$ 4. $62\dfrac{1}{2}$ 5. $135\dfrac{71}{1,000}$ 6. $27\dfrac{1}{4}$

Exercise E 1. 85.4   2. 74.4   3. 2.9   4. .054   5. 1.062   6. 1.060   7. 103.058

Review Exercises 1. (a) tenths   (b) thousandths   (c) hundreds   (d) hundredths
2. (a) 2   (b) 3   (c) 4   (d) 7   (e) 1
3. (a) fifty-seven and fifty-four hundredths
(b) one thousand, six hundred fifty-two and fifty-nine thousandths
(c) nine tenths   (d) five hundredths   4. (a) 23.52   (b) 5.027   (c) .143
5. (a) $\dfrac{73}{100}$ (b) $\dfrac{1}{5}$ (c) $203\dfrac{123}{1,000}$ (d) $59\dfrac{19}{200}$
6. (a) .54   (b) .42   (c) .70   (d) .04

Post-Test 1. (a) thousandths   (b) tenths   2. (a) 7   (b) 3   (c) 2   (d) 9
3. (a) two hundred eighty-nine and fifty-seven thousandths
(b) two hundred seventy-five thousandths   4. (a) 6.052   (b) .0047
5. (a) $\dfrac{21}{200}$ (b) $6\dfrac{23}{100}$ 6. (a) 27.6   (b) 8.4   (c) 29.0

Chapter 13

Section 13.1 1. 1.78   2. 1.096   3. 28.2   4. 144.37   5. 127.799   6. 13.183
7. 32.93   8. 11.02   9. 60.626   10. 99.305   11. 3.93   12. 384.59

Section 13.2 1. .08   2. .882   3. 3.6   4. 11.17   5. 9.379   6. 1.217   7. 21.07
8. 1.02   9. $28.96   10. $58.96
11. This Payment: $18.19 Bal forward: $670.84
12. This Payment: $29.89 Bal forward: $989.19.

Exercise A 1. .22   2. 12.479   3. .48   4. .725   5. 13.56   6. 11.75   7. 2.15
8. 28.82   9. 17.07   10. 183.82

Review Exercises 1. 1.26   2. 23.48   3. 26.176   4. 41.68   5. .08   6. 7.52
7. 8.996   8. 5.3

Post-Test 1. 87.573   2. 32.387   3. 179.99   4. 114.073   5. 95.933
6. 109.89   7. 584.47   8. 156.02

Chapter 14

Section 14.1 1. .0255   2. 92.1087   3. .081   4. 20.93616   5. .00318   6. 109.8
7. .54   8. 44.5936   9. .00021   10. .0000065

Section 14.2 1. 27.507   2. 86.4   3. 40.75   4. .000213   5. 17.3   6. 3,200
7. 1,590   8. 1,500   9. .3   10. 3

Exercise A 1. .0195   2. .516   3. .051   4. 1,750   5. 30   6. 35.1   7. .00266
8. .3   9. 700   10. 1,500

Review Exercises   **1.** 24.98206   **2.** .000741   **3.** .00819   **4.** 140.64   **5.** .000375
   6. 1756.1   **7.** 7.05   **8.** 37,200

Post-Test   **1.** 2.16284   **2.** .2373   **3.** .3125   **4.** 101.7   **5.** .000036   **6.** 159
   7. 317,590   **8.** 2752.6

Chapter 15

Section 15.1   **1.** 2.7   **2.** 2.012   **3.** .25   **4.** .004   **5.** .6   **6.** 2.73   **7.** 1.91
   8. .03   **9.** .01   **10.** .89

Section 15.2   **1.** 1.92   **2.** 1.2535   **3.** .0892   **4.** .01352   **5.** 5.9   **6.** .63   **7.** .4
   8. .049

Section 15.3   **1.** 3.65   **2.** .012   **3.** 400   **4.** 83.64   **5.** 45.752   **6.** 505.33   **7.** .44
   8. 26.67   **9.** 1.87   **10.** 6335.40   **11.** 7.06   **12.** 2.01

Exercise A   **1.** 12.3   **2.** .02   **3.** 12.6   **4.** .173   **5.** 2.4   **6.** 1.23   **7.** 1.5
   8. 413.3   **9.** 1826.1   **10.** 1.0

Review Exercises   **1.** .125   **2.** .013   **3.** .25   **4.** .66   **5.** .18   **6.** .185   **7.** .00173
   8. .5   **9.** 12.5   **10.** 0.02   **11.** .8   **12.** 8.2   **13.** 26.7

Post-Test   **1.** 1.024   **2.** .00235   **3.** .003   **4.** 7.04   **5.** 1.24   **6.** .71   **7.** .97
   8. 5.77   **9.** 4.09   **10.** 0.11

Chapter 16

Section 16.1   **1.** .04   **2.** .0231   **3.** $28.58   **4.** $28.94   **5.** 4.4 seconds
   6. .19 pound   **7.** $366.40   **8.** $5.67   **9.** 9 gallons   **10.** about $221.40
   11. 2-pound can   **12.** 15 in a box   **13.** $623.46
   14. About 16.4 miles per gallon.

Section 16.2   **1.** .25   **2.** 3.8   **3.** $.58\overline{3}$ **4.** $5.\overline{4}$ **5.** $.\overline{857142}$ **6.** 5.495
   7. $\dfrac{5}{6}$ **8.** 17.4625   **9.** $18\dfrac{18}{25}$ **10.** $12\dfrac{5}{12}$

Exercise A   **1.** $.625_{}$ **2.** $.\overline{2}$ **3.** 2.75   **4.** $5.111\ldots = 5.\overline{1}$ **5.** $3.\overline{857142}$ **6.** 5.375
   7. $12.0\overline{6}$ **8.** $.4\overline{857142}$

Review Exercises   **1.** .745   **2.** $227.63   **3.** $463.86   **4.** $5.13   **5.** $7.50   **6.** $70.25
   7. $.\overline{7}$ **8.** .3125   **9.** 2.375   **10.** 8.9465   **11.** $\dfrac{41}{45}$ **12.** $17\dfrac{2}{3}$

Post-Test   **1.** .721   **2.** .91   **3.** 549.8 miles   **4.** $35.27   **5.** $3.73   **6.** 176 gallons
   7. $390.49   **8.** 265.4 miles   **9.** 15.4 miles   **10.** 15.7495   **11.** $6\dfrac{8}{15}$
   12. .138 to the nearest thousandth

Unit Three Post-Test   **1.** (a) $\dfrac{13}{1,000}$ (b) $7\dfrac{4}{5}$ (c) $107\dfrac{49}{1,000}$ **2.** (a) 6.1   (b) 7.2   (c) 18.0
   3. (a) 337.58   (b) 4.7   (c) .924   (d) .254   (e) 65.073   (f) 13.05
   (g) 1.9   (h) .00712   **4.** $35.82   **5.** $240.18   **6.** 9 gallons

UNIT 4
ratio, proportion, and percent

UNIT FOUR PRETEST

This test is taken *before* you read the material to see what you already know. If you cannot work each sample problem, you need to read the indicated chapters carefully. Answers to this test can be found on pages 431–432. You may wish to read each chapter even when you are familiar with the information, but you may omit reading those chapters for which you answered *each* sample problem correctly.

CHAPTER 17

1. My stainless steel table service includes 24 spoons, 16 forks, and 8 knives. Write, in lowest terms, the ratio of
   (a) forks to spoons
   (b) spoons to knives
   (c) knives to forks

2. Twenty-six applicants were given a typing proficiency test. Because they could type at least 60 words per minute, 17 were then interviewed for the job. Write the ratio of the number of applicants who were *not* interviewed to the number who were interviewed.

3. If a car is driven 500 kilometers in 20 hours, what is the rate in kilometers per hour (the ratio of kilometers to hours)?

4. Simplify each ratio and reduce to lowest terms.

   (a) $\dfrac{3\frac{1}{5}}{2\frac{2}{3}}$
   (b) $\dfrac{4.5}{.09}$

5. Which of the following are equal ratios?

   (a) $\dfrac{4}{6}$ and $\dfrac{10}{15}$
   (b) $\dfrac{9}{10}$ and $\dfrac{10}{11}$

6. Express, in lowest terms, the ratio of
   (a) 40 weeks to 1 year
   (b) 4 dimes to 5 nickels

7. Find the missing value in each proportion.

   (a) $5:6 = 30:N$
   (b) $\dfrac{4}{N} = \dfrac{2}{7}$
   (c) $\dfrac{N}{8} = \dfrac{7}{3}$

8. In a bread recipe the ratio of flour to milk is $4:3$. If 5 cups of flour are used, how many cups of milk should be used?

CHAPTER 18

9. Rewrite the ratio form of each percent using a % symbol.

   (a) $\dfrac{9}{100}$ (b) $\dfrac{67}{100}$ (c) $\dfrac{103}{100}$

10. Rewrite each of the following as a percent in ratio form.

    (a) 83%   (b) 7%   (c) 151%

11. Write $\dfrac{4}{5}$ as a percent using % notation.

12. Write $\dfrac{1}{2}$% as a ratio in lowest terms.

13. Change 26.5% to an equivalent decimal.

14. Rewrite 4.3 as an equivalent percent using % notation.

15. 31 is 40% of what number?

16. Find 64% of 750.

17. Chad's house is worth $70,000. How much would he receive if the house were completely destroyed by fire and it was insured for 90% of its value?

17

ratio and proportion

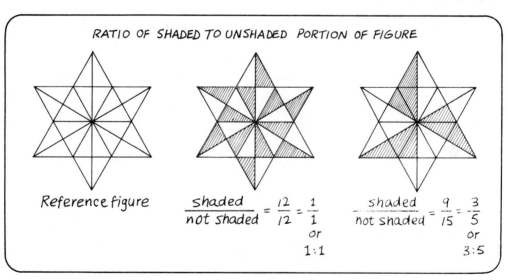

RATIO OF SHADED TO UNSHADED PORTION OF FIGURE

Reference figure

$$\frac{shaded}{not\ shaded} = \frac{12}{12} = \frac{1}{1}$$
or
1:1

$$\frac{shaded}{not\ shaded} = \frac{9}{15} = \frac{3}{5}$$
or
3:5

This section introduces ratios. After its study you should be able to answer questions such as the following:

1. Write a ratio to describe
   a. The comparison of 225 foreign cars to 400 domestic cars.
   b. The rate, in gallons lost per day, of a leaky faucet that loses 14 gallons of water in a week.

2. Determine whether $\frac{2}{5}$ and $\frac{3}{7}$ are equal ratios.

Definition   A **ratio** is the comparison of two numbers by division.

	A pair of numbers can describe a rate.		A pair of numbers may be used to compare two quantities.
Rate	GUM GUM (25¢) (25¢) (25¢) 2 packs for 75¢	Comparison	♥♥ ♥ ♥♥ ★★★
Ratio	$\frac{2}{75}$	Ratio of ♥ to ★	$\frac{5}{3}$
Read the ratio	2 for 75	Read the ratio	5 to 3

A ratio can be written in three different ways.   The ratio of ♥ to ★ in the above comparison can be written as follows:

> 5 to 3,   Written-out form.
> 5 : 3,   Colon form (colon is short for ÷).
> $\frac{5}{3}$,   Fraction form.

Order   Do you think 3 : 5 represents the same ratio of ♥ to ★? It does not.   Although it relates to the same quantities, 3 : 5 compares the number of stars to the number of hearts; that is, ★ to ♥.

> The two numbers that make up a ratio must be written in the same order that the quantities appear in the statement of comparison.

EXAMPLE 1:   If there are 17 females and 12 males in a class, what is the ratio of males to females?

> Statement of comparison: males to females
>
> Ratio: 12 to 17
>
> or   12 : 17
>
> or $\frac{12}{17}$

Again, notice that the numbers of the ratio appear in the same order as the corresponding words appear in the statement of comparison.

EXAMPLE 2:   Eunice Cox has a total of 15 children enrolled in her Sunday school class.   One Sunday morning 2 children were absent.   What was the ratio of the number of children absent to the number of children present?

Statement of comparison: children absent to children present

The ratio cannot be written immediately from the numbers given in the example because the problem did not state how many children were present.   However, a simple subtraction will provide that number:

$$
\begin{array}{rl}
15 & \text{Total number of children enrolled.} \\
-\ 2 & \text{Number of children absent.} \\
\hline
13 & \text{Number of children present.}
\end{array}
$$

Statement of comparison: children absent to children present

Ratio: 2 to 13

or $2:13$

or $\dfrac{2}{13}$

EXAMPLE 3:   Eighteen families live in my building. They own a total of 45 cars. What is the ratio of the number of cars to the number of families in my building?

Statement of comparison: cars to families

Ratio: 45 to 18

or $45:18$

or $\dfrac{45}{18}$

You already know how to reduce the ratio $\dfrac{45}{18}$.

$$\frac{45}{18} = \frac{5}{2}$$

or $5:2$

or 5 to 2

It would be difficult to close your eyes and picture 45 cars parked in front of 18 apartments. It should be easier to picture 5 cars in front of 2 apartments. This is why, if possible, most ratios are reduced to lowest terms: Reducing a ratio makes the comparison easier to understand.

EXAMPLE 4:   Express, in lowest terms, the ratio of voters who actually cast a vote to the persons registered to vote in an election when 4,328 votes were cast and there were 21,640 persons registered.

Statement of comparison: voters to persons registered

Ratio: $\dfrac{4{,}328}{21{,}640}$

Reducing by 2's: $\dfrac{4{,}328}{21{,}640} = \dfrac{2{,}164}{10{,}820} = \dfrac{1{,}082}{5{,}410} = \dfrac{541}{2{,}705}$

Reducing by 541: $\dfrac{541}{2{,}705} = \dfrac{1}{5}$

This ratio means that *one out of five* persons registered actually voted. Reducing made both numbers much smaller and therefore much easier to compare.

EXAMPLE 5:   A pitcher gives up 2 earned runs in $3\frac{1}{2}$ innings of pitching.

What is the ratio, in lowest terms, of earned runs to the number of innings pitched?

Statement of comparison: earned runs to innings pitched

Ratio: $\qquad 2$ to $3\frac{1}{2}$

$\qquad$ or $\quad 2:3\frac{1}{2}$

$\qquad$ or $\quad \dfrac{2}{3\frac{1}{2}}$

To simplify a ratio that contains a fraction, perform the division that the fraction indicates.

$$\frac{2}{3\frac{1}{2}} = 2 \div 3\frac{1}{2}$$

$$= \frac{2}{1} \div \frac{7}{2}$$

$$= \frac{2}{1} \times \frac{2}{7}$$

$$= \frac{4}{7}$$

The ratio of earned runs to the number of innings pitched, in lowest terms, is 4 to 7, or $4:7$, or $\dfrac{4}{7}$. This means that giving up 2 earned runs in $3\frac{1}{2}$ pitched innings is the same as giving up 4 runs in 7 innings.

EXAMPLE 6:   The capacity of the fresh-food section of a refrigerator is 13.5 cubic feet and that of the freezer section is 6.75 cubic feet. What is the ratio, in reduced form, of the freezer capacity to the fresh food capacity?

Statement of comparison: freezer to fresh food

$\qquad$ Ratio: $\qquad$ 6.75 to 13.5

$\qquad$ or $\quad 6.75:13.5$

$\qquad$ or $\quad \dfrac{6.75}{13.5}$

To simplify a ratio containing decimals, move the decimals in the numerator and denominator the same number of places to the right in order to eliminate the decimals and to keep the value of the ratio unchanged.

I shall move the decimal points *one place*.

$$\frac{6.75}{13.5} = \frac{6.75}{13.5}$$

$$= \frac{67.5}{135}$$

But this did not remove the decimal from the numerator, so I shall move the decimal points *two places* from their original positions.

$$\frac{6.75}{13.5} = \frac{6.75}{13.50} \qquad \text{Annex a zero.}$$

$$= \frac{675}{1,350}$$

$$= \frac{135}{270} = \frac{27}{54} \qquad \text{Reducing by 5 twice.}$$

$$= \frac{9}{18} \qquad \text{Reducing by 3.}$$

$$= \frac{1}{2} \qquad \text{Reducing by 9.}$$

The ratio of the freezer capacity to the fresh food capacity is $1:2$. This means that the refrigerator has twice as much room for fresh food as for frozen food.

EXAMPLE 7:   Use the capacity figures from Example 6 to write the ratio, in lowest terms, of the freezer capacity to the total capacity of the refrigerator.

Because the total capacity was not given, I must determine that before I can write a statement of comparison.

$$
\begin{array}{ll}
13.5 \text{ cubic feet} & \text{Fresh food capacity.} \\
+\ 6.75 \text{ cubic feet} & \text{Freezer capacity.} \\
\hline
20.25 \text{ cubic feet} & \text{Total capacity.}
\end{array}
$$

Statement of comparison: freezer to total

Ratio: 6.75 to 20.25

or   6.75 : 20.25

or $\dfrac{6.75}{20.25}$

In order to eliminate the decimals from the ratio I must move the decimals two places to the right.

$$\frac{6.75}{20.25} = \frac{6.75}{20.25}$$

$$= \frac{675}{2,025}$$

$$= \frac{135}{405} = \frac{27}{81} \qquad \text{Reducing by 5.}$$

$$= \frac{3}{9} \qquad \text{Reducing by 9.}$$

$$= \frac{1}{3} \qquad \text{Reducing by 3.}$$

Therefore, the reduced ratio of the freezer capacity to the total capacity is $1:3$.

EXAMPLE 8:   Express the ratio of 4 months to 2 years.

Let me warn you that this example is not so simple as it may first appear. If you try to compare the quantities by thinking only of the numbers given, 4 is greater than 2.

$$4 > 2$$

But this is an incorrect comparison because 4 months is less than 2 years:

$$4 \text{ months} < 2 \text{ years}$$

To avoid making an incorrect comparison of quantities of like or similar measure, use the following technique.

When comparing *like* quantities by a ratio, express the quantities in the same unit of measure, if possible.

Here, months and years are like quantities because they are both measures of time.

Because year is the larger unit of time, I shall change years to months:

$$1 \text{ year} = 12 \text{ months}$$

$$2 \text{ years} = 24 \text{ months}$$

Statement of comparison: 4 months to 2 years

Ratio (like quantities):   4 months to 24 months

Ratio: $4 : 24$

Reduced: $1 : 6$

EXAMPLE 9:   Write the comparison of 6 nickels to 3 quarters as a ratio.

Because it is usually easier to change the larger unit to the smaller unit, I shall change 3 quarters to the number of nickels that has the same value.

$$6 \text{ nickels} = 6 \text{ nickels}$$

$$3 \text{ quarters} = 15 \text{ nickels}$$

$\dfrac{6 \text{ nickels}}{15 \text{ nickels}}$ Same unit.

$\dfrac{6}{15}$ Without labels.

$\dfrac{2}{5}$ Reduced.

This ratio can also be written in terms of the smaller unit cents instead of nickels.

$$6 \text{ nickels} = 30\text{¢}$$

$$3 \text{ quarters} = 75\text{¢}$$

$\dfrac{30\text{¢}}{75\text{¢}}$ Same unit.

$\dfrac{30}{75}$ Without labels.

$\dfrac{2}{5}$ Reduced.

Notice that the reduced form is $\frac{2}{5}$ whether I compare nickels to nickels or cents to cents.

EXAMPLE 10: Write the comparison of 4 dimes to 10 nickels as a ratio in lowest terms.

As in the previous example, this comparison can be made by changing dimes to nickels or by changing both quantities to cents. I shall show both methods.

$$4 \text{ dimes} = 8 \text{ nickels} \qquad\qquad 4 \text{ dimes} = 40\cancel{c}$$
$$10 \text{ nickels} = 10 \text{ nickels} \qquad\qquad 10 \text{ nickels} = 50\cancel{c}$$

$\dfrac{4 \text{ dimes}}{10 \text{ nickels}}$	Statement of comparison.	$\dfrac{4 \text{ dimes}}{10 \text{ nickels}}$
$\dfrac{8 \text{ nickels}}{10 \text{ nickels}}$	Same unit.	$\dfrac{40\cancel{c}}{50\cancel{c}}$
$\dfrac{8}{10}$	Without labels.	$\dfrac{40}{50}$
$\dfrac{4}{5}$	Reduced.	$\dfrac{4}{5}$

Although both $8 : 10$ and $40 : 50$ are correct comparisons of 4 dimes to 10 nickels, there is only one correct reduced ratio, $4 : 5$.

When finding the ratio between quantities, observe the following rules:

1. Express the quantities in the same unit of measure whenever possible.
2. Write the ratio without labels.
3. Simplify the ratio by reducing the final answer to lowest terms.

Rate A very useful ratio is one that expresses a rate that can be reduced so that the denominator is 1. One or more of the following examples of rate should be familiar to you.

A driving rate:

$$\frac{156 \text{ miles}}{3 \text{ hours}} = \frac{52 \text{ miles}}{1 \text{ hour}}, \quad \text{or} \quad 52 \text{ miles per hour}$$

A turntable rate:

$$\frac{90 \text{ revolutions}}{2 \text{ minutes}} = \frac{45 \text{ revolutions}}{1 \text{ minute}}, \quad \text{or} \quad 45 \text{ revolutions per minute}$$

A sales rate:

$$\frac{\$1.48}{4 \text{ pounds}} = \frac{\$.37}{1 \text{ pound}}, \quad \text{or} \quad 37\cancel{c} \text{ per pound}$$

A typing rate:

$$\frac{315 \text{ words}}{5 \text{ minutes}} = \frac{63 \text{ words}}{1 \text{ minute}}, \quad \text{or} \quad 63 \text{ words per minute}$$

A punting rate:

$$\frac{252 \text{ yards}}{6 \text{ punts}} = \frac{42 \text{ yards}}{1 \text{ punt}}, \quad \text{or} \quad 42 \text{ yards per punt}$$

EXAMPLE 11:   My kitchen faucet leaked 10 liters of water in 5 days. What was the rate of loss in liters per day?

Statement of comparison: liters per day

Given measures: 10 liters in 5 days

Reduced ratio: $2:1$

The rate of loss was 2 liters of water per day.

EXAMPLE 12:   A copier makes 225 copies in 3 minutes. What is the machine's rate in copies per minute?

Statement of comparison: copies per minute

Given measures: 225 copies in 3 minutes

Reduced ratio: $75:1$

The machine's rate of producing copies is 75 copies per minute.

Try to work these problems *before* looking at the solutions. For each example write the ratio(s) described in lowest terms.

Examples	Solutions
Write three forms of the ratio of Becky's age to Matthew's age if Becky is 3 years old and Matthew is 5 years old.	Comparison: Becky's age to Matthew's age 3 to 5 or $3:5$ or $\dfrac{3}{5}$
Write three forms of the following ratio: the number of months having 6 or more letters in their name to the number of months having 5 or fewer letters.	Comparison: months with 6 or more letters to months with 5 or fewer 7 to 5 or $7:5$ or $\dfrac{7}{5}$
A football team played 12 games, lost 3, and played to no ties. Write three forms of the ratio of their wins to their losses.	wins to losses $(12-3)$ to 3 9 to 3 3 to 1 Reduced.

Examples	Solutions
	or $\quad 3:1$
	or $\quad \dfrac{3}{1}$
	Note: Do *not* rewrite $\dfrac{3}{1}$ as 3 because a ratio is the comparison of *two* numbers.
Jerry planted 15 tomato (T), 12 squash (S), 6 pepper (P), and 12 cucumber (C) plants. Write each of the following ratios as a fraction in lowest terms:	
(a) squash to tomato plants;	(a) $\dfrac{S}{T} = \dfrac{12}{15} = \dfrac{4}{5}$
(b) tomato to pepper plants;	(b) $\dfrac{T}{P} = \dfrac{15}{6} = \dfrac{5}{2}$
(c) cucumber to squash plants;	(c) $\dfrac{C}{S} = \dfrac{12}{12} = \dfrac{1}{1}$
	Note: A ratio of $1:1$ is always the comparison of equal quantities.
(d) pepper to cucumber plants.	(d) $\dfrac{P}{C} = \dfrac{6}{12} = \dfrac{1}{2}$
When buying or selling stock, stock analysts consider the price-to-earnings (P.E.) ratio. If the price of a certain stock is \$36 and its earnings are \$4, what is the P.E. ratio of that stock?	$\dfrac{\text{price}}{\text{earnings}} = \dfrac{36}{4} = \dfrac{9}{1}$
Maude can type, on the average, 58 address labels in 29 minutes. What is her average typing rate expressed in labels per minute?	labels : minutes $58:29$ $2:1$ Maude's average typing rate is 2 labels per minute.
What is the ratio of 8 eggs to 2 dozen eggs?	1 dozen eggs $= 12$ eggs 2 dozen eggs $= 24$ eggs $\dfrac{8 \text{ eggs}}{2 \text{ dozen eggs}}$ $\dfrac{8 \text{ eggs}}{24 \text{ eggs}}$ $\qquad$ Same unit. $\dfrac{8}{24}$ $\qquad$ Without labels. $\dfrac{1}{3}$ $\qquad$ Reduced.

Examples	Solutions
A small refrigerator in the Brevard family's camper has a capacity of 4.9 cubic feet. Their kitchen refrigerator has a 14.7-cubic-foot capacity. Write a ratio, in lowest terms, comparing the capacities of camper to kitchen refrigerators.	camper to kitchen $$4.9 : 14.7$$ or $\dfrac{4.9}{14.7}$ Move each decimal *one place* in order to remove the decimals. $$\frac{4.9}{14.7} = \frac{4.9}{14.7}$$ $$= \frac{49}{147} = \frac{7}{21} = \frac{1}{3}$$
A $6\dfrac{1}{2}$-foot-tall man casts a $4\dfrac{1}{2}$-foot shadow. What is the ratio, in lowest terms, of the man's height to the length of his shadow?	man to shadow $$6\frac{1}{2} \text{ to } 4\frac{1}{2}$$ or $\dfrac{6\frac{1}{2}}{4\frac{1}{2}}$ Use the rule for division of fractions to simplify and reduce. $$\frac{6\frac{1}{2}}{4\frac{1}{2}} = 6\frac{1}{2} \div 4\frac{1}{2}$$ $$= \frac{13}{2} \div \frac{9}{2}$$ $$= \frac{13}{2} \times \frac{2}{9}$$ $$= \frac{13}{9}$$

Now try these—no hints given.

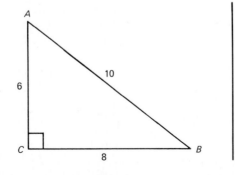

Examples	Solutions
In the right triangle above, what is the ratio of	
(a) the longest side to the shortest side?	(a) $\dfrac{5}{3}$
(b) side AC to side CB?	(b) $\dfrac{3}{4}$
(c) side AB to the sum of all three sides?	(c) $\dfrac{5}{12}$
What is the ratio of caffeine to aspirin in a pain reliever that contains 195 milligrams of aspirin and 65 mg of caffeine in each tablet?	$\dfrac{1}{3}$
What is the ratio of 4 weeks to 12 days?	$\dfrac{7}{3}$

If you had difficulty with the last three examples, please talk with an instructor or read this section again *before* you attempt the following exercises.

Exercises
Part I

Answers to these exercises can be found on page 432.

1. Write three forms of the ratio of the number of letters in Sybil Douthat's first name to the number of letters in her last name.

2. Roger has 23 cars, 7 boats, and 16 planes in his collection of plastic models. Write a ratio for each of the following comparisons:
   (a) cars to planes   **(b)** planes to cars
   (c) boats to planes   **(d)** cars to boats

3. Jamie flipped a penny 100 times. It landed heads up 52 times. Write, in lowest terms, the ratio of times it landed heads up to the number of times it landed tails up.

4. What is the reduced ratio of 50 minutes to 2 hours?

5. What is the reduced ratio of 6 dimes to 2 quarters?

6. What is the ratio, in lowest terms, of 9 months to 2 years?

7. A college football team won 8 out of 10 games played. There were no tie games. Write a ratio, in reduced form, to represent each of the following comparisons:
   (a) wins to games played
   (b) wins to losses
   (c) losses to games played

8. There are 8 males and 7 females in Darryl's industrial arts class. Write a ratio to represent each of the following comparisons:
   (a) males to females
   (b) males to the entire class
   (c) females to the entire class

9. If Mona is $5\frac{1}{2}$ feet tall and casts a $2\frac{1}{2}$-foot shadow, what is the reduced ratio of her height to the length of her shadow?

10. A tree 4.05 meters tall casts a shadow that measures 2.7 meters. Write the ratio, in lowest terms, of the length of the tree's shadow to its height.

11. Marvin was able to drive 156 miles on 4 gallons of gasoline. What was his rate of fuel consumption expressed in miles per gallon—that is, the reduced ratio of miles to gallons used?

12. Write a reduced ratio to represent the rate, in gallons lost per day, of a leaky faucet that loses 84 gallons of water in a week.

Exercises Part II

1. Write three forms of the ratio of the number of letters in Gabriel Perboyre's last name to the number of letters in his first name.

2. Rick's grandmother has 3 dogs, 4 birds, and 7 cats. Write a ratio for each of the following comparisons:
   (a) dogs to cats   **(b)** cats to dogs
   (c) birds to cats   **(d)** birds to dogs

3. Before deciding which coat to buy, Joshua tried on 17 coats that fit and 8 that did not fit. Write the ratio of the number of coats that fit Joshua to the number of coats that he tried on.

4. What is the reduced ratio of 2 quarters to 8 dimes?

5. Jocelyn drove 220 miles in 4 hours. What was her average speed—that is the reduced ratio of the distance traveled to the time?

6. Thomas typed 260 words in 5 minutes. What was his typing rate—that is, the reduced ratio of words typed to time?

SECTION 17.2
PROPORTIONS

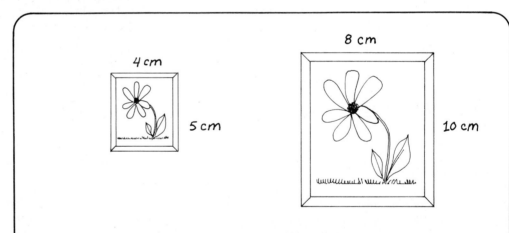

The same negative can be enlarged to many different sizes, but the subject of the picture will keep it's shape because the dimensions are kept proportional.

$$\text{width} \rightarrow \frac{4}{5} \qquad = \qquad \frac{8}{10} \leftarrow \text{length}$$

$$5 \times 8 \quad = \quad 4 \times 10$$
$$40 \quad = \quad 40$$

After studying this section you should be able to find the missing term to make the following a proportion:

$$\frac{3}{8} = \frac{?}{20}$$

Equal Ratios Because ratios can be expressed in fraction form, the rules you learned for working with fractions will be true for ratios also. Equal ratios make the same comparison or describe the same ratio.

In Section 6.6 you learned that two fractions are equal if their cross products are equal. This rule is true for ratios and will be extended in the next section of this chapter.

EXAMPLE 1:   Are $\frac{12}{30}$ and $\frac{14}{35}$ equal ratios?

$$\frac{12}{30} \overset{?}{=} \frac{14}{35} \longrightarrow \begin{array}{l} 30 \times 14 \\ 12 \times 35 \end{array}$$

$$30 \times 14 \overset{?}{=} 12 \times 35$$

$$420 = 420$$

Yes, $\frac{12}{30}$ and $\frac{14}{35}$ are equal ratios:

$$\frac{12}{30} = \frac{14}{35}$$

EXAMPLE 2:   There are 198 girls and 270 boys at Cross High School. There are 260 girls and 325 boys at Macedonia High School. Is the ratio of girls to boys at Cross equal to the ratio of girls to boys at Macedonia?

Cross High: $\dfrac{\text{girls}}{\text{boys}} = \dfrac{198}{270}$

Macedonia High: $\dfrac{\text{girls}}{\text{boys}} = \dfrac{260}{325}$

$$\frac{198}{270} \overset{?}{=} \frac{260}{325}$$

$$270 \times 260 \overset{?}{=} 198 \times 325$$

$$70,200 \neq 64,350$$

No, the ratio of girls to boys at Cross is not equal to the ratio of girls to boys at Macedonia:

$$\frac{198}{270} \neq \frac{260}{325}$$

Proportion In mathematics a pair of equal ratios has a special importance, and has therefore been given a special name: proportion. A **proportion** is the expressed equality of two ratios.

In simplest form a proportion is composed of whole numbers. The order in which the numbers are written is most important.

Means/Extremes   The **means** are the second and third terms in a proportion. The **extremes** are the first and fourth terms in a proportion.

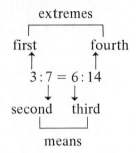

In fraction form, $3:7 = 6:14$ becomes

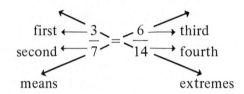

Hint: Remember the <u>*means*</u> as the <u>m</u>iddle two terms (second and third terms) and the *extremes* as the two most separated terms (first and fourth terms).

    The new vocabulary of proportions causes a rewording of the rule that two ratios are equal if their cross products are equal.

Proportion Rule

In any given proportion the product of the means is equal to the product of the extremes.

    EXAMPLE 3:   Do the ratios $\dfrac{8}{12}$ and $\dfrac{6}{9}$ form a proportion?

$$\frac{8}{12} \overset{?}{=} \frac{6}{9}$$

product of means $\longleftarrow 12 \times 6 \overset{?}{=} 8 \times 9 \longrightarrow$ product of extremes

$$72 = 72$$

Yes, the ratios form a proportion because the product of the means is equal to the product of the extremes; that is, $\dfrac{8}{12} = \dfrac{6}{9}$.

    EXAMPLE 4:   Find the missing term to make the following a proportion:

$$\frac{5}{4} = \frac{?}{6}$$

In order to help you find this answer in an organized fashion I shall introduce a few rules of algebra.

    First, I put an N in place of the missing term. This N is called a **variable** or an **unknown**. When I find a number value for N I shall have the answer to this example.

$$\frac{5}{4} = \frac{N}{6}$$

Next, I use the proportion rule:

$$\text{product of means} = \text{product of extremes}$$

$$4 \times N = 5 \times 6$$

$$4 \times N = 30$$

You have seen that in formulas the multiplication of a number by a variable written without any sign of operation. So from now on I shall write expressions such as $4 \times N$ as $4N$.

$$4N = 30$$

In algebra a rule of equality states that you can divide both sides of any equation by the same number (other than zero). How do I decide what number to use for such division? I divide both sides of the equation by the number multiplying the variable N so that only one N remains.

$$4N = 30 \qquad \text{Divide both sides of the equation by 4.}$$

$$\frac{\overset{1}{\cancel{4}N}}{\underset{1}{\cancel{4}}} = \frac{30}{4}$$

$$N = 7\frac{2}{4}, \quad \text{or} \quad 7\frac{1}{2}$$

Therefore, the missing term in the given proportion is $7\frac{1}{2}$.

EXAMPLE 5:   Find the missing term in the proportion $3 : ? = 12 : 20$. First, I replace the missing term with N:

$$3 : N = 12 : 20$$

Then I apply the proportion rule. When writing the product of the means, I shall write "N times 12" as "12 times N." Both multiplications have the same answer. I choose the form $12N$ because such a multiplication is generally written with the number before the variable.

$$12N = 3 \times 20$$

$$12N = 60$$

To find the value of one N (or N) I must divide both sides of the equation by 12, the number multiplying the variable.

$$\frac{\overset{1}{\cancel{12}N}}{\underset{1}{\cancel{12}}} = \frac{60}{12}$$

$$N = 5$$

Notice that the solution consists of five equations. You should always have five separate equations when you are finding the missing term in a given proportion.

EXAMPLE 6:   Find the first term of a proportion if the second term is 36, the third term is 5, and the fourth term is 9.

$$\frac{N}{36} = \frac{5}{9} \qquad \text{First equation: Write the proportion with } N \text{ in place of the missing term.}$$

$36 \times 5 = 9N$	Second equation: The product of the means equals the product of the extremes.
$180 = 9N$	Third equation: Perform the indicated multiplication of two numbers (here 36 and 5).
$\dfrac{180}{9} = \dfrac{\overset{1}{\cancel{9}}N}{\underset{1}{\cancel{9}}}$	Fourth equation: Divide both sides by the number that is multiplying N (here it is 9).
$20 = N$	Fifth equation: Perform the divisions and show the answer for N.

Now it is your turn. Find the answer to each example before looking at the solution.

Examples	Solutions
Are $\dfrac{6}{15}$ and $\dfrac{8}{20}$ equal ratios?	$\dfrac{6}{15} \overset{?}{=} \dfrac{8}{20}$ $15 \times 8 = 6 \times 20$ $120 = 120$ Yes!
Are $\dfrac{11}{17}$ and $\dfrac{9}{14}$ equal ratios?	$\dfrac{11}{17} \overset{?}{=} \dfrac{9}{14}$ $17 \times 9 = 11 \times 14$ $153 \neq 154$ No, because their cross products are not equal.
In the proportion $\dfrac{2}{3} = \dfrac{4}{6}$, name (a) the means; (b) the extremes.	(a) Means: 3 and 4. (b) Extremes: 2 and 6.
In the proportion $5:3 = 20:12$, name (a) the means; (b) the extremes.	(a) Means: 3 and 20. (b) Extremes: 5 and 12.
Find the value of N in each of the following proportions. (a) $\dfrac{3}{N} = \dfrac{2}{6}$	(a) $\dfrac{3}{N} = \dfrac{2}{6}$ Write the proportion. $2N = 3 \times 6$ Apply proportion rule. $2N = 18$ Multiply two numbers.

Examples	Solutions
	$\dfrac{\overset{1}{\cancel{2}N}}{\underset{1}{\cancel{2}}} = \dfrac{18}{2}$ Show division of both sides by 2.
	$N = 9$ Write answer to each division.
(b) $\dfrac{N}{1} = \dfrac{7}{3}$	(b) $\dfrac{N}{1} = \dfrac{7}{3}$ Proportion.
	$1 \times 7 = 3N$ Proportion rule.
	$7 = 3N$ Multiplication answer.
	$\dfrac{7}{3} = \dfrac{3N}{3}$ Division of both sides by 3.
	$2\dfrac{1}{3} = N$ Answers to divisions.
(c) $\dfrac{30}{5} = \dfrac{18}{N}$	(c) $\dfrac{30}{5} = \dfrac{18}{N}$
	$5 \times 18 = 30N$
	$90 = 30N$
	$\dfrac{90}{30} = \dfrac{30N}{30}$
	$3 = N$
(d) $\dfrac{N}{12} = \dfrac{3}{4}$	(d) $\dfrac{N}{12} = \dfrac{3}{4}$
	$12 \times 3 = 4N$
	$36 = 4N$
	$\dfrac{36}{4} = \dfrac{4N}{4}$
	$9 = N$

Now try these—no hints given.

Are $\dfrac{7}{28}$ and $\dfrac{5}{19}$ equal ratios?	No.
Is $\dfrac{5}{100} = \dfrac{3}{60}$ a proportion?	Yes.
Find the value of N in each of the following proportions.	
(a) $\dfrac{1}{9} = \dfrac{4}{N}$	(a) 36

Examples	Solutions
(b) $\dfrac{20}{N} = \dfrac{5}{4}$	(b) 16
(c) $\dfrac{N}{7} = \dfrac{2}{3}$	(c) $4\dfrac{2}{3}$

If you did not answer the last five questions correctly, please talk with an instructor before beginning the exercises that follow.

Exercises
Part I

Answers to these exercises can be found on page 432.

1. Which of the following are equal ratios?

   (a) $\dfrac{2}{3}$ and $\dfrac{3}{4}$ (b) $\dfrac{12}{18}$ and $\dfrac{10}{15}$

   (c) $\dfrac{4}{7}$ and $\dfrac{3}{5}$ (d) $\dfrac{5}{7}$ and $\dfrac{20}{28}$

2. Identify the means and the extremes in each proportion.

   (a) $4:5 = 12:15$ (b) $\dfrac{8}{20} = \dfrac{4}{10}$

3. Use the proportion rule to decide if the following are proportions.

   (a) $21:60 = 35:100$ (b) $\dfrac{11}{12} = \dfrac{13}{14}$

Find the missing value in each proportion.

4. $N:21 = 3:7$ 5. $\dfrac{2}{5} = \dfrac{N}{20}$ 6. $\dfrac{2}{N} = \dfrac{3}{8}$

7. $4:7 = 20:N$ 8. $\dfrac{6}{9} = \dfrac{26}{N}$ 9. $\dfrac{4}{5} = \dfrac{12}{N}$

10. $\dfrac{N}{8} = \dfrac{3}{5}$

Part II

1. Which of the following are equal ratios?

   (a) $\dfrac{4}{5}$ and $\dfrac{12}{15}$ (b) $\dfrac{4}{5}$ and $\dfrac{5}{6}$

2. Name the means in the proportion $\dfrac{4}{10} = \dfrac{6}{15}$.

3. Name the extremes in the proportion $\dfrac{4}{6} = \dfrac{10}{15}$.

4. Use the proportion rule to decide if the following are proportions.

   (a) $\dfrac{10}{11} = \dfrac{12}{13}$ (b) $27:60 = 18:40$

5. Find the missing value in each of the following proportions.

   (a) $N:3 = 12:18$ (b) $\dfrac{5}{7} = \dfrac{8}{N}$

   (c) $\dfrac{2}{N} = \dfrac{3}{8}$ (d) $4:8 = N:6$

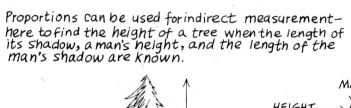

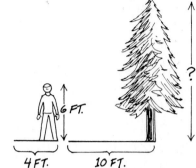

Mastering the steps required to solve a given proportion, as you did in the previous section, is only half the job of mastering the study of proportion. The important part is setting up the proportion. This section will deal with word problems that can be solved by setting up and solving a proportion.

EXAMPLE 1: Telita can type, on the average, 9 lines of handwritten notes in 1 minute. At this rate, about how long should it take her to type a 78-line, handwritten report?

This can be recognized as a proportion problem because two ratios are described and a proportion is the expressed equality of two ratios. Both rates described in the problem involve a number of handwritten lines and the number of typing minutes required.

First, write the labels for this comparison on the left and then show a proportion format to be filled in according to the labels.

$$\frac{\text{handwritten lines}}{\text{minutes of typing}} \qquad - = -$$

The problem expresses the rate at which Telita usually types. Write this rate in one of the open ratios that are part of the proportion format.

$$\frac{\text{handwritten lines}}{\text{minutes of typing}} \qquad \frac{9}{1} = -$$

Now use the information from the second sentence to fill in the second ratio. Be careful to put 78 in the numerator because it is the number of handwritten lines, and to use N in the denominator to represent the missing value: how many minutes it should take Telita to type those 78 lines.

$$\frac{\text{handwritten lines}}{\text{minutes of typing}} \qquad \frac{9}{1} = \frac{78}{N}$$

The problem is half finished!

You can now use the technique of solving a proportion that you learned in the second section of this chapter. Be sure to show all five steps of your solution.

$$\frac{9}{1} = \frac{78}{N}$$

$$1 \times 78 = 9N$$

$$78 = 9N$$

$$\frac{78}{9} = \frac{\cancel{9}^1 N}{\cancel{9}_1}$$

$$8\frac{6}{9} = N$$

$$\text{or} \quad 8\frac{2}{3}$$

Therefore, it should take Telita $8\frac{2}{3}$ minutes (or 8 minutes, 40 seconds) to type the 78-line report. Because people are not machines, the rate at which they can perform skilled jobs is subject to variation. That is why the example asked about how long it should take.

Although $8\frac{2}{3}$ is the exact value, and hence the only value, for N in the proportion, I would answer that it would probably take Telita about 9 minutes to type that report.

EXAMPLE 2:   My recipe for Creole Jambalaya calls for 2 cups of cubed leftover ham and serves 8. If I plan to serve 12, how much ham will I need?

Label the parts of the ratio described in the problem, then fill in the proportion format.

$$\frac{\text{cups of ham}}{\text{servings}} \qquad \underline{\quad} = \underline{\quad}$$

$$\frac{\text{cups of ham}}{\text{servings}} \qquad \frac{2}{8} = \underline{\quad}$$

From the recipe (information given).

$$\frac{\text{cups of ham}}{\text{servings}} \qquad \frac{2}{8} = \frac{N}{12}$$

For my particular plans (information needed).

Solve the resulting proportion.

$$\frac{2}{8} = \frac{N}{12}$$

$$8N = 2 \times 12$$

$$8N = 24$$

$$\frac{\cancel{8}^1 N}{\cancel{8}_1} = \frac{24}{8}$$

$$N = 3$$

Therefore, I need 3 cups of cubed ham to make enough Creole Jambalaya to serve 12 persons.

EXAMPLE 3: At a going-out-of-business sale all record albums in the store were priced at 4 albums for $9. At this rate, how much would 11 albums cost?

In setting up the ratio labels for each proportion I always use the two quantities in the order that they are first mentioned in the problem. So for this problem I would write $\frac{\text{albums}}{\text{dollars}}$. But you could reverse the order of the labels and write $\frac{\text{dollars}}{\text{albums}}$. The important step, as far as order is concerned, is writing both ratios of the proportion in the same order that you wrote the labels. For this problem I shall use the $\frac{\text{dollars}}{\text{albums}}$ labeling.

$\dfrac{\text{dollars}}{\text{albums}}$ $\dfrac{\quad}{\quad} = \dfrac{\quad}{\quad}$ Information given.

                               Information needed.

$\dfrac{\text{dollars}}{\text{albums}}$ $\dfrac{9}{4} = \dfrac{N}{11}$

$$4N = 9 \times 11$$

$$4N = 99$$

$$\frac{\overset{1}{\cancel{4}}N}{\underset{1}{\cancel{4}}} = \frac{99}{4}$$

$$N = 24\frac{3}{4}$$

If you cannot rewrite $24\frac{3}{4}$ dollars as $24.75 without doing the division, then divide 99 by 4 to two decimal places, because N represents money in this problem.

$$\frac{99}{4} = 4\overline{)99.00}^{\,24.75}$$

For each of the following problems fill in the terms of a proportion, then solve the resulting proportion.

Examples	Solutions
If a copy machine can produce 70 copies of an original in 1 minute, how long will it take to produce 245 copies of that original?	$\dfrac{\text{copies}}{\text{minutes}}$ $\dfrac{\quad}{\quad} = \dfrac{\quad}{\quad}$ Information given. Information needed.

Examples	Solutions
	$$\frac{70}{1} = \frac{245}{N}$$ $$1 \times 245 = 70N$$ $$245 = 70N$$ $$\frac{245}{70} = \frac{\overset{1}{\cancel{70}}N}{\underset{1}{\cancel{70}}}$$ $$3\frac{35}{70} = N$$ or $3\frac{1}{2}$ minutes
If a recipe calls for flour and sugar to be combined in the ratio of 3 : 2, how many cups of flour are needed when 3 cups of sugar are used?	$\dfrac{\text{flour}}{\text{sugar}}$ Given. $$— = —$$ Needed. $$\frac{3}{2} = \frac{N}{3}$$ $$2N = 3 \times 3$$ $$2N = 9$$ $$\frac{\overset{1}{\cancel{2}}N}{\underset{1}{\cancel{2}}} = \frac{9}{2}$$ $$N = 4\frac{1}{2} \text{ cups of flour}$$
At an ice-cream parlor 4 quarts of ice cream were used to make 50 sundaes. How many quarts are needed to make 175 sundaes?	$\dfrac{\text{quarts}}{\text{sundaes}}$ Given. $$— = —$$ Needed. $$\frac{4}{50} = \frac{N}{175}$$ $$50N = 4 \times 175$$ $$50N = 700$$ $$\frac{50N}{50} = \frac{700}{50}$$ $$N = 14 \text{ quarts}$$

Examples	Solutions
If a 5-foot-tall girl casts a 4-foot shadow, how tall is a signpost that casts a 14-foot shadow?	$\dfrac{\text{height}}{\text{shadow}}$ $\underset{\uparrow}{\underline{\quad}} = \underset{\uparrow}{\underline{\quad}}$ Girl. / Signpost. $$\frac{5}{4} = \frac{N}{14}$$ $$4N = 5 \times 14$$ $$4N = 70$$ $$\frac{4N}{4} = \frac{70}{4}$$ $$N = 17\frac{2}{4}$$ or $17\frac{1}{2}$ feet high

Now try these—no hints given.

If a tableful of books is sale priced at 4 books for $5.00, how much will 10 books cost?	$12.50
It took Malcolm 4 hours to read the first 180 pages of a novel. If he continues reading at the same rate, how long will it take him to read 405 pages?	9 hours
The ratio of a person's weight on the earth compared to his or her weight on the moon is 6:1. How much would a 144-pound person weigh on the moon?	24 pounds
If 40 pounds of fertilizer cover 4,000 square feet of lawn, how many square feet will 15 pounds cover?	1,500 square feet

If you were in any way unsure about your solutions to the last four problems, please work with an instructor before attempting the exercises that follow.

Exercises Part I

Answers to these exercises can be found on page 432.

1. Stan can type 240 words in 5 minutes. At this rate, how long will it take him to type a 1,680-word report?

2. If 20 pounds of grass seed will cover 800 square feet of soil, how many square feet will 45 pounds cover?

3. Sherry saves $20 each month. How long will it take her to save $170 if she continues saving at this rate?

4. An investment of $2,550 earned $306 in one year. How much would have to be invested at the same rate in order to earn $648?

5. Eric scored 110 points in 6 basketball games. If he continues to score at this rate, how many points can he be expected to score in 21 games?

6. If Keith's shadow is 3 feet long when that of a 13-foot signpost is 7 feet long, how tall is Keith?

7. If 3 cans of potato sticks sell for $1.26, how many cans can you buy for $6.30?

8. The ratio of a person's weight on Mars compared to his or her weight on Earth is 2:5. How much would Pedro weigh on Mars if his weight on earth is 175 pounds?

9. A recipe calls for 2 cups of milk and 5 cups of flour. How many cups of milk are needed when 20 cups of flour are used?

10. It took Takisha 2 hours to paint a 300-square-foot area of a barn. Working at this rate, how long will it take her to paint 1,000 square feet?

Part II

1. A 6-foot-tall man casts a 4-foot shadow. How tall is a tree that casts an 18-foot shadow?

2. The ratio of a person's weight on the earth compared to his or her weight on the moon is 6:1. How much would a 132-pound person weigh on the moon?

3. A baseball team won 5 of its first 9 games. How many games would you expect them to win of their first 36 games if they continue to play with the same degree of success?

4. A recipe calls for 2 cups of sugar and 3 cups of flour. How many cups of flour are needed when 8 cups of sugar are used?

5. Gregory uses about 3 gallons of paint to do 2 rooms. How many gallons will he need to paint 8 rooms?

REVIEW MATERIAL

**Review Skill
Ratios
Containing
Fractions**

Simplify a ratio containing one or two fractions by using the division rule and reducing.

For example, simplify $\dfrac{7\frac{1}{2}}{1\frac{2}{3}}$.

$$\frac{7\frac{1}{2}}{1\frac{2}{3}} = 7\frac{1}{2} \div 1\frac{2}{3}$$

$$= \frac{15}{2} \div \frac{5}{3}$$

$$= \frac{15}{2} \times \frac{3}{5}$$

$$= \frac{9}{2} \quad \text{or} \quad 9:2$$

Exercise A Answers to this exercise can be found on page 432.

Write each of the following as a ratio in lowest terms.

1. $\dfrac{2\frac{1}{2}}{5}$

2. $\dfrac{1}{2}$ to 2

3. $\dfrac{3}{4}$ to $\dfrac{5}{8}$

4. $\dfrac{3\frac{1}{3}}{10}$

5. $4 : \dfrac{3}{8}$

6. $\dfrac{2}{3} : \dfrac{4}{15}$

7. $2\dfrac{2}{3}$ to $1\dfrac{1}{15}$

8. $\dfrac{2\frac{1}{2}}{3\frac{1}{3}}$

9. $\dfrac{\frac{3}{5}}{1\frac{4}{5}}$

10. $1\dfrac{3}{4}$ to $3\dfrac{1}{2}$

Exercise B No answers are given.

Write each of the following as a ratio in lowest terms.

1. $\dfrac{6}{2\frac{2}{5}}$

2. $\dfrac{1}{3}$ to 2

3. $22 : 2\dfrac{1}{5}$

4. $\dfrac{\frac{3}{4}}{\frac{9}{8}}$

5. $3\dfrac{1}{4}$ to $2\dfrac{7}{16}$

Review Skill
Ratios
Containing
Decimals

Simplify a ratio containing one or two decimals by moving each decimal point the same number of places to the right in order to make the numerator and denominator whole numbers. You may have to annex zeros to one of the numbers.

EXAMPLE 1:   Simplify $\dfrac{1.01}{1.02}$.

Move each decimal *two* places to the right.

$$\dfrac{1.01}{1.02} = \dfrac{101}{102}$$

EXAMPLE 2:   Simplify $\dfrac{1.01}{10.2}$.

Move each decimal *two* places to the right.

$$\dfrac{1.01}{10.20} = \dfrac{101}{1020}$$

EXAMPLE 3:   Simplify $\dfrac{5}{7.5}$.

Move each decimal *one* place to the right.

$$\frac{5.0}{7.5} = \frac{50}{75}$$

Reduce to lowest terms.

$$\frac{50}{75} = \frac{10}{15} = \frac{2}{3}$$

Exercise C Answers to this exercise can be found on page 432.

Write each of the following as a ratio in lowest terms.

1. $\dfrac{.7}{.9}$ 2. $\dfrac{1.7}{2}$ 3. $\dfrac{.05}{.17}$

4. $\dfrac{3}{1.1}$ 5. $\dfrac{.007}{.008}$ 6. $\dfrac{1.2}{2.4}$

7. $\dfrac{.12}{2.4}$ 8. $\dfrac{.006}{.009}$ 9. $\dfrac{.17}{3.4}$

10. $\dfrac{3}{2.4}$ 11. $\dfrac{2.5}{4.5}$ 12. $\dfrac{1.5}{12}$

13. $\dfrac{.08}{.32}$ 14. $\dfrac{.8}{.32}$ 15. $\dfrac{.007}{.21}$

Exercise D No answers are given.

Write each of the following as a ratio in lowest terms.

1. $\dfrac{.4}{.5}$ 2. $\dfrac{2.9}{3}$ 3. $\dfrac{.09}{.11}$

4. $\dfrac{.04}{3.2}$ 5. $\dfrac{.4}{3.2}$ 6. $\dfrac{.009}{.018}$

7. $\dfrac{1.2}{2}$ 8. $\dfrac{6}{1.5}$

Review Skill
Ratios
Comparing Like
Quantities

Whenever a ratio compares like quantities, you must change these quantities to the same unit of measure *before* reducing.

For example, write a simplified ratio comparing 30 hours to 2 days.

Hours and days are like quantities because they are both measures of time. Express each in the same unit of time.

1 day = 24 hours

2 days = 48 hours

Therefore,

30 hours to 2 days

becomes   30 hours : 48 hours

$$\frac{30}{48} = \frac{5}{8} \qquad \text{Reduced.}$$

Answers to this exercise can be found on page 432.

Write the simplified ratio for each of the following.

1. 1 nickel to 2 dimes
2. 14 doughnuts to 2 dozen doughnuts
3. 5 days to 1 week
4. 16 pennies to 4 nickels
5. 40 hours to 2 days
6. 15 minutes to 1 hour
7. 2 dimes to 2 quarters
8. 1 year to 14 months
9. 2 weeks to 14 days
10. 6 nickels to 1 quarter
11. 6 $5 bills to 4 $10 bills
12. 18 months to 3 years
13. 3 quarters to 8 dimes
14. 8 $1 bills to 2 $20 bills
15. 1 dozen eggs to 15 eggs
16. 12 nickels to 6 dimes
17. 2 minutes to 2 hours
18. 70 years to 1 century
19. 4 dimes to 13 nickels
20. 30 seconds to 1 minute

Exercise F No answers are given.

Write the simplest ratio for each of the following.

1. 11 days to 2 weeks
2. 7 nickels to 7 dimes
3. 4 $5 bills to 3 $20 bills
4. 2 years to 30 months
5. 20 pennies to 1 quarter
6. 3 quarters to 9 dimes
7. 2 dozen doughnuts to 16 doughnuts
8. 8 nickels to 2 quarters
9. 9 individual cans to 2 six-packs
10. 50 years to 1 century

**Review Skill
Ratios Requiring
One Calculation**

Type 1   Parts are given and *addition* is necessary to determine *how many in all*.

EXAMPLE 1:   A college football team won 7, lost 3, and played to no ties. Write the ratio of the team's wins to the number of games played that season.

Calculation required:

$$\begin{array}{r} 7 \quad \text{wins} \\ + \ 3 \quad \text{losses} \\ \hline 10 \quad \text{games played} \end{array}$$

Comparison: wins to games played

Ratio: 7 : 10

Type 2   Total and one part are known so *subtraction* is required to find the *remaining part*.

EXAMPLE 2:   In a season of 10 games a college football team lost only 1 game and tied none. Write the ratio of the team's wins to its losses.

Calculation required:

$$\begin{array}{r} 10 \quad \text{games played} \\ - \ 1 \quad \text{loss} \\ \hline 9 \quad \text{wins} \end{array}$$

Comparison: wins to losses

Ratio: 9 : 1

Exercise G   Answers to this exercise can be found on page 432.

Solve each of the following by performing the required calculation before writing the ratio in lowest terms.

1.   Randy planted 4 apple trees and 3 pear trees. Write the ratio comparing the number of pear trees to the total number of fruit trees he planted.
2.   Edna used only gold ornaments and red ornaments on her Christmas tree. If she has 72 ornaments and 24 of those are gold, write a reduced ratio comparing Edna's gold to red ornaments.
3.   A college basketball team won 23 out of 30 games played. There were no tie games. Write the ratio of losses to wins.
4.   For a day when 6 students are absent and 22 present, write the reduced ratio of those in attendance to the total number in the class.
5.   Ted's music appreciation class consists of 12 females and 19 males. Write the ratio of males to the total number in the class.
6.   Mr. Johnston received 19 applications for a clerical vacancy. Three applicants were not qualified for the position. Write a ratio comparing the number of qualified applicants to the total number of applicants.
7.   There were 49 children riding a school bus and there were 3 vacant places. Write a ratio comparing the number of passengers to the total capacity of the bus.
8.   Sheldon answered 42 questions correctly and 8 questions incorrectly on a test. Write and reduce the ratio of his correct answers to the number of questions on the test.

Exercise H   No answers are given.

Write the ratio described in each of the following.

1.   Bernard uses only silver ornaments and blue ornaments to decorate his Christmas tree. If he has 38 silver and 42 blue ornaments, write the reduced ratio of his silver ornaments to his entire collection of ornaments.
2.   A college basketball team won 26, lost 4, and tied 1 game. Write the ratio of wins to games played that season.
3.   Mildred's math class consists of 29 students. When 5 are absent what is the ratio of students present to class enrollment?
4.   Ira has 7 female and 9 male cousins. Write the ratio comparing the number of his female cousins to the total number of Ira's cousins.

**Review Skill
Solving
a Proportion**

The five-equation approach to solving a proportion is outlined below in finding the missing value in the proportion $6:N = 5:7$.

$$\frac{6}{N} = \frac{5}{7}$$

First equation: Write the proportion so that ratios are written as fractions.

$$5N = 6 \times 7$$

Second equation: Show that the product of the means equals the product of the extremes.

$$5N = 42$$

Third equation: Simplify the multiplication.

$$\frac{\overset{1}{\cancel{5}}N}{\underset{1}{\cancel{5}}} = \frac{42}{5}$$

Fourth equation: Divide both sides by the number multiplying N (here it is 5).

$$N = 8\frac{2}{5}$$

Fifth equation: Perform the division and show the value for N.

Exercise I Answers to this exercise can be found on page 432.

Find the missing value in each of the following proportions.

1. $4:N = 6:8$

2. $\dfrac{N}{7} = \dfrac{8}{14}$

3. $\dfrac{3}{5} = \dfrac{N}{20}$

4. $\dfrac{N}{12} = \dfrac{3}{4}$

5. $1:N = 3:7$

6. $5:3 = N:18$

7. $\dfrac{7}{2} = \dfrac{4}{N}$

8. $N:20 = 4:5$

9. $\dfrac{N}{7} = \dfrac{2}{3}$

10. $2:9 = 4:N$

11. $\dfrac{N}{8} = \dfrac{2}{7}$

12. $4:N = 6:15$

Exercise J No answers are given.

Solve each of the following proportions.

1. $\dfrac{5}{7} = \dfrac{20}{N}$

2. $7:6 = N:3$

3. $9:11 = 5:N$

4. $7:10 = N:20$

5. $\dfrac{N}{2} = \dfrac{7}{8}$

6. $8:N = 3:2$

**Review Skill
Setting Up
a Proportion**

The procedure for setting up a proportion in order to solve a problem is illustrated below.

If a copying machine can produce 70 copies of an original in one minute, how long will it take to produce 175 copies of that original?

Information given:	70 copies in 1 minute
First ratio:	70 : 1
Information needed:	175 copies in N minutes
Second ratio:	175 : N

The second ratio must be written in the same order that the first ratio was written.

$$\frac{70}{1} = \frac{175}{N}$$
$$1 \times 175 = 70N$$
$$175 = 70N$$
$$\frac{175}{70} = \frac{70N}{70}$$

$$2\frac{35}{70} = N$$

$$\text{or} \quad 2\frac{1}{2} = N$$

Therefore, the machine will take $2\frac{1}{2}$ minutes to produce 175 copies.

Exercise K   Answers to this exercise can be found on page 432.

Set up and solve the proportions described in the following problems.

1. Bette makes leather belts as a hobby. She finds that she can make 4 belts in 6 hours. How many belts should she expect to make in 15 hours?
2. On a certain map the scale is 3 inches to 50 miles. The route Dennis plans to travel measures 9 inches on the map. How many miles does that represent?
3. If 6 jawbreakers cost 45¢, how much do 10 jawbreakers cost?
4. Eunice drove 8 miles in 10 minutes. At that rate how far should she be able to travel in 45 minutes?
5. One factory inspector can check 12 machine parts in 8 minutes. At that rate how many parts can the inspector check in 36 minutes?
6. It takes Arnie 2 hours to ride his bike 9 miles. If he can keep up the pace, how far can he ride in 5 hours?
7. How tall is a pole that casts a 9-foot shadow when a 4-foot-tall child casts a 3-foot shadow?
8. If 2 shirts cost $27, how many of this type shirt can be bought for $108?
9. How many pounds of ham are needed for 54 servings if an 8-pound shankless ham contains 36 servings?
10. Fruit juice and ginger ale can be mixed in a 2:3 ratio to make punch. How much ginger ale would be used with 5 quarts of fruit juice?

Exercise L   No answers are given.

Set up and solve the proportions described in the following problems.

1. If 6 pieces of bubble gum cost 44¢, how much do 9 pieces of bubble gum cost?
2. It takes Glenna 2 hours to ride her bike 9 miles. If she keeps up this pace, how far can she ride in 3 hours?
3. If 2 shirts cost $23, how many of this type shirt can be bought for $92?
4. Tamara drove 12 miles in 10 minutes. At that rate, how long should it take her to travel 54 miles?
5. One factory inspector can check 12 machine parts in 8 minutes. How long should it take him to check 39 parts?

REVIEW EXERCISES

Refer to the section listed if you have difficulty. Answers to these exercises can be found on page 433.

Section 17.1

1. Write three forms of the ratio of Catherine's age to Stephen's age if Stephen is 9 and Catherine is 7 years old.

2. Henry planted 20 corn, 15 collard, and 12 tomato plants. Write each ratio as a fraction in lowest terms.
   (a) corn to collard plants
   (b) tomato to corn plants
   (c) corn to tomato plants
   (d) tomato to collard plants

3. Eight of the Cainhoy cheerleaders came to practice on Thursday and two were absent. Write the reduced ratio comparing the number of absent cheerleaders to the total number on the Cainhoy squad.

4. Thirty-six of the 88 keys on a standard piano keyboard are black. Write the reduced ratio of the number of black keys to the number of white keys.

5. Simplify each of the following ratios.

   (a) $\dfrac{5\frac{1}{2}}{3\frac{2}{3}}$

   (b) $\dfrac{1.2}{.04}$

Write, in lowest terms, the ratio that compares each of the following.

6. 5 dimes to 2 quarters

7. 2 hours to 100 minutes

8. 6 eggs to one dozen eggs

9. On a trip to Corpus Christi, Anne traveled 396 miles on 12 gallons of gasoline. Write a reduced ratio indicating her miles per gallon for that trip.

10. Harvey scored 14 hits in 35 times at bat. Write a reduced ratio to represent his hits to his total times at bat.

Section 17.2

11. Are $\dfrac{4}{10}$ and $\dfrac{6}{15}$ equal ratios?

12. Do the ratios $\dfrac{8}{12}$ and $\dfrac{10}{15}$ form a proportion?

13. Name the means in the proportion $2:7 = 6:21$.

14. Name the extremes in the proportion $\dfrac{3}{5} = \dfrac{6}{10}$.

15. Find the missing term in each of the following proportions.

   (a) $\dfrac{N}{5} = \dfrac{9}{15}$

   (b) $\dfrac{2}{N} = \dfrac{7}{9}$

   (c) $3:10 = N:20$

   (d) $\dfrac{5}{8} = \dfrac{6}{N}$

Section 17.3

16. Jeannie can decorate 5 cookies in 2 minutes. How long should it take her to decorate 70 cookies?

17. On a certain map the scale is $\dfrac{1}{2}$ inch to 10 miles. If Jaye's route measures 5 inches on that map, how far should she plan to drive?

18. How tall is a pole that casts a 6.3-foot shadow when a 5-foot-tall woman casts a 3.5-foot shadow?

19. If 5 pairs of socks cost $14, how many pairs can be purchased for $42?

20. My favorite recipe for chocolate chip cookies calls for sugar and flour in the ratio of 3:4. If I use 6 cups of flour for an extra large batch, how much sugar should I use?

CHAPTER POST-TEST

This test should determine if you have mastered, the topics in Chapter 17. Answers to this test can be found on page 433.

1. T.E. planted 18 corn, 12 okra, and 8 tomato plants. Write each ratio as a fraction in lowest terms.
   (a) tomato to okra plants   (b) corn to okra plants
   (c) tomato to corn plants   (d) corn to tomato plants

2. There are 52 white and 36 black keys on a standard piano keyboard. Write the reduced ratio of the number of white keys to the total number of keys on a piano.

3. Simplify the ratio $\dfrac{2\frac{1}{2}}{7}$.

4. Write, in lowest terms, the ratio that compares each of the following.
   (a) 12 days to 2 weeks   (b) 6 nickels to 2 quarters

5. Sissy types 189 words in 3 minutes. Write a reduced ratio to show her typing rate in words per minute.

6. Find the missing term in each of the following proportions.
   (a) $\dfrac{8}{N} = \dfrac{4}{26}$ (b) $7:4 = 6:N$

7. If 3 pencils cost 45¢, how much do 5 pencils cost?

8. Ginger ale and fruit juice can be mixed in a 3:2 ratio to make punch. How much ginger ale would be used with 15 pints of fruit juice?

18
percent

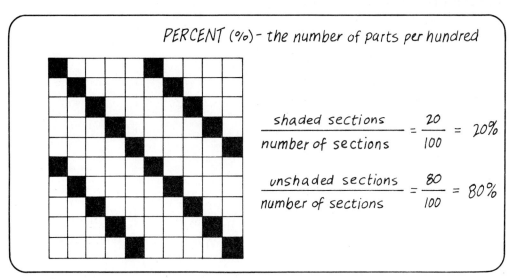

PERCENT (%) – the number of parts per hundred

$$\frac{\text{shaded sections}}{\text{number of sections}} = \frac{20}{100} = 20\%$$

$$\frac{\text{unshaded sections}}{\text{number of sections}} = \frac{80}{100} = 80\%$$

Definition   This section introduces the meaning of percent and then discusses how ratios and decimals can be written as percents.

A **percent** is a ratio whose second term is 100. It is a comparison *by one hundred*, a rate *per hundred*.

EXAMPLE 1:   Write a ratio to describe the shaded portion of the hundreds square shown.

$$\frac{17}{100}$$ Number of shaded parts.
Total number of equal parts.

401

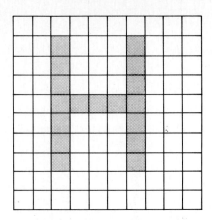

This same comparison of $\dfrac{17}{100}$ can be written as 17%. The percent symbol, %, means *by the hundred*, or *per hundred*.

Notice that $\dfrac{17}{100} = 17\%$. Each expression represents *seventeen per hundred*.

In ratio form a percent must always have the term 100 expressed in the denominator. In % form, only the first term of the ratio is written and the % symbol is written in place of the 100.

Examples	Solutions
Rewrite the ratio form of each percent using a % symbol.	
(a) $\dfrac{95}{100}$	(a) 95%
(b) $\dfrac{3}{100}$	(b) 3%
(c) $\dfrac{140}{100}$	(c) 140%
(d) $\dfrac{33\frac{1}{3}}{100}$	(d) $33\frac{1}{3}\%$
(e) $\dfrac{50}{100}$	(e) 50%
Rewrite each of the following as a percent in ratio form.	
(a) 23%	(a) $\dfrac{23}{100}$
(b) 200%	(b) $\dfrac{200}{100}$
(c) $12\frac{1}{2}\%$	(c) $\dfrac{12\frac{1}{2}}{100}$

Examples	Solutions
(d) 9%	(d) $\dfrac{9}{100}$
(e) 75%	(e) $\dfrac{75}{100}$

The previous examples were easy because the ratios already had 100 as their second term. The following discussion deals with the procedure for writing any ratio as a percent. In order to do this I shall refer to an illustration using the hundreds square.

Step 1: I divided the square into four equal parts and shaded one of them to represent the ratio $\dfrac{1}{4}$.

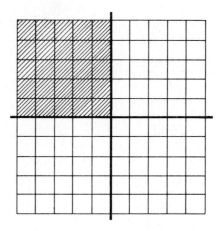

Step 2: I shaded 25 of the 100 small squares that make up the hundreds square in order to represent the ratio $\dfrac{25}{100}$.

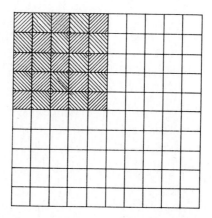

The shaded portion in Step 1 appears to be the same as the shaded portion in Step 2. I can check this by using the proportion rule: If the product of the means is equal to the product of the extremes, then the ratios representing the shaded portions are equal.

$$\frac{1}{4} \overset{?}{=} \frac{25}{100}$$

$$4 \times 25 \overset{?}{=} 1 \times 100$$

$$100 = 100$$

Yes, $\frac{1}{4}$ and $\frac{25}{100}$ are just different names for the same ratio. When I need to express $\frac{1}{4}$ as a percent it is helpful to write the ratio as $\frac{25}{100}$ and then rewrite $\frac{25}{100}$ as 25%.

$$\frac{1}{4} = \frac{25}{100} = 25\%$$

Changing Fractions to Percents

EXAMPLE 2:   Express $\frac{3}{5}$ as a percent.

First, write an equal ratio for $\frac{3}{5}$ that has 100 as the denominator, either by the method of building fractions (Section 6.4) or by solving a proportion (Section 17.1).

Building-fraction method:

$$\frac{3}{5} = \frac{?}{100}$$

Because $100 \div 5 = 20$, I shall use 20 to build.

$$\frac{3}{5} = \frac{3 \times 20}{5 \times 20}$$

$$= \frac{60}{100} = 60\%$$

Proportion-solving method:

$$\frac{3}{5} = \frac{N}{100}$$

$$5N = 3 \times 100$$

$$5N = 300$$

$$\frac{5N}{5} = \frac{300}{5}$$

$$N = 60$$

Hence, $\frac{60}{100} = 60\%$.

By either method, $\frac{3}{5} = \frac{60}{100}$. Using the percent symbol, I write $\frac{3}{5} = 60\%$.

EXAMPLE 3:   Write $\frac{5}{8}$ as a percent.

I shall use the proportion-solving method to change $\frac{5}{8}$ to a ratio having 100 as its second term.

Do you know why I chose not to use the building-fraction method? That is right, $100 \div 8$ does not come out to a whole number, and that makes the building-fraction method too complicated. So I shall use the proportion-solving method for ease of operation.

$$\frac{5}{8} = \frac{N}{100}$$

$$8N = 5 \times 100$$

$$8N = 500$$

$$\frac{8N}{8} = \frac{500}{8}$$

$$N = 62\frac{4}{8}, \quad \text{or} \quad 62\frac{1}{2}$$

Hence, $\dfrac{5}{8} = \dfrac{62\frac{1}{2}}{100}$, or $62\frac{1}{2}\%$.

EXAMPLE 4:   Write 2 as a percent.

First, I write 2 as a ratio, filling in the understood denominator: $\dfrac{2}{1}$.

Then I use the building-fraction method to express $\dfrac{2}{1}$ as an equal ratio having 100 as denominator.

$$\frac{2}{1} = \frac{?}{100}$$

Because $100 \div 1 = 100$, I shall use 100 to build.

$$\frac{2}{1} = \frac{2 \times 100}{1 \times 100}$$

$$= \frac{200}{100}$$

Hence, $\dfrac{2}{1} = \dfrac{200}{100}$, or 200%.

Changing Percents to Fractions

Now, as is so frequently done in mathematics, let me reverse the procedure.

EXAMPLE 5:   Write 85% as a ratio in lowest terms.

$$85\% = \frac{85}{100} \qquad \text{Reduce by 5.}$$

$$= \frac{17}{100}$$

Hence, in lowest terms, $85\% = \dfrac{17}{20}$.

EXAMPLE 6:   Write 300% as a fraction in lowest terms.

$$300\% = \frac{300}{100} \qquad \text{Reduce by 100.}$$

$$= \frac{3}{1}, \quad \text{or} \quad 3$$

Hence, in lowest terms, $300\% = 3$.

EXAMPLE 7:   Write $33\frac{1}{3}\%$ as a ratio in lowest terms.

$$33\frac{1}{3}\% = \frac{33\frac{1}{3}}{100}$$

This obviously cannot be reduced by dividing both terms of the fraction by a common factor.

$$\frac{33\frac{1}{3}}{100} = \frac{\frac{100}{3}}{100} = \frac{100}{3} \div 100$$

$$= \frac{100}{3} \div \frac{100}{1}$$

$$= \frac{\overset{1}{\cancel{100}}}{3} \times \frac{1}{\underset{1}{\cancel{100}}}$$

$$= \frac{1}{3}$$

Hence, $33\frac{1}{3}\% = \frac{1}{3}$.

EXAMPLE 8:   Write $\frac{1}{2}\%$ as a ratio in lowest terms.

Be careful—this expression is smaller than 1%!

$$\frac{1}{2}\% = \frac{\frac{1}{2}}{100} = \frac{1}{2} \div 100$$

$$= \frac{1}{2} \div \frac{100}{1}$$

$$= \frac{1}{2} \times \frac{1}{100}$$

$$= \frac{1}{200}$$

Hence, $\frac{1}{2}\% = \frac{1}{200}$.

Examples	Solutions
Express each ratio as a percent. Write each answer in % notation.	
(a) $\dfrac{7}{10}$	(a) $\dfrac{7}{10} = \dfrac{7 \times 10}{10 \times 10} = \dfrac{70}{100} = 70\%$
(b) $\dfrac{4}{5}$	(b) $\dfrac{4}{5} = \dfrac{80}{100} = 80\%$
(c) 4	(c) $4 = \dfrac{4}{1} = \dfrac{400}{100} = 400\%$
(d) $\dfrac{13}{20}$	(d) $\dfrac{13}{20} = \dfrac{65}{100} = 65\%$
(e) $\dfrac{4}{9}$	(e) Because $4 \div 9$ does not come out to a whole-number answer, use the proportion-solving method.

$$\frac{4}{9} = \frac{N}{100}$$

$$9N = 4 \times 100$$

$$9N = 400$$

$$\frac{9N}{9} = \frac{400}{9}$$

$$N = 44\frac{4}{9}$$

Hence, $\dfrac{4}{9} = 44\dfrac{4}{9}\%.$

| (f) $\dfrac{3}{4}$ | (f) $\dfrac{3}{4} = \dfrac{75}{100} = 75\%.$ |

Write each percent as a ratio in lowest terms.	
(a) 50%	(a) $50\% = \dfrac{50}{100} = \dfrac{1}{2}$
(b) $\dfrac{2}{3}\%$	(b) $\dfrac{2}{3}\% = \dfrac{\frac{2}{3}}{100} = \dfrac{2}{3} \div 100$

$$= \frac{2}{3} \div \frac{100}{1}$$

$$= \frac{2}{3} \times \frac{1}{\underset{50}{\cancel{100}}}$$

$$= \frac{1}{150}$$

Examples	Solutions
(c) 122%	(c) $122\% = \dfrac{122}{100} = \dfrac{61}{50}$
(d) 20%	(d) $20\% = \dfrac{20}{100} = \dfrac{1}{5}$
(e) 37%	(e) $37\% = \dfrac{37}{100}$
(f) $12\frac{1}{2}\%$	(f) $12\frac{1}{2}\% = \dfrac{12\frac{1}{2}}{100} = 12\frac{1}{2} \div 100$
	$= \dfrac{25}{2} \div \dfrac{100}{1}$
	$= \dfrac{\overset{1}{\cancel{25}}}{2} \times \dfrac{1}{\underset{4}{\cancel{100}}}$
	$= \dfrac{1}{8}$

Now try these—no hints given.

Write $\dfrac{2}{5}$ as a percent using % notation.	40%
Write 65% as a ratio in lowest terms.	$\dfrac{13}{20}$
Write $\dfrac{6}{7}$ as a percent using % notation.	$85\dfrac{5}{7}\%$
Write $5\frac{1}{4}\%$ as a ratio in lowest terms.	$\dfrac{21}{400}$

Up to this point you have encountered only whole numbers, mixed numbers, and fractions in your study of percent. The following problems will involve decimals.

Some of the examples you just finished involved changing percents to fractions, such as $47\% = \dfrac{47}{100}$.

But $\dfrac{47}{100}$ can also be written as .47; therefore,

$$47\% = \dfrac{47}{100} = .47$$

Changing Percents to Decimals Because percent is defined as a ratio whose second term is 100, 47% can be changed to .47 by direct application of the rule for division by powers of ten (from Section 15.2): To divide by 100, move the decimal *two places to the left*.

> To change a percent to a decimal, move the decimal *two places to the left* and drop the percent symbol.
>
> $$45\% = .45$$
>
> $$190\% = 1.90$$
>
> $$9.2\% = .092$$

EXAMPLE 9:   Change 14% to an equivalent decimal.

When a decimal is not shown in a given number it is understood to belong to the right of the rightmost digit.

Thus, $14\% = 14.\%$

and $14.\% = .14$

move decimal | drop the
two places to left | percent symbol

Answer: $14\% = .14$

Examples	Solutions
Change each percent to an equivalent decimal.	
(a) 64%	(a) $64\% = 64\% = .64$
(b) 12.5%	(b) $12.5\% = 12.5\% = .125$
(c) 2%	(c) $2\% = 2\% = .02$
(d) .8%	(d) $.8\% = .8\% = .008$
(e) 137%	(e) $137\% = 137\% = 1.37$
(f) 95%	(f) $95\% = 95\% = .95$
(g) .25%	(g) $.25\% = 00.25\% = .0025$
(h) 62%	(h) $62\% = 62\% = .62$

Changing Decimals to Percents

If you want to convert the decimal form of a number to % notation, reverse the procedure you just practiced.

$$.47 = \frac{47}{100} = 47\%$$

Note that the decimal is moved *two places to the right* and the % symbol is placed directly to the right of the number.

> To change a decimal to a percent, move the decimal *two places to the right* and annex the percent symbol.
>
> $$.65 = 65\%$$
>
> $$1.25 = 125\%$$
>
> $$.015 = 1.5\%$$

EXAMPLE 10:   Rewrite 9.5 as a percent in % notation.

Move the decimal two places to the right and annex the % symbol

$$9.5 = 9.50\% = 950.\%$$

However, most whole numbers are written without the decimal point. Therefore, 950.% = 950%. Therefore, 9.5 = 950%.

Examples	Solutions
Rewrite each decimal as an equivalent percent in % notation.	
(a) .77	(a) .77 = 77%
(b) .5	(b) .5 = 50%
(c) 1.9	(c) 1.9 = 190%
(d) .085	(d) 0.085 = 8.5%
(e) 8	(e) 8 = 800.%, or 800%
(f) .99	(f) .99 = 99%
(g) 25	(g) 25 = 2,500.% or 2,500%
(h) .30	(h) .30 = 30%

Because you cannot perform any calculation with an expression that includes the percent symbol, 98%, for example, must be rewritten as $\dfrac{98}{100}$ or .98 whenever you need to use it operationally.

You should feel comfortable with the following procedures that change the form, but not the value, of a percent expression.

Ratio Form to % Notation

1. If the denominator is 100, drop the 100 and annex the % symbol.

EXAMPLE 11:

$$\frac{3}{100} = 3\%$$

$$\frac{15\frac{2}{3}}{100} = 15\frac{2}{3}\%$$

2. If the denominator is other than 100, but will divide 100 evenly, change the given fraction to an equivalent fraction with a denominator of 100, then drop the 100 and annex the % symbol to the new numerator.

EXAMPLE 12:

$$\frac{8}{5} = \frac{160}{100} = 160\%$$

$$\frac{7}{10} = \frac{70}{100} = 70\%$$

If the denominator is other than 100, and will not divide 100 evenly, use the proportion-solving method to change the given fraction to a ratio with 100 as its second term.

EXAMPLE 13:

$$\frac{5}{6} = \frac{N}{100}$$

$$6N = 5 \times 100$$

$$6N = 500$$

$$\frac{6N}{6} = \frac{500}{6}$$

$$N = 83\frac{1}{3}$$

Hence, $\dfrac{5}{6} = \dfrac{83\frac{1}{3}}{100}$.

Drop the 100 denominator and annex the % symbol.

$$\frac{5}{6} = \frac{83\frac{1}{3}}{100} = 83\frac{1}{3}\%$$

% Notation to Ratio Form to Lowest Terms Drop the % symbol, indicate 100 as denominator, and simplify the resulting fraction. For example,

EXAMPLE 14:

$$33\% = \frac{33}{100}$$

$$9\% = \frac{9}{100}$$

$$2\frac{1}{2}\% = \frac{2\frac{1}{2}}{100} = 2\frac{1}{2} \div 100$$

$$= \frac{5}{2} \div \frac{100}{1}$$

$$= \frac{\overset{1}{\cancel{5}}}{2} \times \frac{1}{\underset{20}{\cancel{100}}}$$

$$= \frac{1}{40}$$

EXAMPLE 15:

$$104\% = \frac{104}{100} = \frac{52}{50} = \frac{26}{25}$$

$$6.5\% = \frac{6.5}{100}$$

$$= \frac{6.5}{100} \times \frac{10}{10} = \frac{65}{1,000}$$

$$= \frac{65}{1,000} = \frac{13}{200}$$

% Notation to Decimal Form

Drop the % symbol and indicate 100 as denominator by moving the decimal point *two places* to the *left*.

EXAMPLE 16:

$$25\% = .25$$

$$12.5\% = .125$$

$$.02\% = .0002$$

$$4\% = .04$$

Decimal to % Notation

Move the decimal point *two places* to the *right* and annex the % symbol. For example,

EXAMPLE 17:

$$.98 = 98\%$$

$$.005 = .5\%$$

$$7 = 700\%$$

$$9.3 = 930\%$$

Exercises
Part I

Answers to these exercises can be found on page 433.

1. Rewrite the ratio form of each percent using a % symbol.

   (a) $\dfrac{73}{100}$ (b) $\dfrac{7}{100}$ (c) $\dfrac{129}{100}$ (d) $\dfrac{12\frac{1}{2}}{100}$

2. Rewrite each of the following as a percent in ratio form. Do *not* reduce.

   (a) 59% (b) $8\frac{1}{2}\%$ (c) 4% (d) 150%

3. Write each ratio as a percent using % notation.

   (a) $\dfrac{2}{5}$ (b) $\dfrac{7}{8}$ (c) 3 (d) $\dfrac{9}{20}$

4. Write each percent as a ratio in lowest terms.

   (a) 58% (b) $\frac{1}{3}\%$ (c) 30% (d) 145%

5. Change each percent to an equivalent decimal.
   (a) 48% (b) 11.9% (c) 6% (d) .05%

6. Rewrite each decimal as an equivalent percent using % notation.
   (a) .81 (b) 4.3 (c) .007 (d) 12

Exercises
Part II

No answers are given.

1. Rewrite the ratio form of each percent using a % symbol.

   (a) $\dfrac{13\frac{3}{4}}{100}$ (b) $\dfrac{97}{100}$

2. Rewrite each of the following as a percent in ratio form. Do *not* reduce.
   (a) 9%   (b) 83%

3. Write each ratio as a percent using % notation.

   (a) $\dfrac{3}{8}$ (b) 4

4. Write each percent as a ratio in lowest terms.

   (a) $6\frac{2}{3}\%$ (b) 44%

5. Change each percent to an equivalent decimal.
   (a) 34%   (b) 8%

6. Rewrite each decimal as an equivalent percent using % notation.
   (a) .005   (b) .71

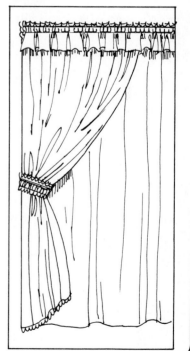

SAVE 25% ON ALL CURTAINS AND SHEERS

SALE   5.99

60 X 54" or 60 X 63"
each, reg.   7.99

Semisheer voile panel
in eggshell, white, rose,
or copper. Polyester and
cotton blend is machine
washable.

60 X 84" panel, reg. 8.99
                    Sale 6.49
118 X 84", reg. 17.99
                    Sale 13.49
Valance, reg. 7.99
                    sale 5.99

This section uses proportions to solve basic percent problems.

Any basic percent problem can be broken down into three types of numbers and compared by the *percent proportion:*

$$\frac{A}{W} = \frac{P}{100}$$

where

A = the *amount* being compared,

W = the *whole* to which something is being compared, and

P = the *percent*, or the comparison as based on a whole of 100.

EXAMPLE 1:   Find the value of A when $W = 65$ and $P = 80$.
First, write the percent proportion.

$$\frac{A}{W} = \frac{P}{100}$$

Second, wherever the problem gave a number value for a variable (letter) in the proportion, write that number in place of the letter. Here, write 65 in place of W and 80 in place of P.

$$\frac{A}{W} = \frac{P}{100}$$

$$\frac{A}{65} = \frac{80}{100}$$

Now you are back to something familiar. Solve the proportion that resulted from the substitutions.

$$\frac{A}{W} = \frac{80}{100}$$

$$65 \times 80 = 100A$$

$$5,200 = 100A$$

$$\frac{5,200}{100} = \frac{100A}{100}$$

$$52 = A$$

EXAMPLE 2:   Find P when $W = 400$ and $A = 10$.

$\dfrac{A}{W} = \dfrac{P}{100}$ Write the percent proportion.

$\dfrac{10}{400} = \dfrac{P}{100}$ Substitute number values given for W and A.

$400P = 10 \times 100$ Solve the proportion.

$400P = 1,000$

$\dfrac{400P}{400} = \dfrac{1,000}{400}$

$P = 2\dfrac{1}{2}, \quad \text{or} \quad 2.5$

EXAMPLE 3:   Find W when $A = 27$ and $P = 150$.

$$\frac{A}{W} = \frac{P}{100}$$ Write the percent proportion.

$$\frac{27}{W} = \frac{150}{100}$$ Substitute values for A and P.

$$150W = 27 \times 100$$ Solve the proportion.

$$150W = 2{,}700$$

$$\frac{150W}{150} = \frac{2{,}700}{150}$$

$$W = 18$$

Try the following to test your ability to solve the percent proportion for one unknown value.

Examples	Solutions
Find W when $P = 55$ and $A = 44$.	$$\frac{A}{W} = \frac{P}{100}$$ $$\frac{44}{W} = \frac{55}{100}$$ $$55W = 44 \times 100$$ $$55W = 4{,}400$$ $$\frac{55W}{55} = \frac{4{,}400}{55}$$ $$W = 80$$
Find A when $W = 5$ and $P = 50$.	$$\frac{A}{W} = \frac{P}{100}$$ $$\frac{A}{5} = \frac{50}{100}$$ $$5 \times 50 = 100A$$ $$250 = 100A$$ $$\frac{250}{100} = \frac{100A}{100}$$ $$2\frac{50}{100} = A$$ or, in lowest terms, $$2\frac{1}{2} = A$$
Find P when $A = 9$ and $W = 4$.	$$\frac{A}{W} = \frac{P}{100}$$

Examples	Solutions
	$\dfrac{9}{4} = \dfrac{P}{100}$
	$4P = 9 \times 100$
	$4P = 900$
	$\dfrac{4P}{4} = \dfrac{900}{4}$
	$P = 225$

Now try these—no hints given.

Find A when $P = 5$ and $W = 40$.	$A = 2$
Find W when $A = 14$ and $P = 7$.	$W = 200$
Find P when $A = 3$ and $W = 12$.	$P = 25$

If the last three solutions are not clear to you, please study the examples in this section again or consult an instructor before continuing.

Applications In order to apply the percent proportion to a word problem, it is necessary to be able to identify A, W, and P based on the statement of a problem.

The easiest of the three to identify is the percent, P, because it is always followed by either the word *percent* or the symbol %. The whole, W, is the basis of comparison and frequently follows the word *of*. The amount, A, represents the part being compared to the whole.

EXAMPLE 4:   Identify A, W, and P in each of the following.

(a)   Find 23% of 97.

P — indicated by %

W — follows the word *of*

Answer: $P = 23$

$W = 97$

A = the amount you are looking for: the number being compared to 97

(b)   5% of what number is 8?

P — indicated by %

W — follows *of*

Answer: $P = 5$

W = the value to be found

$A = 8$

(c) If 4 out of 24 students in a class are absent, what percent of the class is absent?

Answer: P = the value to be found

$W = 24$ The whole class; followed *of*.

$A =\ \ 4$ The number of students absent.

(d) 72 is what percent of 18?

Answer: P = the value to be found

$W = 18$ Follows *of* in the question.

$A = 72$ Amount compared to 18.

Notice that the value of A is larger than the value of W in this particular problem. Do not let this bother you—a percent can be larger than 100.

In each of the following examples, identify the amount, A; the whole, W; and the percent, P.

Examples	Solutions
12% of what number is 40?	Look for P first—indicated here by %. $P = 12$ Next, look for W—follows *of*— $W = W$ that is, the value to solve for. Finally identify A—the only quantity not yet labeled. $A = 40$
Find what percent 10 is of 8.	*What percent* means that P will remain in variable form until the proportion is solved: $P = P$. $W = 8$ Follows *of*. $A = 10$
What percent of 80 is 15?	$P = P$ Variable to solve for. $W = 80$ Follows *of*. $A = 15$
11% of a number is 6. Find that number.	$P = 11$ $W = W$ *Of* a number: value to solve for. $A = 6$
Now try these—no hints given.	
97 is what percent of 210?	$P = P$ $W = 210$ $A = 97$

Examples	Solutions
86% of 420 is what number?	$P = 86$ $W = 420$ $A = A$
Find 62% of 895.	$P = 62$ $W = 895$ $A = A$
7% of some number is 20. Find that number.	$P = 7$ $W = W$ $A = 20$

You should understand each of the preceding before continuing to work in this section of the chapter.

You are now fully equipped to solve word problems involving percents. First, identify P, W, and A; then substitute these values in the percent proportion; and, finally, solve the percent proportion for the unknown value.

EXAMPLE 5:   8% of a number is 96. Find that number.

$$P = 8 \qquad \text{Identify } P, W, \text{ and } A.$$

$$W = W$$

$$A = 96$$

$$\frac{A}{W} = \frac{P}{100} \qquad \begin{array}{l}\text{Substitute these values in the} \\ \text{percent proportion.}\end{array}$$

$$\frac{96}{W} = \frac{8}{100}$$

$$8W = 96 \times 100$$

$$8W = 9{,}600$$

$$\frac{8W}{8} = \frac{9{,}600}{8}$$

$$W = 1{,}200$$

The number to be found is 1,200.

EXAMPLE 6:   A house is worth $60,000. How much would the owner receive if the house were completely destroyed by fire and it was insured for 90% of its value?

$$P = 90$$

$$W = 60{,}000$$

$$A = A$$

$$\frac{A}{W} = \frac{P}{100}$$

$$\frac{A}{60,000} = \frac{90}{100}$$

$$60,000 \times 90 = 100A$$

$$5,400,000 = 100A$$

$$\frac{5,400,000}{100} = \frac{100A}{100}$$

$$54,000 = A$$

The insurance company would have to pay the owner $54,000.

You may have been taught to do percent problems such as this one by direct manipulation. I shall work this problem by that method.

According to the numbers of the example above, you must find the amount for which the house was insured: 90% of $60,000.

As with fractions, the word *of* signals the operation of multiplication. Change 90% to decimal form and multiply by $60,000.

$$
\begin{array}{r}
60,000 \\
\times \quad\quad .90 \\
\hline
00\ 000 \\
54\ 00\ 00 \\
\hline
54,000.00 \quad \text{dollars}
\end{array}
$$

You can use this method in any percent problem when W and P have been assigned values and you are looking for the value of A.

Because this type of percent problem can be solved by a one-step multiplication, you may want to perform this step on a calculator, using the percent key.

> The **percent key** changes a percent notation (such as 75%) to its equivalent decimal form while it is performing the indicated operation.

Return to Example 6 and find 90% of $60,000 using a calculator.
Make the following calculator entries:

$$60,000 \ \boxed{\times} \ 90 \ \boxed{\%}$$

and the answer shows immediately as 54,000. You do *not* have to depress $\boxed{=}$, but you do have to reread the problem in order to label the answer as $54,000.

If your calculator's percent key does not perform this function as described, read the instruction sheet that came with your calculator one more time in order to learn how to use its $\underline{\%}$ key.

If you plan to use a calculator for this type problem you should practice the following examples with your calculator.

Examples	Solutions
Find 12% of 850.	Enter: $850 \ \boxed{\times} \ 12 \ \boxed{\%}$
	Answer: 102
Find 7% of $8,900.	Enter: $8,900 \ \boxed{\times} \ 7 \ \boxed{\%}$
	Answer: $623

Examples	Solutions
Find 75% of $19.88.	Enter: 105 ☒ 4 ⟦%⟧ Answer: $14.91
What is 4% of 105?	Enter: 105 ☒ 4 ⟦%⟧ Answer: 4.2
A blazer originally marked $47.80 is on sale at 35% off. What is the dollar amount of savings?	Find: 35% of $47.80 Enter: 47.80 ☒ 35 ⟦%⟧ Answer: $16.73

EXAMPLE 7: Forty persons from the Adult Learning Lab took the GED (General Educational Development) test last year. If 85% of those tested passed the test, how many persons from the lab received a GED certificate last year?

$$P = 85$$
$$W = 40$$
$$A = A$$
$$\frac{A}{W} = \frac{P}{100}$$
$$\frac{A}{40} = \frac{85}{100}$$
$$40 \times 85 = 100A$$
$$3,400 = 100A$$
$$\frac{3,400}{100} = \frac{100A}{100}$$
$$34 = A$$

Therefore, 34 persons passed the GED test.

This problem can also be done by the direct multiplication method because the unknown value is A.

$$85\% \times 40 = A \qquad 85\% \ of \ 40$$

Change 85% to decimal form

$$85\% = .85$$

and multiply to find the value of A.

$$\begin{array}{r} 40 \\ \times \ .85 \\ \hline 200 \\ 320 \\ \hline 34.00 \end{array}$$

By either method $A = 34$.

Solve each of the following examples by using the percent proportion. If you need to consult the solution column before you have arrived at the final answer, uncover only enough to get you working again. Then compare your entire solution to the one provided.

Examples	Solutions
15 is what percent of 75?	$P = P$ $\qquad$ $\dfrac{A}{W} = \dfrac{P}{100}$ $W = 75$ $A = 15$ $\qquad$ $\dfrac{15}{75} = \dfrac{P}{100}$ $\qquad\qquad$ $75P = 15 \times 100$ $\qquad\qquad$ $75P = 1{,}500$ $\qquad\qquad$ $\dfrac{75P}{75} = \dfrac{1{,}500}{75}$ $\qquad\qquad$ $P = 20$ 15 is 20% of 75.
A man earning $5.50 an hour is given a 6% pay increase. How much will this raise his hourly wage?	$P = 6$ $\qquad$ $\dfrac{A}{W} = \dfrac{P}{100}$ $W = 5.50$ $A = A$ $\qquad$ $\dfrac{A}{5.50} = \dfrac{6}{100}$ $\qquad\qquad$ $6 \times 5.50 = 100A$ $\qquad\qquad$ $33.00 = 100A$ $\qquad\qquad$ $\dfrac{33.00}{100} = \dfrac{100A}{100}$ $\qquad\qquad$ $.33 = A$ His hourly wage will be increased by $.33.
If your rent is $190 per month and it is increased by 8%, how much is the increase?	$P = 8$ $\qquad$ $\dfrac{A}{W} = \dfrac{P}{100}$ $W = 190$ $A = A$ $\qquad$ $\dfrac{A}{190} = \dfrac{8}{100}$ $\qquad\qquad$ $190 \times 8 = 100A$ $\qquad\qquad$ $1{,}520 = 100A$ $\qquad\qquad$ $\dfrac{1{,}520}{100} = \dfrac{100A}{100}$ $\qquad\qquad$ $15.20 = A$ Your increase in monthly rent is $15.20.
23 people drop out of a karate class after the first week. What percent dropped out if the class numbered 28 at the beginning?	$P = P$ $\qquad$ $\dfrac{A}{W} = \dfrac{P}{100}$ $W = 28$ $A = 23$ $\qquad$ $\dfrac{23}{28} = \dfrac{P}{100}$

Examples	Solutions
	$28P = 23 \times 100$
	$28P = 2{,}300$
	$\dfrac{28P}{28} = \dfrac{2{,}300}{28}$
	$P = 82\dfrac{1}{7}$
	The drop-out rate was $82\dfrac{1}{7}\%$.
Gwen paid 28% of her gross annual salary of $18,300 for income tax. How much did she pay that year for income tax?	$P = 28$ $\dfrac{A}{W} = \dfrac{P}{100}$
	$W = 18{,}300$ $\dfrac{A}{18{,}300} = \dfrac{28}{100}$
	$A = A$ $18{,}300 \times 28 = 100A$
	$\dfrac{512{,}400}{100} = \dfrac{100A}{100}$
	$5{,}124 = A$
	For that year Gwen's income tax was $5,124.
What percent of 4,500 is 150?	$P = P$ $\dfrac{A}{W} = \dfrac{P}{100}$
	$W = 4{,}500$ $\dfrac{150}{4{,}500} = \dfrac{P}{100}$
	$A = 150$ $4{,}500P = 150 \times 100$
	$4{,}500P = 15{,}000$
	$\dfrac{4{,}500P}{4{,}500} = \dfrac{15{,}000}{4{,}500}$
	$P = 3\dfrac{1}{3}$
	150 is $3\dfrac{1}{3}\%$ of 4,500.
$4\dfrac{1}{2}$ is 30% of what number?	$P = 30$ $\dfrac{A}{W} = \dfrac{P}{100}$
	$W = W$
	$A = 4\dfrac{1}{2}$ $\dfrac{4\dfrac{1}{2}}{W} = \dfrac{30}{100}$

Examples	Solutions
	$$30W = 4\frac{1}{2} \times 100$$ $$30W = \frac{9}{2} \times \frac{100}{1}$$ $$30W = 450$$ $$\frac{30W}{30} = \frac{450}{30}$$ $$W = 15$$ $4\frac{1}{2}$ is 30% of 15.

If you had difficulty with any of the above examples, please talk with an instructor before attempting the exercises that follow.

Exercises
Part I

Answers to these exercises can be found on page 433.

Solve each of the following problems by using the percent proportion.

1. 27 is 60% of what number?
2. 9 is 8% of what number?
3. Find 28% of 7,400.
4. Find $7\frac{1}{2}$% of $80.
5. 5 is what percent of 2?
6. 12 is what percent of 60?
7. Robin earns $14,000 per year. If her state income tax is 2% of her salary, how much tax must she pay?
8. Jason bought a $595 stereo and received an 8% discount for paying cash. How much money did Jason save by paying cash?
9. After an accident, Lenore's car—insured for 60% of its value—was declared a total loss. The insurance company sent Lenore a check for $3,120. What was the value of Lenore's car?
10. After making a credit card payment, Ginger still had to pay $1\frac{1}{2}$% interest on $400 unpaid balance that month. What was the amount of interest she paid that month?
11. What percent of the class was absent on a day when 14 of the class of 35 students were absent?
12. Fran's house was insured for 95% of its assessed value. For what amount was Fran's house insured when it was assessed at $60,000?
13. 6% of a number is 18. Find that number.
14. What percent of 20 is 17?
15. Find 98% of 500.

16. A furniture salesperson gave Jeremiah a discount of 5% because of a few scratches on the piece he had selected. How much did Jeremiah save if the table he bought was marked to sell for $180?

1. Patty's grocery bill is $62.54. If a sales tax of 4% is added to that amount, how much tax will Patty be charged?
2. A house is insured for 90% of its value. What was the value of that house before it was completely destroyed by fire if the owner received $52,200 from the insurance company?
3. $\frac{1}{2}$ is what percent of 3?
4. 15 is what percent of 75?
5. Find what percent 3 is of 5.
6. 7% of some number is 20. Find that number.
7. A man earning $5.50 an hour is given a 6% pay increase. By how much will his hourly wage rise?
8. A sweater originally selling for $25 was sale priced at $18. The sale price was what percent of the original price?

REVIEW MATERIAL

**Review Skill
Changing
Fractions
to Percents**

In order to express a fraction as a percent, write an equal ratio that has 100 as the denominator. Then change to % notation.

EXAMPLE 1: Express $\frac{2}{5}$ as a percent.

$$\frac{2}{5} = \frac{N}{100}$$

$$5N = 2 \times 100$$

$$5N = 200$$

$$\frac{5N}{5} = \frac{200}{5}$$

$$N = 40$$

Hence, $\frac{2}{5} = \frac{40}{100}$ and $\frac{40}{100} = 40\%$.

EXAMPLE 2: Express $\frac{1}{8}$ as a percent.

$$\frac{1}{8} = \frac{N}{100}$$

$$8N = 1 \times 100$$

$$8N = 100$$

$$\frac{8N}{8} = \frac{100}{8}$$

$$N = 12\frac{1}{2}$$

Hence, $\dfrac{1}{8} = \dfrac{12\frac{1}{2}}{100}$ and $\dfrac{12\frac{1}{2}}{100} = 12\frac{1}{2}\%$.

Exercise A Answers to this exercise can be found on page 433.

Express each of the following ratios as a percent using % notation.

1. $\dfrac{1}{4}$ 2. $\dfrac{2}{3}$ 3. $\dfrac{7}{10}$ 4. $\dfrac{1}{2}$ 5. $\dfrac{5}{8}$ 6. $\dfrac{3}{4}$ 7. $\dfrac{5}{6}$

8. $\dfrac{3}{10}$ 9. $\dfrac{9}{20}$ 10. $\dfrac{11}{25}$ 11. $\dfrac{3}{8}$ 12. $\dfrac{4}{5}$ 13. $\dfrac{13}{20}$ 14. $\dfrac{43}{50}$

Exercise B No answers are given.

Express each of the following ratios as a percent using % notation.

1. $\dfrac{9}{10}$ 2. $\dfrac{1}{3}$ 3. $\dfrac{7}{8}$ 4. $\dfrac{11}{50}$ 5. $\dfrac{1}{6}$ 6. $\dfrac{17}{20}$ 7. $\dfrac{7}{25}$

Review Skill Applying the Percent Proportion

To solve a percent problem, identify A, W, and P, then solve the percent proportion for the missing value.

$$\text{Percent proportion: } \dfrac{A}{W} = \dfrac{P}{100}$$

The easiest of these to identify is the percent, P, because it is always followed by the word *percent* or the symbol %.

The whole, W, is the basis of comparison and frequently follows the word *of*.

The amount, A, represents the part being compared to the whole.

EXAMPLE: If 4 of 24 students in a class are absent, what percent of the class is absent?

$$P = \text{the value to be found}$$
$$W = 24$$
$$A = 4$$

$$\dfrac{A}{W} = \dfrac{P}{100}$$

$$\dfrac{4}{24} = \dfrac{P}{100}$$

$$24P = 4 \times 100$$

$$24P = 400$$

$$\dfrac{24P}{24} = \dfrac{400}{24}$$

$$P = 16\frac{2}{3}$$

$$\frac{P}{100} = \frac{16\frac{2}{3}}{100} = 16\frac{2}{3}\%$$

Hence, $16\frac{2}{3}\%$ of the class is absent.

Exercise C   Answers to this exercise can be found on page 433.

1.   Reggie paid 28% of his gross annual salary of $18,300 for income tax. How much did he pay that year for income tax?
2.   7% of a number is 105. Find that number.
3.   9 is 60% of what number?
4.   5 is what percent of 2?
5.   Ashley earns $17,000 per year. If her state income tax is 2% of her salary, how much tax must she pay?
6.   What percent of 20 is 14?
7.   For cash and carry, Jeff received a 10% discount when he bought his new desk. How much did the desk originally cost if Jeff paid only $135?
8.   Find 5% of 645.
9.   Find 82% of 75.
10. A dress originally priced at $25 was reduced by $7. The markdown was what percent of the original price?

Exercise D   No answers are given.

1.   Christy earns $9,745 per year. If her state income tax is 10% of her salary, how much state tax does she pay?
2.   Find 8% of 500.
3.   9 is what percent of 12?
4.   24 is 60% of what number?
5.   An appliance salesperson gave Charlotte a 4% discount for cash. How much did Charlotte save on an appliance marked to sell for $450?

REVIEW EXERCISES

Refer to the section listed if you have difficulty. Answers to these exercises can be found on pages 433–434.

Section 18.1   1.   Rewrite the ratio form of each percent using a % symbol.

(a) $\frac{65}{100}$ (b) $\frac{5}{100}$ (c) $\frac{112}{100}$ (d) $\frac{11\frac{3}{4}}{100}$ (e) $\frac{12.5}{100}$

2.   Rewrite each of the following as a percent in ratio form—do *not* reduce.

(a) 115%   (b) 92%   (c) $33\frac{1}{3}\%$ (d) $5\frac{1}{2}\%$ (e) 4%

3. Write each ratio as a percent using % notation.

(a) $\dfrac{3}{5}$ (b) $\dfrac{5}{8}$ (c) 7   (d) $\dfrac{15}{100}$ (e) $\dfrac{6}{20}$

4. Write each percent as a ratio in lowest terms.

(a) 56%   (b) $\dfrac{2}{3}$%   (c) 60%   (d) 8%   (e) 125%

5. Change each percent to an equivalent decimal.
(a) 38%   (b) 14.2%   (c) 8%   (d) .5%   (e) 125%

6. Rewrite each decimal as an equivalent percent using % notation.
(a) .986   (b) 1.5   (c) .008   (d) .05   (e) 2

Section 18.2 Use the percent proportion $\left(\dfrac{A}{W} = \dfrac{P}{100}\right)$ to solve each problem in this section of the review.

7. Find W when $P = 12$ and $A = 15$.
8. Find A when $W = 40$ and $P = 60$.
9. Find P when $A = 7$ and $W = 2$.
10. 12% of what number is 30?
11. Find what percent 10 is of 8.
12. Find 45% of 60.
13. If John's rent of $260 per month is increased by 15%, by how much is his rent raised?
14. Eight swimmers passed their lifesaving test the first time they took it. If 12 swimmers took the test, what percent of them passed on the first attempt?
15. Genevieve's health insurance paid for 80% of her prescription medicine one year. If the insurance paid her $125 that year, what was the total cost of her prescriptions?

16. What is 45% of 980?
17. Find 18% of $6,700.
18. Luggage was sale priced at a 37% reduction. What was the savings on a suitcase originally marked for $52?

CHAPTER POST-TEST

This test should determine if you have mastered the topics in Chapter 18. Answers to this test can be found on page 434.

Rewrite the ratio form of each percent using a % symbol.

1. $\dfrac{82}{100}$ 2. $\dfrac{3\frac{1}{2}}{100}$ 3. $\dfrac{5}{100}$

Rewrite each of the following as a percent in ratio form—do *not* reduce.

4. 99%   5. 115%   6. 13.5%

Write each ratio as a percent using % notation.

7. $\dfrac{3}{20}$

8. $\dfrac{1}{8}$

9. $\dfrac{4}{5}$

Write each percent as a ratio in lowest terms.

10. 52%

11. $\dfrac{5}{6}\%$

12. 6%

Change each percent to an equivalent decimal.

13. 7%

14. 91%

15. 15.4%

Rewrite each decimal as an equivalent percent using % notation.

16. .44

17. .003

18. 2.8

Use the percent proportion to solve the following problems.

19. Find what percent 8 is of 5.
20. 16% of what number is 72.
21. Find 33% of $8,700.
22. Record albums were on special for 10% off the marked price. How much would be saved on an album marked at $12.90?
23. During the flu season 17 of a company's 85 employees were out sick one day. What percent of their workers were not at work that day?
24. Geraldo's dental insurance covered 70% of his oral surgery expenses. If his insurance paid $336, what was the total charge for Geraldo's oral surgery?

UNIT FOUR
RATIO, PROPORTION, AND PERCENT
REVIEW AND POST-TEST

Ratio A ratio is the comparison of two numbers by division. The ratio of 7 to 9 can be written as

$$7 \text{ to } 9 \quad \text{or} \quad 7:9 \quad \text{or} \quad \dfrac{7}{9}$$

Ratios should be simplified whenever possible.

The ratio $\dfrac{10}{18}$ simplifies to $\dfrac{5}{9}$ by reducing by 2.

The ratio $\dfrac{3\frac{1}{2}}{7}$ simplifies to $\dfrac{1}{2}$ by applying the rule for division of fractions.

The ratio $\dfrac{.18}{3.6}$ simplifies to $\dfrac{1}{20}$ after moving both decimal points two places to the right.

When a ratio compares like quantities, express the quantities in the same unit of measure whenever possible.

$$\frac{12 \text{ days}}{2 \text{ weeks}} = \frac{12 \text{ days}}{14 \text{ days}} = \frac{12}{14} = \frac{6}{7}$$

Rate Rate is a special ratio that compares unlike quantities. Many rates appear as ratios having 1 for their second term.

$$45 \text{ miles per hour is a rate expressing } \frac{45 \text{ miles}}{1 \text{ hour}} \quad \text{or} \quad 45:1$$

Proportion A proportion is the expressed equality of two ratios, such as $\frac{4}{6} = \frac{8}{12}$. In any proportion, the product of the means is equal to the product of the extremes.

$$\frac{4}{6} = \frac{8}{12}$$

product of means $\qquad$ $6 \times 8 = 4 \times 12$ $\qquad$ product of extremes

$$48 = 48$$

Knowing how to solve a proportion for a missing value is a valuable skill.

$$\frac{N}{3} = \frac{7}{10}$$

$$3 \times 7 = 10N$$

$$21 = 10N$$

$$\frac{21}{10} = \frac{10N}{10}$$

$$2\frac{1}{10} = N \quad \text{or} \quad 2.1 = N$$

Percent Percent is a special ratio having 100 as its second term.

$$99\% = \frac{99}{100} \quad \text{or} \quad 99:100$$

Because percent is used so frequently in everyday business, it is important that you be familiar with the following notations and know how to change from one form to another without changing the value of the percent.

Ratio form to % notation:

$$\frac{7}{100} = 7\%$$

% notation to ratio form in lowest terms:

$$28\% = \frac{28}{100} = \frac{14}{50} = \frac{7}{25}$$

% notation to decimal form:

$$79\% = 79\% = .79$$

Decimal form to % notation:

$$.53 = .53\% = 53\%$$

A fraction can be changed to % notation by expressing it first as a ratio with 100 as its second term.

$$\frac{7}{20} = \frac{?}{100} \qquad\qquad \frac{2}{3} = \frac{?}{100}$$

Use fraction-building method   Use proportion-solving method

$$\frac{7}{20} = \frac{7 \times 5}{20 \times 5} \qquad\qquad \frac{2}{3} = \frac{N}{100}$$

$$= \frac{35}{100} = 35\% \qquad\qquad 3N = 2 \times 100$$

$$3N = 200$$

$$\frac{3N}{3} = \frac{200}{3}$$

$$N = 66\frac{2}{3}$$

$$\frac{2}{3} = \frac{N}{100} = \frac{66\frac{2}{3}}{100} = 66\frac{2}{3}\%$$

Percent Proportion Any basic percent problem can be broken down into three types of numbers and compared by the *percent proportion*:

$$\frac{A}{W} = \frac{P}{100}$$

where A = the *amount* being compared,
$\quad W$ = the *whole* to which something is being compared, and
$\quad P$ = the *percent*, or the comparison based on a whole of 100.

Before working the Unit Four Post-Test, work (or rework)

Chapter 17 Review Exercises, page 398
Chapter 18 Review Exercises, page 426

Study the sections for which you cannot work the sample problems.

UNIT FOUR POST-TEST

Answers to this test can be found on page 434.

1. Manny has 14 rock albums, 8 country albums, and 6 classical albums. Write, in lowest terms, the ratio of his (a) classical to rock albums, (b) country to classical albums, (c) rock to country albums.

2. As part of a political campaign, Stan made 30 telephone calls one evening. Eight of the numbers he was given never answered. Write the ratio of the number of answered calls to the number that went unanswered and reduce to lowest terms.

3. If a truck is driven 354 miles in 6 hours, what is the rate in miles per hour (the ratio of miles to hours)?

4. Simplify each ratio and reduce to lowest terms.

(a) $\dfrac{4\frac{1}{3}}{6\frac{1}{2}}$ (b) $\dfrac{.06}{4.2}$

5. Are $\dfrac{6}{10}$ and $\dfrac{9}{15}$ equal ratios?

6. Express, in lowest terms, the ratio of
   (a) 85 minutes to 2 hours   (b) 4 dimes to 3 quarters

7. Find the missing value in each proportion.

(a) $N:9 = 2:3$ (b) $\dfrac{3}{5} = \dfrac{N}{7}$

8. In a bread recipe the ratio of flour to milk is $4:3$. If 5 cups of milk are used, how many cups of flour should be used?

9. Rewrite the ratio form of each percent using a % symbol.

(a) $\dfrac{6}{100}$ (b) $\dfrac{71}{100}$ (c) $\dfrac{7\frac{1}{4}}{100}$

10. Rewrite each of the following as a percent in ratio form.
    (a) 81% $= \dfrac{81}{100}$ (b) 3%   (c) 103%

11. Write $\dfrac{3}{5}$ as a percent using % notation.

12. Write $5\frac{1}{3}\%$ as a ratio in lowest terms.

13. Change 4.5% to an equivalent decimal.

14. Rewrite .92 as an equivalent percent using % notation.

15. Find 62% of $450.

16. 210 is 40% of what number?

17. Ron's computer is worth $6,500. How much would he receive if his computer were stolen and it was insured for 75% of its value?

18. 7 is what percent of 21?

19. Perrit's insurance paid 90% of his outpatient expenses. If the insurance company paid $450 toward his outpatient bill, what were the total charges?

20. In a class of 20 students, what percent of the class is present on a day when 3 students are absent?

ANSWERS

Unit Four
Pretest
1. (a) $2:3$ (b) $3:1$ (c) $1:2$ 2. $9:17$
3. $25:1$ or 25 kilometers per hour   4. (a) $6:5$ (b) $50:1$ 5. (a)
6. (a) $\dfrac{10}{13}$ (b) $\dfrac{8}{5}$ 7. (a) 36   (b) 14   (c) $18\frac{2}{3}$ 8. $3\frac{3}{4}$ cups

Answers   431

9. (a) 9% (b) 67% (c) 103%   **10.** (a) $\dfrac{83}{100}$ (b) $\dfrac{7}{100}$ (c) $\dfrac{151}{100}$

11. 80%   **12.** $\dfrac{1}{200}$ **13.** .265   **14.** 430%   **15.** $77\dfrac{1}{2}$ **16.** 480

17. $63,000

Chapter 17

Section 17.1 **1.** $\dfrac{5}{7}$, 5 to 7, 5:7   **2.** (a) 23:16 (b) 16:23 (c) 7:16 (d) 23:7
3. 13:12   **4.** 5:12   **5.** 6:5   **6.** 3:8   **7.** (a) 4:9 (b) 4:1 (c) 5:9
8. (a) 8:7 (b) 8:15 (c) 7:15   **9.** 11:2   **10.** 2:3
11. 39:1 or 39 miles per gallon   **12.** 12:1 or 12 gallons per day

Section 17.2 **1.** (b) and (d) are equal ratios
2. (a) means: 5 and 12; extremes: 4 and 15
(b) means: 20 and 4; extremes: 8 and 10
3. (a) Yes, because 2,100 = 2,100. (b) No, because 156 ≠ 154.   **4.** 9
5. 8   **6.** $5\dfrac{1}{3}$ **7.** 35   **8.** 39   **9.** 15   **10.** $4\dfrac{4}{5}$

Section 17.3 **1.** 35 minutes   **2.** 1,800 square feet   **3.** $8\dfrac{1}{2}$ months   **4.** $5,400

5. 385 points   **6.** $5\dfrac{4}{7}$ feet   **7.** 15 cans   **8.** 70 pounds   **9.** 8 cups

10. $6\dfrac{2}{3}$ hours

Exercise A **1.** $\dfrac{1}{2}$ **2.** 1:32   **3.** 6:5   **4.** 1:3   **5.** 32:3   **6.** 5:2   **7.** 5:2   **8.** $\dfrac{3}{4}$

9. 1:3   **10.** 1:2

Exercise C **1.** 7:9   **2.** 17:20   **3.** 5:17   **4.** 30:11   **5.** 7:8   **6.** 1:2   **7.** 1:20
8. 2:3   **9.** 1:20   **10.** 5:4   **11.** 5:9   **12.** 1:8   **13.** 1:4   **14.** 5:2
15. 1:30

Exercise E **1.** 1:4   **2.** 7:12   **3.** 5:7   **4.** 4:5   **5.** 5:6   **6.** 1:4   **7.** 2:5
8. 6:7   **9.** 1:1   **10.** 6:5   **11.** 3:4   **12.** 1:2   **13.** 15:16   **14.** 1:5
15. 4:5   **16.** 1:1   **17.** 1:60   **18.** 7:10   **19.** 8:15   **20.** 1:2

Exercise G **1.** 3:7   **2.** 1:2   **3.** 7:23   **4.** 11:14   **5.** 19:31   **6.** 16:19
7. 49:52   **8.** 21:25

Exercise I **1.** $5\dfrac{1}{3}$ **2.** 4   **3.** 12   **4.** 9   **5.** $2\dfrac{1}{3}$ **6.** 30   **7.** $1\dfrac{1}{7}$ **8.** 16   **9.** $4\dfrac{2}{3}$

10. 18   **11.** $2\dfrac{2}{7}$ **12.** 10

Exercise K **1.** 10 belts   **2.** 150 miles   **3.** 75¢   **4.** 36 miles   **5.** 54 parts

6. $22\dfrac{1}{2}$ miles   **7.** 12 feet   **8.** 8 shirts   **9.** 12 pounds   **10.** $7\dfrac{1}{2}$ quarts

Review Exercises 1. 7 to 9, 7:9, $\frac{7}{9}$ 2. (a) $\frac{4}{3}$ (b) $\frac{3}{5}$ (c) $\frac{5}{3}$ (d) $\frac{4}{5}$ 3. $\frac{1}{5}$ 4. 9:13

5. (a) $\frac{3}{2}$ (b) $\frac{30}{1}$ 6. 1:1 7. 6:5 8. 1:2 9. 33:1 10. 2:5

11. Yes. 12. Yes. 13. 7 and 6 14. 3 and 10

15. (a) 3 (b) $2\frac{4}{7}$ (c) 6 (d) $9\frac{3}{5}$ 16. 28 minutes 17. 100 miles

18. 9 feet 19. 15 pairs 20. $4\frac{1}{2}$ cups

Post-Test 1. (a) $\frac{2}{3}$ (b) $\frac{3}{2}$ (c) $\frac{4}{9}$ (d) $\frac{9}{4}$ 2. 13:22 3. $\frac{5}{14}$

4. (a) 6:7 (b) 3:5 5. 63:1 6. (a) 52 (b) $3\frac{3}{7}$ 7. 75¢

8. $22\frac{1}{2}$ pints

Chapter 18

Section 18.1 1. (a) 73% (b) 7% (c) 129% (d) $12\frac{1}{2}$%

2. (a) $\frac{59}{100}$ (b) $\frac{8\frac{1}{2}}{100}$ (c) $\frac{4}{100}$ (d) $\frac{150}{100}$

3. (a) 40% (b) $87\frac{1}{2}$% (c) 300% (d) 45%

4. (a) $\frac{29}{50}$ (b) $\frac{1}{300}$ (c) $\frac{3}{10}$ (d) $\frac{29}{20}$

5. (a) .48 (b) .119 (c) .06 (d) .0005

6. (a) 81% (b) 430% (c) .7% (d) 1,200%

Section 18.2 1. 45 2. 112.5 3. 2,072 4. 6 5. 250 6. 20 7. $280
8. $47.60 9. $5,200 10. $6 11. 40 12. $57,000 13. 300
14. 85 15. 490 16. $9

Exercise A 1. 25% 2. $66\frac{2}{3}$% 3. 70% 4. 50% 5. $62\frac{1}{2}$% 6. 75% 7. $83\frac{1}{3}$%

8. 30% 9. 45% 10. 44% 11. $37\frac{1}{2}$% 12. 80% 13. 65%
14. 86%

Exercise C 1. $5,124 2. 1,500 3. 15 4. 250 5. $340 6. 70 7. $150
8. 32.25 9. 61.5 10. 28

Review Exercises 1. (a) 65% (b) 5% (c) 112% (d) $11\frac{3}{4}$% (e) 12.5%

2. (a) $\frac{115}{110}$ (b) $\frac{92}{100}$ (c) $\frac{33\frac{1}{3}}{100}$ (d) $\frac{5\frac{1}{2}}{100}$ (e) $\frac{4}{100}$

3. (a) 60% (b) $62\frac{1}{2}$% (c) 700% (d) 15% (e) 30%

4. (a) $\dfrac{14}{25}$ (b) $\dfrac{2}{300}$ (c) $\dfrac{3}{5}$ (d) $\dfrac{2}{25}$ (e) $\dfrac{5}{4}$

5. (a) .38   (b) .142   (c) .08   (d) .005   (e) 1.25

6. (a) 98.6%   (b) 150%   (c) .8%   (d) 5%   (e) 200%   7. 125   8. 24

9. 350   10. 250   11. 125   12. 27   13. $39   14. $66\dfrac{2}{3}$

15. $156.25   16. 441   17. $1,260   18. $19.24

Post-Test   1. 82%   2. $3\dfrac{1}{2}\%$ 3. 5%   4. $\dfrac{99}{100}$ 5. $\dfrac{115}{100}$ 6. $\dfrac{13.5}{100}$ 7. 15%

8. $12\dfrac{1}{2}\%$ 9. 80%   10. $\dfrac{13}{25}$ 11. $\dfrac{1}{120}$ 12. $\dfrac{3}{50}$ 13. .07   14. .91

15. .154   16. 44%   17. .3%   18. 280%   19. 160   20. 450

21. $2,871   22. $1.29   23. 20   24. $480

Unit Four Post-Test   1. (a) 3:7   (b) 4:3   (c) 7:4   2. 11:4   3. 59:1 or 59 miles per hour

4. (a) $\dfrac{2}{3}$ (b) $\dfrac{1}{70}$ 5. Yes.   6. (a) 17:24   (b) 8:15   7. (a) 6   (b) $4\dfrac{1}{5}$

8. $6\dfrac{2}{3}$ 9. (a) 6%   (b) 71%   (c) $7\dfrac{1}{4}\%$ 10. (a) $\dfrac{81}{100}$ (b) $\dfrac{3}{100}$ (c) $\dfrac{103}{100}$

11. 60%   12. $\dfrac{4}{75}$ 13. .045   14. 92%   15. $279   16. 525

17. $4,875   18. $33\dfrac{1}{3}$ 19. $500   20. 85

UNIT 5
measurement

UNIT FIVE PRETEST

This test is given before you read the material to see what you already know. You may need to refer to the tables starting on page 439.

    If you cannot work each sample problem, you need to read the indicated chapters carefully. Answers to this test can be found on page 479.

Chapter 19   1. Change 840 grams to ounces.

                  2. Change 162 meters to kilometers.

                  3. Change 16 centimeters to millimeters.

Chapter 20   4. Change 360 seconds to minutes.

                  5. Change 60 ounces to pounds and ounces.

                  6. Add: 6 lb 8 oz

                                    +2 lb 9 oz

                  7. Divide: 3 $\overline{\smash{)}\,7\text{ ft 3 in.}}$

19

measurement and the metric system

Introduction How long is it? How much does it weigh? Before you can answer these questions you need to answer an additional question: Compared with what?

Measurement involves two components: (1) the **amount** and (2) the **unit** of comparison.

For example, how long is this line segment?

I first think, compared with what?

For the moment, let us call this distance a havin.

437

1 havin

Compared to a havin, the line segment is 6 havins long.

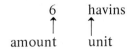

6   havins
↑   ↑
amount   unit

Saying that the segment is 6 havins long means that the length is made up of a standard havin repeated 6 times.

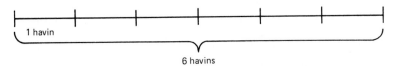

1 havin

6 havins

I made up the unit havin, but the idea of measurement is the same in any system. Measurement is a process of describing a quantity by comparing it with a standard unit. The amount is the number of times that the unit is repeated.

Had I decided to compare the length of the segment to this length,

1 whoin

the line segment would be 1 standard whoin, repeated 3 times, or 3 whoins.

1 whoin

3 whoins

I made up whoins also. A true measurement system is not made up by one or two individuals, but is agreed upon by a large group of people.

Worldwide standards are necessary today because of international travel, business, and science. There is one system that is used by 95% of the world—the **metric system**.

If you have studied the English system, you will find that the metric system is superior in organization and easier to work with.

Metric System of Prefixes The metric system uses one standard table of prefixes for the units of length, volume, and weight to relate amounts within each type of measurement.

Metric System Prefixes

Prefix	Symbol	Relation to Unit
kilo	k	1 kilo = 1,000 units
hecto	h	1 hecto = 100 units
deka	da	1 deka = 10 units
Greater than 1 unit		Less than 1 unit
deci	d	10 deci = 1 unit
centi	c	100 centi = 1 unit
milli	m	1,000 milli = 1 unit

This table of relationships is based on powers of 10.

There are other prefixes used in the metric system, but the most commonly used of the prefixes shown are kilo, centi, and milli. You should memorize these.

$$\boxed{\begin{array}{l} 1 \text{ kilo} = 1{,}000 \text{ units} \\ 100 \text{ centi} = 1 \text{ unit} \\ 1{,}000 \text{ milli} = 1 \text{ unit} \end{array}}$$

Length   The basic unit of length in the metric system is the **meter**. Compared with a unit of length in the English system,

$$1 \text{ meter (m)} = 3.3 \text{ feet (ft)}$$

Thus, a meter is longer than a yard.

Let us see what makes the metric system easier to use and learn. Compare the way in which the different units within each system are related.

English System (length)
_____

1 foot (ft) = 12 inches (in.)
1 yard (yd) = 3 feet (ft)
1 mile (mi) = 1,760 yards (yd) or 5,280 feet (ft)
_____

Metric System (length)
_____

1 kilometer (km) = 1,000 meters (m)
100 centimeters (cm) = 1 meter
1,000 millimeters (mm) = 1 meter
_____

In the English system you see 12's, 3's, and 5,280's in the conversion formulas. But in the metric system you see only powers of 10 and the prefixes just discussed. There are fewer fractions in the metric system and conversions are easier to remember because of the prefix organization.

Weight or Mass   Let us compare the units of weight. In the metric system, the basic unit of weight is the **gram**. For comparison,

$$1 \text{ ounce (oz)} \cong 28 \text{ grams (g)}$$

English System (weight)
_____

1 pound (lb) = 16 ounces (oz)
1 ton = 2,000 pounds (lb)
_____

Notice the 16's and 2,000's. In the metric system the same prefixes are used for weight that were used for length, and the relationships are the same—they are based on powers of 10.

Metric System (weight)

1 kilogram (kg) = 1,000 grams (g)
100 centigrams (cg) = 1 gram
1,000 milligrams (mg) = 1 gram

Volume   Volume measures how much space the object occupies. The English system has a new set of numbers for the conversions with volume. The metric system uses the same prefixes as for the length and weight and the basic unit of volume is the **liter**. For comparison,

$$1 \text{ liter } (\ell) \cong 1.06 \text{ quarts (qt)}$$

English System (volume)

1 quart (qt) = 2 pints (pt)
1 gallon (gal) = 4 quarts (qt) or 231 cubic inches (cu in.)

Metric System (volume)

1 kiloliter (kl) = 1,000 units
100 centiliters (cl) = 1 unit
1,000 milliliters (ml) = 1 unit

Also used in the metric system is a cubic centimeter (cc)—a cube with each edge measuring 1 centimeter. In fact, the system of volume is designed so that

$$\boxed{1 \text{ cubic centimeter (cc)} = 1 \text{ milliliter (ml)}}$$

As I stated in the introduction to this section, all measurements depend on standards of comparison. In the metric system these standards are: the meter for length, the gram for weight, and the liter for volume.

It is sometimes hard to convince someone who learned the English system first that the metric system is easier. But stop and look back at the way the English system is organized. It looks like a system put together by three different committees of people who never met.

The metric system is clearly very organized within each type of measurement, and the same prefix organization is used for each type. It is the system used in science, engineering, and medicine, and is the system generally used in most of the world. In the United States, it is the system used in science, and soon will be used commonly in business.

Conversions   To learn to use the metric system, you need to know more than how to convert
and Unit   back to the English system. The main objective of this section is for you to learn
Fractions   how to use unit fractions for conversions in any system or between any two systems.
        To convert between the metric and the English systems use these useful conversion formulas.

$$\boxed{\begin{aligned}
1 \text{ inch (in.)} &= 2.54 \text{ centimeters (cm)} \\
3.3 \text{ feet (ft)} &= 1 \text{ meter (m)} \\
.621 \text{ miles (mi)} &= 1 \text{ kilometer (km)} \\
1 \text{ ounce (oz)} &= 28 \text{ grams (g)} \\
1 \text{ pound (lb)} &= 454 \text{ grams (g)} \\
1.06 \text{ quarts (qt)} &= 1 \text{ liter } (\ell)
\end{aligned}}$$

Unit Fractions   The important fact to remember in this section is the unique property of 1: The product of a number and 1 is the number itself.

$$5 \times 1 = 5$$

Different forms of 1 were used to build fractions

$$\frac{2}{3} = \frac{2}{3} \times \frac{4}{4}$$
$$= \frac{8}{12}$$

The fraction $\frac{4}{4}$ equals 1, and the two fractions $\frac{2}{3}$ and $\frac{8}{12}$ are equal, although they have different forms.

Remember: Although the form of a number can change when it is multiplied by 1, its value remains unchanged.
The conversion formula

$$3.3 \text{ ft} = 1 \text{ m}$$

can be changed into two **unit fractions**:

$$\frac{3.3 \text{ ft}}{1 \text{ m}} = 1 \quad \text{and} \quad \frac{1 \text{ m}}{3.3 \text{ ft}} = 1$$

Conversion is then a matter of choosing the fraction that will cancel the unwanted unit, leaving the unit desired.

EXAMPLE 1:   Change 6 feet to meters.
Multiply 6 feet by the unit fraction that has feet in the denominator to cancel the feet, leaving the unit meters in the numerator.

$$6 \text{ ft} = 6 \text{ ft} \times \frac{1 \text{ m}}{3.3 \text{ ft}} \qquad \text{Cancels feet, leaving meters.}$$

$$= \frac{6 \text{ m}}{3.3}$$

$$= 1.8 \text{ m} \qquad \text{To the nearest tenth.}$$

Unit Fraction Conversions

The following summarizes the conversion procedure using unit fractions:

1. Use the conversion formula that relates the two units of the problem to form a unit fraction.
2. Multiply by the unit fraction that has the old unit (the one you wish to cancel) in the denominator and the new unit in the numerator.

EXAMPLE 2:   Change 5 inches to centimeters.

Because 1 inch = 2.54 centimeters, there are two unit fractions.

$$\frac{1 \text{ in.}}{2.54 \text{ cm}} = 1 \quad \text{and} \quad \frac{2.54 \text{ cm}}{1 \text{ in.}} = 1$$

Because I want to cancel inches, I use the unit fraction that has inches in the denominator and centimeters in the numerator.

$$5 \text{ in.} = 5 \,\cancel{\text{in.}} \times \frac{2.54 \text{ cm}}{1 \,\cancel{\text{in.}}}$$

$$= 12.7 \text{ cm}$$

EXAMPLE 3:   Change 10 centimeters to inches.

This time use

$$\frac{1 \text{ in.}}{2.54 \text{ cm}} = 1$$

to cancel centimeters and leave inches.

$$10 \text{ cm} = 10 \,\cancel{\text{cm}} \times \frac{1 \text{ in.}}{2.54 \,\cancel{\text{cm}}}$$

$$= \frac{10 \text{ in.}}{2.54}$$

$$= 3.9 \text{ in.} \qquad \text{To the nearest tenth.}$$

EXAMPLE 4:   1,000 kilometers equal how many miles?

I use the formula .621 miles = 1 kilometer.

To change from kilometers to miles, I need to cancel kilometers, leaving miles. I use the conversion formula

$$\frac{.621 \text{ mi}}{1 \text{ km}} = 1$$

Therefore,

$$1,000 \text{ km} = 1,000 \,\cancel{\text{km}} \times \frac{.621 \text{ mi}}{1 \,\cancel{\text{km}}}$$

$$= 621 \text{ mi}$$

EXAMPLE 5:   Change 20 pounds to grams.

Because 1 pound = 454 grams, I shall use

$$\frac{454 \text{ g}}{1 \text{ lb}} = 1$$

to cancel pounds and leave grams.

$$20 \text{ lb} = 20 \cancel{\text{lb}} \times \frac{454 \text{ g}}{1 \cancel{\text{lb}}}$$

$$= 9{,}080 \text{ g}$$

Now you try these.

Examples	Solutions
3 liters equal how many quarts?	First, find the conversion formula that relates liters and quarts. $$1.06 \text{ qt} = 1 \text{ }\ell$$ Next, use the unit fraction that will cancel liters and leave quarts. $$\frac{1.06 \text{ qt}}{1 \text{ }\ell} = 1 \qquad \begin{array}{l}\text{Leave quarts.}\\ \text{Cancel liters.}\end{array}$$ Therefore, $$3 \text{ }\ell = 3 \text{ }\cancel{\ell} \times \frac{1.06 \text{ qt}}{1 \text{ }\cancel{\ell}}$$ $$= 3.18 \text{ qt}$$
Change 84 grams to ounces.	Because 1 oz = 28 g $$84 \text{ g} = 84 \cancel{\text{g}} \times \frac{1 \text{ oz}}{28 \cancel{\text{g}}}$$ $$= \frac{84 \text{ oz}}{28}$$ $$= 3 \text{ oz}$$

No hints this time!

Change 6 inches to centimeters.	15.24 cm
Change 5.589 miles to kilometers.	9 km

Ready for the exercises?

Exercises Part I

Answers to these exercises can be found on page 479.

1. Change 10 inches to centimeters.
2. Change 12.7 centimeters to inches.
3. Change 6.21 miles to kilometers.
4. Change 5 meters to feet.
5. Change 5.3 quarts to liters.

Exercises Part II

1. Change 12 inches to centimeters.
2. Change 13.86 centimeters to inches.
3. Change 420 grams to ounces.

SECTION 19.2
METRIC SYSTEM CONVERSIONS

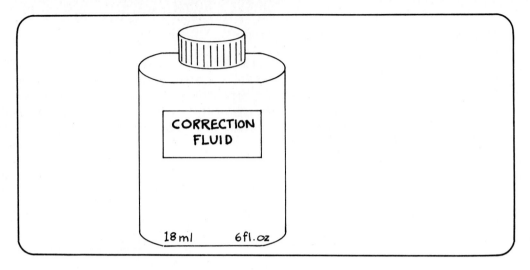

Prefixes and Units

The most common *prefixes* in the metric system are shown in the table below. You should memorize these relationships.

1 kilo = 1,000 units
100 centi = 1 unit
1,000 milli = 1 unit

Conversions in the metric system are simple because one set of prefixes is used with the following *units* for all measurements.

Meters (m) measure length.
Liters (ℓ) measure volume.
Grams (g) measure weight.

Length

Here is a metric ruler, labeled with centimeters; it is 10 centimeters long.

A centimeter is about the thickness of your little finger.

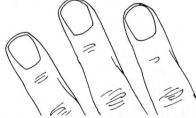

If you divide the centimeter into 10 equal parts, each length is a millimeter, which is about the thickness of a dime.

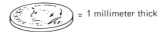
= 1 millimeter thick

The smaller markings not labeled on the metric ruler are millimeters. A meter is 100 centimeters (the ruler above repeated 10 times). A kilometer is 1,000 meters (about .6 mile).

Thus, centimeters and millimeters are used to measure small lengths, whereas a meter is used for greater lengths. Kilometers are used to measure very great distances.

Weight   A gram is very light—a nickel weighs about 5 grams.

5 grams

A milligram is an even lighter amount. An aspirin weighs 300 milligrams.

300 milligrams

A kilogram, however, is heavier. It is similar to the English pound. In fact, 1 kilogram = 2.2 pounds.

Volume   A liter is a little larger than the English quart.

Cola   Cola

1 liter   1 quart

A milliliter is a very small volume—5 milliliters equal 1 teaspoonful.

> In general centi-units and milli-units are used to measure very small amounts, whereas kilo-units are used to measure larger amounts.

Conversions   The following are examples of conversions within the metric system. Before reading these examples, you might want to review how to (mentally) multiply and divide by powers of 10 in Chapters 14 and 15, because that is the only arithmetic necessary in each of the conversion formulas.

EXAMPLE 1:   Change 5 kilometers to meters.
Because 1 kilometer = 1,000 meters, use

$$\frac{1{,}000 \text{ m}}{1 \text{ km}} = 1$$

to cancel kilometers and leave meters as the unit.

Therefore,

$$5 \text{ km} = 5 \text{ km} \times \frac{1{,}000 \text{ m}}{1 \text{ km}}$$

$$= 5 \times 1{,}000 \text{ m}$$

$$= 5000. \text{ m} \qquad \text{Move the decimal point 3 places.}$$

$$= 5{,}000 \text{ m}$$

EXAMPLE 2: Change 600 meters to kilometers.

$$600 \text{ m} = 600 \text{ m} \times \frac{1 \text{ km}}{1{,}000 \text{ m}}$$

$$= \frac{600 \text{ km}}{1{,}000}$$

$$= .600 \text{ km} \qquad \text{Move the decimal point 3 places.}$$

$$= .6 \text{ km}$$

EXAMPLE 3: Change 25 centigrams to grams.
Because 100 centigrams = 1 gram,

$$\frac{1 \text{ g}}{100 \text{ cg}} = 1$$

Therefore,

$$25 \text{ cg} = 25 \text{ cg} \times \frac{1 \text{ g}}{100 \text{ cg}}$$

$$= \frac{25 \text{ g}}{100}$$

$$= .25 \text{ g}$$

I told you it was easy. If you have ever worked with conversions with the English system, the conversion formulas you had to know and the arithmetic necessary to get the answers were much more involved. The metric system uses only powers of 10 and the arithmetic of powers of 10.

EXAMPLE 4: Change 65 centiliters to milliliters.
Because the conversion formulas I gave you only compare to liters, change centiliters to liters and then liters to milliliters with two unit fractions.

$$65 \text{ cl} = 65 \text{ cl} \times \underbrace{\frac{1 \, \ell}{100 \text{ cl}}}_{\substack{\text{Changes} \\ \text{cl to } \ell}} \times \underbrace{\frac{1{,}000 \text{ ml}}{1 \, \ell}}_{\substack{\text{Changes} \\ \ell \text{ to ml}}}$$

$$= \frac{65 \times 1{,}000 \text{ ml}}{100}$$

$$= 650 \text{ ml}$$

EXAMPLE 5:   Change 6,250 centigrams to kilograms.

$$6{,}250 \text{ cg} = 6{,}250 \text{ cg} \times \overbrace{\frac{1 \text{ g}}{100 \text{ cg}}}^{\text{cg to g}} \times \overbrace{\frac{1 \text{ kg}}{1{,}000 \text{ g}}}^{\text{g to kg}}$$

$$= .06250 \text{ kg}$$

$$= .0625 \text{ kg}$$

EXAMPLE 6:   Add 15 centimeters and 12 millimeters.
Before adding, change each to the smaller unit—millimeters.

$$15 \text{ cm} = 15 \text{ cm} \times \frac{1 \text{ m}}{100 \text{ cm}} \times \frac{1{,}000 \text{ mm}}{1 \text{ m}}$$

$$= 150 \text{ mm}$$

Therefore,

$$15 \text{ cm} + 12 \text{ mm} = 150 \text{ mm} + 12 \text{ mm}$$

$$= 162 \text{ mm}$$

Temperature   Time in the metric system is measured the same way as in the English system, but temperature in the metric system is measured in Celsius (*C*) degrees.

This scale was set using two temperatures as reference: the boiling point of water (100°C) and the freezing point of water (0°C).

You are probably familiar with the Fahrenheit (*F*) scale. The drawing compares the two temperature scales.

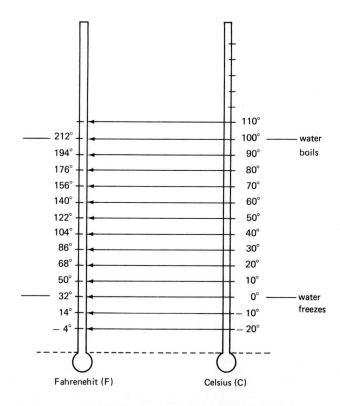

Fahrenehit (F)   Celsius (C)

There are two arithmetic formulas for converting from one scale to the other.

$$F = \frac{9}{5}C + 32$$

$$C = \frac{5}{9}(F - 32)$$

The first formula converts from Celsius to Fahrenheit temperature.

$$F = \frac{9}{5}C + 32$$

Multiply the Celsius temperature (C) by $\frac{9}{5}$,

$$\frac{9}{5}C \qquad \text{Multiplication is indicated by writing them next to each other.}$$

then add 32 to that answer

$$\frac{9}{5}C + 32$$

to get the equivalent Fahrenheit temperature.

$$F = \frac{9}{5}C + 32$$

$$\underbrace{\frac{9}{5}}_{\substack{\text{nine-} \\ \text{fifths} \\ \text{of } C}} \quad \underbrace{+ \; 32}_{\substack{\text{plus} \\ 32}}$$

EXAMPLE 7: Change 25°C to a Fahrenheit temperature. The conversion formula is

$$F = \frac{9}{5}C + 32$$

Take $\frac{9}{5}$ of the Celsius temperature,

$$\frac{9}{5} \times 25 = \frac{9}{\cancel{5}} \times \frac{\cancel{25}^{5}}{1}$$

$$= 45$$

then add 32 to this result.

$$45 + 32 = 77$$

Therefore, 25°C = 77°F.

The formula

$$C = \frac{5}{9}(F - 32)$$

converts a temperature from Fahrenheit to Celsius.

The parentheses group $F - 32$, so this operation should be performed first. Subtract 32 from the Fahrenheit temperature,

$$(F - 32)$$

then multiply the results by $\frac{5}{9}$

$$\frac{5}{9}(F - 32)$$

to obtain the equivalent Celsius temperature.

$$C = \frac{5}{9}(F - 32)$$

EXAMPLE 8: Change 77°F to a Celsius temperature. The conversion formula is

$$C = \frac{5}{9}(F - 32)$$

Subtract 32 from the Fahrenheit temperature,

$$77 - 32 = 45$$

then multiply this result by $\frac{5}{9}$.

$$\frac{5}{9}(45) = \frac{5}{\overset{}{\underset{1}{9}}} \times \frac{\overset{5}{\cancel{45}}}{1}$$

$$= 25$$

Therefore, 77°F = 25°C.

EXAMPLE 9: Change 96°F to a Celsius temperature. Subtract 32 from 96,

$$96 - 32 = 64$$

then find $\frac{5}{9}$ of this result.

$$\frac{5}{9}(64) = \frac{5}{9} \times \frac{64}{1}$$

$$= \frac{320}{9}$$

$$= 35.555\ldots$$

Therefore, 96°F = 35.6°C (to the nearest tenth).

Now it is your turn. Try to work these problems before looking at the solutions.

Examples	Solutions
Change 5 meters to centimeters.	You should have memorized these conversion formulas: $$1 \text{ kilo} = 1{,}000 \text{ units}$$ $$100 \text{ centi} = 1 \text{ unit}$$ $$1{,}000 \text{ milli} = 1 \text{ unit}$$ and these facts: Length is measured in meters. Volume is measured in liters. Weight is measured in grams. Meters and centimeters are related by the formula $$100 \text{ cm} = 1 \text{ m}$$ To change meters to centimeters use the unit fraction that cancels meters and leaves centimeters. $$5 \text{ m} = 5 \text{ m} \times \frac{100 \text{ cm}}{1 \text{ m}}$$ $$= 500 \text{ cm}$$
Change 512.5 milligrams to grams.	$$1{,}000 \text{ mg} = 1 \text{ g}$$ Therefore, $$512.5 \text{ mg} = 512.5 \text{ mg} \times \frac{1 \text{ g}}{1{,}000 \text{ mg}}$$ $$= .5125 \text{ g}$$ $$= .5125 \text{ g}$$
Change 2.5 milliliters to centiliters.	Change milliliters to liters, then liters to centiliters using two unit fractions. $$2.5 \text{ ml} = 2.5 \text{ ml} \times \frac{1 \, \ell}{1{,}000 \text{ ml}} \times \frac{100 \text{ cl}}{1 \, \ell}$$ $$= \frac{2.5 \times 100 \text{ cl}}{1{,}000}$$ $$= .25 \text{ cl}$$
Change 50°C to a Fahrenheit temperature.	$$F = \frac{9}{5} C + 32$$ Take $\frac{9}{5}$ of C,

Examples	Solutions
	$$\frac{9}{5} \times 50 = \frac{9}{\cancel{5}} \times \frac{\cancel{50}^{10}}{1}$$ $$= 90$$ then add 32 to this answer. $$90 + 32 = 122$$ Therefore, $50°C = 122°F$.

Now try these—no hints given.

Change 15.7 meters to kilometers.	.0157 kilometers
Change 51 milligrams to grams.	.051 grams
Change 12 centiliters to milliliters.	120 milliliters

If you had difficulty, please read this section again or talk with an instructor.

Exercises Part I

Answers to these exercises can be found on page 480.

1. Change 6,300 grams to kilograms.
2. Change 1,500 liters to kiloliters.
3. Change 1.23 centimeters to meters.
4. Change .073 kilometers to meters.
5. Change .073 kilograms to grams.
6. Change 5.32 centimeters to millimeters.
7. Change 2 kilograms to centigrams.
8. Change 2.3 millimeters to centimeters.
9. Change 80 meters to kilometers.
10. Change 2.1 kiloliters to liters.

Exercises Part II

1. Change 270 liters to kiloliters.
2. Change .06 kilograms to grams.
3. Change 20 kilograms to centigrams.
4. Change 51 millimeters to centimeters.
5. Change 120 milligrams to centigrams.

REVIEW MATERIAL

Review Skill Conversions in the Metric System

1 kilo = 1,000 units
100 centi = 1 unit
1,000 milli = 1 unit

(a) Change 5 kilograms to grams.

Use the formula

$$1 \text{ kg} = 1,000 \text{ g}$$

Therefore,

$$\frac{1 \text{ kg}}{1,000 \text{ g}} = 1 \quad \text{and} \quad \frac{1,000 \text{ g}}{1 \text{ kg}} = 1$$

Multiply by the unit fraction that cancels kg, leaving g.

$$5 \text{ kg} = 5 \cancel{\text{kg}} \times \frac{1,000 \text{ g}}{1 \cancel{\text{kg}}}$$
$$= 5,000 \text{ g}$$

(b) Change 20 centimeters to millimeters.

Use two unit fractions to change centimeters to meters and meters to millimeters.

$$20 \text{ cm} = 20 \cancel{\text{cm}} \times \frac{1 \cancel{\text{m}}}{100 \cancel{\text{cm}}} \times \frac{1,000 \text{ mm}}{1 \cancel{\text{m}}}$$
$$= 200 \text{ mm}$$

Exercise A   Answers to this exercise can be found on page 480.

1. Change 200 centigrams to grams.
2. Change 4 meters to millimeters.
3. Change 6,520 liters to kiloliters.
4. Change 2 grams to centigrams.
5. Change 50 millimeters to centimeters.

Exercise B   No answers are given.

1. Change 250 centigrams to grams.
2. Change 3 meters to millimeters.
3. Change 7,200 liters to kiloliters.
4. Change 1 kilogram to centigrams.
5. Change 35 millimeters to centimeters.

REVIEW EXERCISES

Refer to the section listed if you have difficulty. Answers to these exercises can be found on page 480.

Section 19.1

1. Change 5 inches to centimeters.
2. Change 99 meters to feet.
3. 10.6 quarts equal how many liters?

Section 19.2

4. Change 600 liters to kiloliters.
5. Change .53 centimeters to meters.
6. Change 1.2 millimeters to meters.
7. Change 5 centimeters to millimeters.

CHAPTER POST-TEST

This test should determine if you have mastered the topics in Chapter 19. Answers to this test can be found on page 480.

1. Change 6 pounds to grams.
2. Change 6.36 quarts to liters.
3. Change 70 liters to kiloliters.
4. Change 1.6 meters to centimeters.
5. Change 6 milligrams to centigrams.

20

measurement and the english system

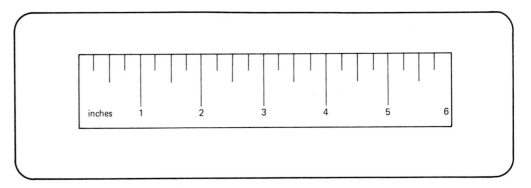

INTRODUCTION

Measurement involves two components: (1) the **amount** and (2) the **unit** of comparison.

For example, the base of my stapler is 5 inches long. The unit of comparison is an inch.

1 Inch

The amount 5 tells me that the base's length is made up of 5 inches.

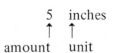

$$5 \quad \text{inches}$$
$$\uparrow \quad \uparrow$$
$$\text{amount} \quad \text{unit}$$

Measurement is a process of describing a quantity by comparing it with a standard unit. The amount is the number of times that the unit is repeated. When I say that the base of my stapler is 5 inches long, I mean that the length of the base is the standard unit of an inch repeated 5 times.

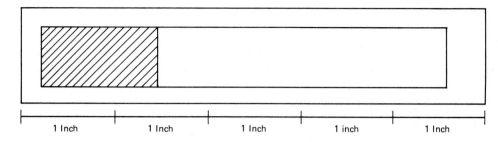

In general, measurements answer questions such as

> How long is it?
>
> How far is it?
>
> How much does it weigh?
>
> How much does it contain?
>
> How large a region is it?

Worldwide standards are necessary today because of international travel, business, and science. There are only two major systems of measurement. In this section, I shall discuss the **English system**. In Section 19.1, I discussed the system used by most of the world—the metric system.

English System

Length or Distance: Inches, feet, yards, and miles are used to measure along a straight line. The relationship of these units to each other and their usual abbreviations are shown.

$$
\begin{array}{l}
1 \text{ foot (ft)} = 12 \text{ inches (in.)} \\
1 \text{ yard (yd)} = 3 \text{ feet (ft)} \\
1 \text{ mile (mi)} = 5{,}280 \text{ feet (ft)}
\end{array}
$$

These three relationships are needed to work with the English systems of length or distances. The figures illustrate the relationships.

Given that along a straight line, this is an inch,

1 foot is the standard inch repeated 12 times along a straight line.

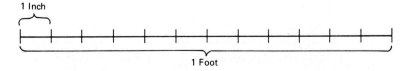

One yard is a foot repeated three times along a straight line.

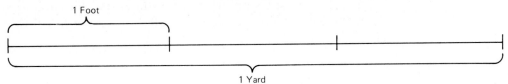

1 Foot

1 Yard

One mile is a foot repeated 5,280 times, or a yard repeated 1,760 times, along a straight line.

Area: Square inches, square feet, square yards, square miles, and acre are used to measure a bounded region on a surface. An inch is used to measure along a straight line. A square inch is used to measure a region.

Consider the region determined by the base of my stapler.

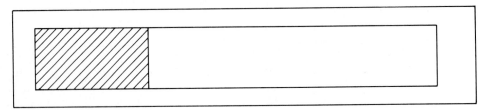

The unit *inch* is used to describe the length and width of the base.

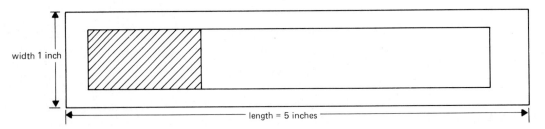

width 1 inch

length = 5 inches

Both measurements are along a straight line. But what is the size of the region?

One standard unit of comparison is a square inch—a region with both length and width measuring 1 inch.

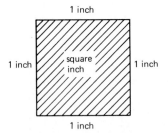

1 inch

1 inch square inch 1 inch

1 inch

To describe the area of the base of my stapler, I can say it is 5 square inches. That is, the region is made up of 5 square inches—1 square inch repeated 5 times over a surface.

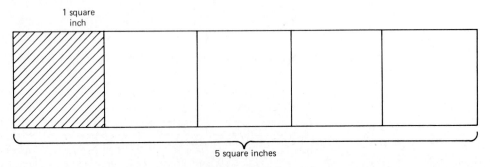

1 square inch

5 square inches

The other units of area are similarly defined. My house is 15,000 square feet large. This means the floor space consists of a square foot, repeated 15,000 times over a surface.

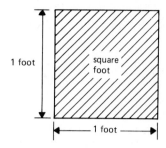

An acre of land is 43,560 square feet—a square foot, repeated 43,560 times over a surface—a very large area!

Time: Seconds, minutes, hours, days, weeks, and years are used to measure time. The relationship of these units to each other and their usual abbreviations are shown.

1 minute (min) = 60 seconds (s)

1 hour (h) = 60 minutes (min)

1 day (d) = 24 hours (h)

1 week (wk) = 7 days (d)

1 year (yr) = 52 weeks (wk) or 365 days (d)

These units are based on a solar day—the interval of time that it takes the earth to turn once on its axis.

Speed: Some of the common units of speed are feet per second (ft/s) and miles per hour (mi/h).

These units describe the amount of distance traveled in one unit of time. For example,

15 feet per second

↑
amount   unit

describes an object traveling 15 feet in distance during 1 second in time. The units of speed are usually written in fractional form, with the unit following the word *per* as the divisor or denominator.

The expression

55 miles per hour

written

55 mi/h

describes an object that takes 1 hr to move the distance of 55 miles.

Volume: Cubic inches, pints, quarts, and gallons are used to measure volume. The relationship of these units to each other and their usual abbreviations are shown.

$$1 \text{ quart (qt)} = 2 \text{ pints (pt)}$$
$$1 \text{ gallon (gal)} = 4 \text{ quarts or } 231 \text{ cubic inches (cu in.)}$$

Inches are used to measure distance in a straight line.

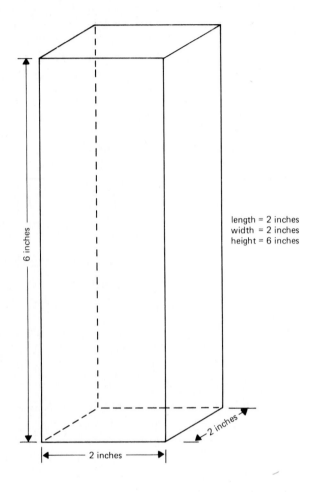

length = 2 inches
width = 2 inches
height = 6 inches

6 inches

2 inches

2 inches

Square inches are used to measure area of a region over a surface.

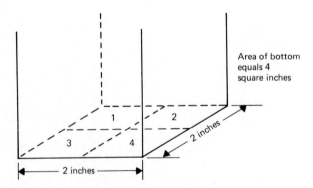

Area of bottom
equals 4
square inches

1   2

3   4

2 inches

2 inches

How much will the carton hold? One standard unit of measure for comparison is a cubic inch—a cube whose length, width, and height each equals 1 inch.

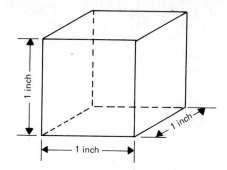

The carton is made up of a cubic inch repeated 24 times in space. Therefore, the carton will hold 24 cubic inches.

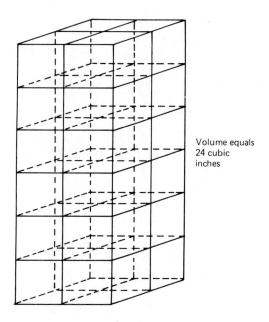

Volume equals 24 cubic inches

A cubic inch repeated 231 times in space equals 1 gallon.

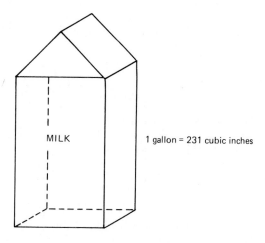

MILK

1 gallon = 231 cubic inches

Weight: Ounces, pounds, and tons are used to measure weight. These units are related to each other as shown.

$$1 \text{ pound (lb)} = 16 \text{ ounces (oz)}$$
$$1 \text{ ton} = 2,000 \text{ pounds (lb)}$$

Weight is a measure of how heavy something is, whereas volume is a measure of the space it occupies. Today most dry materials are sold by weight, rather than by volume.

Conclusion   Measurement is a process of describing a quantity by comparing it with a standard unit. The amount is the number of times the unit is repeated.

I have not covered all units in the English system, nor all types of measurement possible. These examples were only an introduction to the English system.

SECTION 20.2
CONVERSION AND UNIT FRACTIONS

$$96 \text{ inches} = 8 \text{ feet}$$

Conversions   Conversions—changing from one unit to another—are more complicated in the English system than in the metric system (discussed earlier).

For example,

13 yards equal how many inches?

You must remember two facts from Section 20.1:

$$1 \text{ ft} = 12 \text{ in.}$$
$$1 \text{ yd} = 3 \text{ ft}$$

Because each yard is equal to 3 feet, 13 yards would be 3 feet repeated 13 times, or

$$3 \text{ yd} = 13 \times 3 \text{ ft}$$
$$= 39 \text{ ft}$$

Because each foot is equal to 12 inches, 39 feet would be 12 inches repeated 39 times, or

$$39 \text{ ft} = 39 \times 12 \text{ in.}$$
$$= 468 \text{ in.}$$

Therefore,

$$23 \text{ yd} = 468 \text{ in.}$$

Now try the conversion in the other direction. For example,

$$96 \text{ inches equal how many feet?}$$

Because 12 inches equal 1 foot, it is necessary to find the number of 12's in 96 by division.

$$96 \text{ in.} = \frac{96}{12} \text{ ft}$$

$$= 8 \text{ ft}$$

The one fact

$$1 \text{ ft} = 12 \text{ in.}$$

is used to change feet to inches by multiplying by 12, and to change inches to feet by dividing by 12.

Conversions in the English system are complex for two reasons. First, you must memorize (or have a list handy of) each of the conversion equalities, such as

$$1 \text{ ft} = 12 \text{ in.}$$

Second, you must determine which operation—multiplication or division—is appropriate in order to make the conversion.

The first problem is a difficulty of the system. There is no uniformity in the English system. For example, it takes 2 pints to make a quart, but 4 quarts to make a gallon, and it takes 16 ounces to make a pound, but 2,000 pounds to make a ton.

The second problem, however, can be dealt with by using unit fractions.

Unit Fractions   The important fact to remember in this section is the unique property of 1. The product of a number and 1 is the number itself.

$$5 \times 1 = 5$$

Different forms of 1 were used to build fractions

$$\frac{2}{3} = \frac{2 \times 4}{3 \times 4}$$

$$= \frac{8}{12}$$

The fraction $\frac{4}{4}$ equals 1, and the two fractions $\frac{2}{3}$ and $\frac{8}{12}$ are equal, although they have different forms.

Remember: Although the form of a number can change when it is multiplied by 1, its value remains unchanged.

The conversion formula

$$1 \text{ ft} = 12 \text{ in.}$$

can be changed into two **unit fractions**:

$$\frac{1 \text{ ft}}{12 \text{ in.}} = 1 \quad \text{and} \quad \frac{12 \text{ in.}}{1 \text{ ft}} = 1$$

Conversion is then a matter of choosing the fraction that will cancel the unwanted unit, leaving the unit desired. For example,

39 feet equals how many inches?

Multiply by the unit fraction that has feet in the denominator to cancel the feet in 39 feet, leaving the unit inches.

$$39 \text{ ft} = 39 \, \cancel{\text{ft}} \times \frac{12 \text{ in.}}{1 \, \cancel{\text{ft}}}$$

$$= \frac{468 \text{ ft}}{1}$$

$$= 468 \text{ ft}$$

This procedure works because the product of a number and 1 is the number itself (possibly unchanged in form).

Go back and compare this procedure with the thought process used in the first part of this section. I hope that you will agree that the unit fraction conversion is easier.

Unit Fraction Conversions

The following summarizes the conversion procedure using unit fractions:

1. Use the conversion formula that relates the two units of the problem to form a unit fraction.
2. Multiply by the unit fraction that has the old unit (the one you wish to cancel or change) in the denominator and the new unit in the numerator.

This procedure makes it simple to convert units with which you are not even familiar.

EXAMPLE 1: Assume that 1 peck (pk) = 2 gallons (gal). Nine gallons equal how many pecks?

Do I multiply or divide by 2? If you are unfamiliar with pecks, this may be a difficult decision. Unit fractions make the decision for you!

Because 1 peck = 2 gallons, there are two unit fractions:

$$\frac{1 \text{ pk}}{2 \text{ gal}} = 1 \quad \text{and} \quad \frac{2 \text{ gal}}{1 \text{ pk}} = 1$$

I want to cancel gallons and leave pecks, so I use the unit fraction that has gallons in the denominator and pecks in the numerator.

$$\frac{1 \text{ pk}}{2 \text{ gal}} = 1$$

Therefore,

$$9 \text{ gal} = 9 \text{ gal} \times \frac{1 \text{ pk}}{2 \text{ gal}} \qquad \begin{array}{l} \text{Cancels gallons,} \\ \text{leaves pecks.} \end{array}$$

$$= \frac{9 \text{ pk}}{2}$$

$$= 4\frac{1}{2} \text{ pk}$$

EXAMPLE 2: 5 minutes equal how many seconds?

Use the formula, 1 minute = 60 seconds. To cancel minutes, put 1 minute in the denominator of the unit fraction and 60 seconds in the numerator.

$$\frac{60 \text{ s}}{1 \text{ min}} = 1$$

This unit fraction will cancel minutes and leave seconds as the new unit. Therefore,

$$5 \text{ min} = 5 \text{ min} \times \frac{60 \text{ s}}{1 \text{ min}}$$

$$= 300 \text{ s}$$

EXAMPLE 3: 9 pints equal how many quarts?

The conversion formula that relates pints and quarts is

$$1 \text{ qt} = 2 \text{ pt}$$

Use the unit fraction

$$\frac{1 \text{ qt}}{2 \text{ pt}} = 1$$

to cancel pints and leave the answer in quarts.

$$9 \text{ pt} = 9 \text{ pt} \times \frac{1 \text{ qt}}{2 \text{ pt}}$$

$$= \frac{9 \text{ qt}}{2}$$

$$= 4\frac{1}{2} \text{ qt}$$

EXAMPLE 4: Change 19 quarts to gallons and quarts.

Remember that 1 gallon = 4 quarts.

$$19 \text{ qt} = 19 \text{ qt} \times \frac{1 \text{ gal}}{4 \text{ qt}}$$

$$= \frac{19 \text{ gal}}{4}$$

$$= 4 \text{ gal with a remainder of 3 qt}$$

$$= 4 \text{ gal 3 qt}$$

Instead of writing $4\frac{3}{4}$ gallons, you can leave the remainder in the old unit of quarts.

> When converting a smaller unit to a larger unit, you may either leave the answer as a mixed number or use the smaller unit to express the remainder.

EXAMPLE 5:   Write 91 pints as gallons, quarts, and pints.
First, change 91 pints to quarts and pints. Remember that 1 quart = 2 pints.

$$91 \text{ pt} = 91 \text{ pt} \times \frac{1 \text{ qt}}{2 \text{ pt}}$$

$$= \frac{91 \text{ qt}}{2}$$

$$= 45 \text{ qt with a remainder of 1 pt}$$

$$= 45 \text{ qt } 1 \text{ pt}$$

Now change the 45 quarts to gallons and quarts.

$$45 \text{ qt} = 45 \text{ qt} \times \frac{1 \text{ gal}}{4 \text{ qt}}$$

$$= \frac{45 \text{ gal}}{4}$$

$$= 11 \text{ gal with a remainder of 1 qt}$$

Here are the two steps, written together.

$$
\begin{array}{lll}
91 \text{ pt} = & \quad\quad 91\text{ pt} & \text{—— canceled and —} \\
& +45 \text{ qt} \quad 1 \text{ pt} \leftarrow & \text{—— changed to ——} \\
& \overline{\quad 45 \text{ qt} \quad 1 \text{ pt}} \\
\end{array}
$$

$$
\begin{array}{lll}
& \quad 45\text{ qt} \quad 1 \text{ pt} & \text{—— canceled and —} \\
& +11 \text{ gal} \quad 1 \text{ qt} \leftarrow & \text{——— changed to ——} \\
= & \overline{11 \text{ gal} \quad 1 \text{ qt} \quad 1 \text{ pt}} \\
\end{array}
$$

Therefore, 91 pints = 11 gallons 1 quart 1 pint.

In some problems it is necessary to use two conversion formulas and therefore two unit fractions to make the conversion.

EXAMPLE 6:   Change 4 gallons to pints.
Remember that 1 gallon = 4 quarts and that 1 quart = 2 pints.
Change gallons to quarts and quarts to pints with two unit fractions.

changes gal to qt ⌐ changes qt to pt

$$4 \text{ gal} = 4 \text{ gal} \times \frac{4 \text{ qt}}{1 \text{ gal}} \times \frac{2 \text{ pt}}{1 \text{ qt}}$$

$$= 4 \times 4 \times 2 \text{ pt}$$

$$= 32 \text{ pt}$$

EXAMPLE 7:   Change 108 inches to yards.

$$108 \text{ in.} = 108 \text{ in.} \times \frac{1 \text{ ft}}{12 \text{ in.}} \times \frac{1 \text{ yd}}{3 \text{ ft}}$$

where the first fraction *changes in. to ft* and the second *changes ft to yd*.

$$= \frac{108 \text{ yd}}{36}$$

$$= 3 \text{ yd}$$

Now you try these. You may need to refer to Section 20.1 for the conversion formulas.

Examples	Solutions
Change 6 feet to inches.	The conversion formula is $$1 \text{ ft} = 12 \text{ in.}$$ Two unit fractions can be formed. $$\frac{1 \text{ ft}}{12 \text{ in.}} = 1 \quad \text{and} \quad \frac{12 \text{ in.}}{1 \text{ ft}} = 1$$ To cancel feet and leave inches, place the feet in the denominator. $$\frac{12 \text{ in.}}{1 \text{ ft}} = 1$$ Therefore, $$6 \text{ ft} = 6 \text{ ft} \times \frac{12 \text{ in.}}{1 \text{ ft}}$$ $$= 72 \text{ in.}$$
Change 120 seconds to minutes.	Remember that 1 minute = 60 seconds. To cancel seconds, use $$\frac{1 \text{ min}}{60 \text{ s}} = 1$$ Therefore, $$120 \text{ s} = 120 \text{ s} \times \frac{1 \text{ min}}{60 \text{ s}}$$ $$= \frac{120 \text{ min}}{60}$$ $$= 2 \text{ min}$$
Change 9 pints to quarts.	Use the formula $$1 \text{ qt} = 2 \text{ pt}$$

Examples	Solutions
	To cancel pints, use $$\frac{1 \text{ qt}}{2 \text{ pt}} = 1$$ Therefore, $$9 \text{ pt} = 9 \text{ pt} \times \frac{1 \text{ qt}}{2 \text{ pt}}$$ $$= \frac{9 \text{ qt}}{2}$$ $$= 4\frac{1}{2} \text{ qt}$$
57 inches equal how many feet and inches?	Use the formula $$1 \text{ ft} = 12 \text{ in.}$$ Therefore, $$57 \text{ in.} = 57 \text{ in.} \times \frac{1 \text{ ft}}{12 \text{ in.}}$$ $$= \frac{57 \text{ ft}}{12}$$ = 4 ft with a remainder of 9 in. = 4 ft 9 in. Instead of writing a fraction—$4\frac{9}{12}$ or $4\frac{3}{4}$—I leave the remainder in the old unit.
Change 2 yards to inches.	Use the formulas $$1 \text{ yd} = 3 \text{ ft}$$ $$1 \text{ ft} = 12 \text{ in.}$$ Change yards to feet with $$\frac{3 \text{ ft}}{1 \text{ yd}}$$ and feet to inches with $$\frac{12 \text{ in.}}{1 \text{ ft}}$$ Therefore, $$2 \text{ yd} = 2 \text{ yd} \times \frac{3 \text{ ft}}{1 \text{ yd}} \times \frac{12 \text{ in.}}{1 \text{ ft}}$$ $$= \frac{6 \times 12 \text{ in.}}{1}$$ = 72 in.

Examples	Solutions
Now try these—no hints given.	
Change 6 feet to inches.	72 inches
Change 144 inches to feet.	12 feet
Change 15 gallons to pints.	120 pints
Change 70 ounces to pounds and ounces.	4 pounds 6 ounces

Ready for the exercises?

Exercises Part I

Answers to these exercises can be found on page 480.

Perform the indicated conversions.

1. Change 7 yards to feet.
2. Change 84 feet to yards.
3. Change 2 feet to inches.
4. Change 16 inches to feet.
5. Change 72 inches to yards.
6. Change 17 yards to feet.
7. Change 7 miles to feet.
8. Change 89 feet to yards and feet.
9. Change 79 inches to feet and inches.
10. Change 17 feet to yards and feet.
11. Change 29 inches to feet and inches.
12. Change 5 minutes to hours.
13. Change 120 minutes to hours.
14. Change 2 quarts to pints.
15. Change 9 pints to quarts.
16. Change 16 quarts to gallons.
17. Change 18 quarts to gallons and quarts.
18. Change 3 pounds to ounces.
19. Change 92 ounces to pounds and ounces.
20. Change 2 tons to ounces.

Exercises Part II

1. Change 7 feet to inches.
2. Change 84 inches to yards.
3. Change 360 minutes to hours.
4. Change 19 quarts to gallons and quarts.
5. Change 90 ounces to pounds and ounces.

IT'S A BOY! 8 lb 3 oz

Introduction   You have learned arithmetic operations with numbers. When doing arithmetic with numbers of measure—**denominate numbers**—you must consider two components: (1) amount and (2) unit.

Addition   EXAMPLE 1:   Add 3 feet 5 inches and 7 feet 9 inches.

Method One   Change each to inches, the smaller unit.

$$3 \text{ ft } 5 \text{ in.} = \quad 3 \text{ ft} \quad \text{ and } 5 \text{ in.}$$

$$= 3 \text{ ft} \times \frac{12 \text{ in.}}{1 \text{ ft}} \text{ and } 5 \text{ in.} \qquad \text{Change feet to inches.}$$

$$= 36 \text{ in.} \qquad \text{ and } 5 \text{ in.} \qquad \text{Add up the total number of inches.}$$

$$= 41 \text{ in.}$$

$$7 \text{ ft } 9 \text{ in.} = \quad 7 \text{ ft} \quad \text{ and } 9 \text{ in.}$$

$$= 7 \text{ ft} \times \frac{12 \text{ in.}}{1 \text{ ft}} \text{ and } 9 \text{ in.} \qquad \text{Change feet to inches.}$$

$$= 84 \text{ in.} \qquad \text{ and } 9 \text{ in.} \qquad \text{Add up the total number of inches.}$$

$$= 93 \text{ in.}$$

Now adding,

$$3 \text{ ft } 5 \text{ in.} + 7 \text{ ft } 9 \text{ in.} = 41 \text{ in.} + 93 \text{ in.}$$

$$= 134 \text{ in.}$$

This is the answer; however, if I wish the answer to be in the same form as the question, I need to write 134 inches in both feet and inches.

$$134 \text{ in.} = 134 \text{ in.} \times \frac{1 \text{ ft}}{12 \text{ in.}}$$

$$= \frac{134 \text{ ft}}{12}$$

$$= 11 \text{ ft with a remainder of 2 in.}$$

$$= 11 \text{ ft } 2 \text{ in.}$$

Therefore, 3 feet 5 inches + 7 feet 9 inches = 11 feet 2 inches.

Method Two Add feet to feet and inches to inches—that is, add using the same units.

$$3 \text{ ft } \quad 5 \text{ in.}$$
$$+ \; 7 \text{ ft } \quad 9 \text{ in.}$$
$$\overline{10 \text{ ft } \; 14 \text{ in.}}$$

With the answer in this form, 14 inches contains a foot. I shall change 14 inches to feet and inches to simplify the answer.

10 ft 14 in. = 10 ft and 14 in.

$$= 10 \text{ ft and } 14 \text{ in.} \times \frac{1 \text{ ft}}{12 \text{ in.}} \qquad \text{Converting inches to feet and inches.}$$

= 10 ft and 1 ft with a remainder of 2 in.

= <u>10 ft</u> and <u>1 ft</u> 2 in. \qquad Adding the additional foot.

= 11 ft and 2 in.

Notice that the result of both methods is the same. It is always possible to change each denominate number to the smaller unit (Method One) before adding, subtracting, multiplying, or dividing. I prefer Method Two and shall use it for problems in which the denominate numbers include more than one unit.

In the solution of the problem using Method Two, I simplified 10 feet 14 inches to 11 feet 2 inches.

$$10 \text{ ft } \; \cancel{14 \text{ in.}} \; \longrightarrow \; \text{14 in. canceled and changed to} \;\rceil$$
$$+ \; 1 \text{ ft } \quad 2 \text{ in. } \longleftarrow$$
$$= \quad \overline{11 \text{ ft } \quad 2 \text{ in.}} \qquad \text{because 14 in. equals 1 ft 2 in.}$$

Simplifying Units

When denominate numbers include more than one unit, change any smaller unit into a larger unit whenever possible.

Addition of Denominate Numbers

Add like units and simplify the units that are used.

EXAMPLE 2: Add 5 pounds 2 ounces and 2 pounds 15 ounces.
Add like units.

$$5 \text{ lb } \quad 2 \text{ oz}$$
$$+ 2 \text{ lb } \; 15 \text{ oz}$$
$$\overline{7 \text{ lb } \; 17 \text{ oz}}$$

This is the answer, but it is not considered simplified because 17 ounces contains a pound.

Change 17 ounces to pounds and ounces.

$$17 \text{ oz} = 17 \text{ oz} \times \frac{1 \text{ lb}}{16 \text{ oz}}$$

$$= \frac{17 \text{ lb}}{16}$$

= 1 lb with a remainder of 1 oz

= 1 lb 1 oz

Therefore, changing 17 ounces to 1 pound 1 ounce and adding,

$$
\begin{array}{r}
7 \text{ lb } \; \cancel{17 \text{ oz}} \; \text{— canceled and changed to —} \\
+1 \text{ lb } \quad 1 \text{ oz } \longleftarrow \\
\hline
= \quad 8 \text{ lb } \quad 1 \text{ oz} \qquad \text{Simplified answer.}
\end{array}
$$

Therefore, 5 pounds 2 ounces + 2 pounds 15 ounces = 8 pounds 1 ounce.

EXAMPLE 3:   Add 2 gallons 3 quarts 1 pint and 3 gallons 2 quarts 1 pint.

Add like units.

$$
\begin{array}{r}
2 \text{ gal } \; 3 \text{ qt } \; 1 \text{ pt} \\
+3 \text{ gal } \; 2 \text{ qt } \; 1 \text{ pt} \\
\hline
5 \text{ gal } \; 5 \text{ qt } \; 2 \text{ pt}
\end{array}
$$

Simplify the expression starting with the smallest unit of measure.

First, 2 pints contain a quart because 2 pints = 1 quart.

$$
2 \text{ pt} = 2 \cancel{\text{ pt}} \times \frac{1 \text{ qt}}{2 \cancel{\text{ pt}}}
$$

$$
= 1 \text{ qt and } 0 \text{ pt remaining}
$$

Therefore,

$$
\begin{array}{r}
5 \text{ gal } \; 5 \text{ qt } \; \cancel{2 \text{ pt}} \; \text{— 2 pt canceled and changed to —} \\
+ \qquad\quad 1 \text{ qt } \; 0 \text{ pt } \longleftarrow \\
\hline
= \quad 5 \text{ gal } \; 6 \text{ qt} \qquad \text{because 2 pt = 1 qt}
\end{array}
$$

Next, consider the quarts. Because 4 quarts = gallon, 6 quarts contain a gallon. Change 6 quarts to gallons and quarts to simplify 5 gallons 6 quarts.

$$
6 \text{ qt} = 6 \cancel{\text{ qt}} \times \frac{1 \text{ gal}}{4 \cancel{\text{ qt}}}
$$

$$
= 1 \text{ gal with a remainder of } 2 \text{ qt}
$$

$$
= 1 \text{ gal } 2 \text{ qt}
$$

Therefore,

$$
\begin{array}{r}
5 \text{ gal } \; \cancel{6 \text{ qt}} \; \text{— 6 qt canceled and changed to —} \\
+1 \text{ gal } \; 2 \text{ qt } \longleftarrow \\
\hline
= \quad 6 \text{ gal } \; 2 \text{ qt} \qquad \text{because 6 qt = 1 gal 2 qt}
\end{array}
$$

Thus, 5 gallons 5 quarts 2 pints equal 6 gallons 2 quarts. This process is considered necessary to simply denominate numbers with more than one unit. Here are the steps again.

Start with the smallest unit.

$$
\begin{array}{r}
5 \text{ gal } \; 5 \text{ qt } \; \cancel{2 \text{ pt}} \; \text{— 2 pt canceled and changed to —} \\
+ \qquad\quad 1 \text{ qt } \; 0 \text{ pt } \longleftarrow \\
\hline
= \quad 5 \text{ gal } \; 6 \text{ qt}
\end{array}
$$

Next, consider the quarts.

$$
\begin{array}{r}
5 \text{ gal } \; \cancel{6 \text{ qt}} \; \text{— 6 qt canceled and changed to —} \\
+1 \text{ gal } \; 2 \text{ qt } \longleftarrow \\
\hline
= \quad 6 \text{ gal } \; 2 \text{ qt}
\end{array}
$$

Now no unit can be changed into the next larger unit.

EXAMPLE 4:   Add 6 feet 5 inches and 7 feet 9 inches.
Add like units.

$$
\begin{array}{r}
6\ \text{ft} \quad 5\ \text{in.} \\
+\ \underline{7\ \text{ft} \quad 9\ \text{in.}} \\
13\ \text{ft} \quad 14\ \text{in.}
\end{array}
$$

Simplify the expression.

$$
\begin{array}{l}
\quad 13\ \text{ft} \quad \cancel{14\ \text{in.}} \longrightarrow \text{canceled and changed to} \rightharpoondown \\
+\ \underline{\ 1\ \text{ft} \quad 2\ \text{in.}} \leftarrow \\
=\ \ 14\ \text{ft} \quad 2\ \text{in.} \qquad \text{because 14 in. = 1 ft 2 in.}
\end{array}
$$

Therefore, the sum is 14 feet 2 inches.

Subtraction

> **Subtraction of Denominate Numbers**
>
> Subtract like units, borrowing whenever necessary.

EXAMPLE 5:   Subtract 3 feet 5 inches from 7 feet 9 inches.
Subtraction of like units sometimes requires borrowing. In this problem, it does not.

$$
\begin{array}{r}
7\ \text{ft}\ \ 9\ \text{in.} \\
-\underline{3\ \text{ft}\ \ 5\ \text{in.}} \\
4\ \text{ft}\ \ 4\ \text{in.}
\end{array}
$$

EXAMPLE 6:   Subtract 2 minutes 55 seconds from 5 minutes 5 seconds.
It will be necessary to borrow because 55 seconds cannot be subtracted from 5 seconds.

$$
\begin{array}{r}
5\ \text{min}\ \ \ 5\ \text{s} \\
-\underline{2\ \text{min}\ 55\ \text{s}}
\end{array}
$$

Because 1 minute = 60 seconds, borrowing 1 minute adds 60 seconds to the seconds column.

$$
\begin{array}{rl}
4\ \text{min}\ 65\ \text{s} & \text{60 s borrowed plus 5 s} \\
\cancel{5\ \text{min}}\ \ \cancel{5\ \text{s}} & \text{already there equals} \\
-\underline{2\ \text{min}\ 55\ \text{s}} & \text{65 s.} \\
2\ \text{min}\ 10\ \text{s} &
\end{array}
$$

EXAMPLE 7:   Subtract 2 gallons 3 quarts from 5 gallons 1 quart.
Borrow from the gallons in order to subtract.

$$
\begin{array}{rl}
4\ \text{gal}\ 5\ \text{qt} & \text{Because 1 gal = 4 qt, borrowing} \\
\cancel{5\ \text{gal}}\ \ \cancel{1\ \text{qt}} & \text{1 gal adds 4 qt to the 1 qt, giving} \\
-\underline{2\ \text{gal}\ 3\ \text{qt}} & \text{a total of 5 qt.} \\
2\ \text{gal}\ 2\ \text{qt} &
\end{array}
$$

Multiplication

> **Multiplication by a Number**
>
> To multiply a denominate number by a number, multiply each part by the number and simplify, expressing the answer in the units of the problem.

EXAMPLE 8:   Multiply 5 feet 2 inches by 7.
Multiply each part by 7.

$$
\begin{array}{r}
5 \text{ ft} \quad 2 \text{ in.} \\
\times \qquad\quad 7 \\
\hline
35 \text{ ft} \quad 14 \text{ in.}
\end{array}
$$

Now simplify, starting with the smallest unit.

$$
\begin{array}{rl}
& 35 \text{ ft} \quad \cancel{14 \text{ in.}} \text{ — canceled and changed to} \\
+ & \underline{1 \text{ ft} \quad 2 \text{ in.}} \leftarrow \\
= & 36 \text{ ft} \quad 2 \text{ in.} \qquad \text{because 14 in. = 1 ft 2 in.}
\end{array}
$$

Therefore, the product in simplified form is 36 feet 2 inches. It is not necessary to change 36 feet to yards because the units in the problem involve only feet and inches.

Division

Division by a Number

To divide a denominate number by a number, divide each part by the number, carrying any remainder to the next unit.

EXAMPLE 9:   Divide 3 minutes 16 seconds by 2.
Divide each part by 2.

$$
\begin{array}{r}
1 \text{ min} \qquad\qquad \\
2 \overline{)\, 3 \text{ min} \quad 16 \text{ s}} \\
\underline{2 \text{ min}} \qquad\qquad \\
1 \text{ min} = 60 \text{ s} \\
\underline{\phantom{1 \text{ min} =}} \\
76 \text{ s}
\end{array}
$$

Remaining minute must be converted to seconds and added to seconds in order to continue the division.

Now divide 2 into 76 seconds.

$$
\begin{array}{r}
1 \text{ min} \quad 38 \text{ s} \\
2 \overline{)\, 3 \text{ min} \quad 16 \text{ s}} \\
\underline{2 \text{ min}} \qquad\qquad \\
1 \text{ min} = 60 \text{ s} \\
\underline{} \\
76 \text{ s} \\
\underline{76 \text{ s}} \\
0
\end{array}
$$

EXAMPLE 10:   Divide 2 gallons 3 quarts 1 pint by 3.
First, 3 will not divide 2 gallons.

$$
\begin{array}{r}
0 \text{ gal} \qquad\qquad\qquad \\
3 \overline{)\, 2 \text{ gal} \quad 3 \text{ qt} \quad 1 \text{ pt}} \\
\underline{0 \text{ gal}} \qquad\qquad\qquad \\
2 \text{ gal} = \quad 8 \text{ qt} \\
\underline{} \\
11 \text{ qt}
\end{array}
$$

The 2 gal must be changed to quarts to continue the division.

$$
2 \text{ gal} = 2 \text{ gal} \times \frac{4 \text{ qt}}{1 \text{ gal}}
$$

$$
= 8 \text{ qt}
$$

Now there are 11 quarts total.

Now divide 3 into 11 qt.

$$
\begin{array}{r}
\;\;0\text{ gal}\quad\;\;3\text{ qt}\phantom{\;\;1\text{ pt}} \\ \hline
3\,\big|\;\;2\text{ gal}\quad\;\;3\text{ qt}\quad 1\text{ pt} \\
\;\;\underline{0\text{ gal}}\phantom{\quad\;\;3\text{ qt}\quad 1\text{ pt}} \\
\;\;2\text{ gal} =\;\underline{8\text{ qt}}\phantom{\quad 1\text{ pt}} \\
\phantom{3\,|2\text{ gal} =}\;11\text{ qt}\phantom{\quad 1\text{ pt}} \\
\phantom{3\,|2\text{ gal} =\;1}\underline{9\text{ qt}}\phantom{\quad 1\text{ pt}}
\end{array}
$$

2 qt = $\dfrac{4\text{ pt}}{5\text{ pt}}$ The 2 qt is changed into pints and added to the pints already there.

Finally, divide 3 into 5 pt.

$$
\begin{array}{r}
\;\;0\text{ gal}\quad\;\;3\text{ qt}\quad 1\tfrac{2}{3}\text{ pt} \\ \hline
3\,\big|\;\;2\text{ gal}\quad\;\;3\text{ qt}\quad 1\text{ pt} \\
\;\;\underline{0\text{ gal}} \\
\;\;2\text{ gal} =\;\underline{8\text{ qt}} \\
\phantom{3\,|2\text{ gal} =}\;11\text{ qt} \\
\phantom{3\,|2\text{ gal} =\;1}\underline{9\text{ qt}}
\end{array}
$$

2 qt = $\dfrac{4\text{ pt}}{5\text{ pt}}$

$\dfrac{3\text{ pt}}{2\text{ pt}}$ 3 divides 5 pt only once with a remainder of 2 pt.

Because 3 will not divide the remaining 2 pints, write the quotient as a mixed number,

$$\text{quotient} \longrightarrow 1\tfrac{2}{3}\text{ pt} \begin{array}{l} \longleftarrow \text{ remainder} \\ \longleftarrow \text{ divisor} \end{array}$$

Therefore, 2 gallons 3 quarts 1 pint $\div$ 3 = 3 quarts $1\tfrac{2}{3}$ pints. Notice that the 0 gallon is dropped.

Now it is your turn.

Examples	Solutions
Simplify 5 gallons 9 quarts.	Start with the smaller unit and try to write it using the next larger unit. $$9\text{ qt} = 9\text{ qt} \times \frac{1\text{ gal}}{4\text{ qt}}$$ $$= \frac{9\text{ qt}}{4}$$ $$= 2\text{ gal }1\text{ qt}$$ Therefore, $$\begin{array}{r} 5\text{ gal }\;9\text{ qt} \longrightarrow \\ +\,2\text{ gal }\;1\text{ qt} \longleftarrow \\ \hline 7\text{ gal }\;1\text{ qt} \end{array}$$ Simplified answer.

Examples	Solutions
Simplify 6 feet 19 inches to feet and inches.	This expression needs to be simplified because 19 inches contain a foot. $$19 \text{ in.} = 19 \text{ in.} \times \frac{1 \text{ ft}}{12 \text{ in.}}$$ $$= \frac{19 \text{ ft}}{12}$$ $$= 1 \text{ ft } 7 \text{ in.}$$ Therefore, I change 19 inches into 1 foot 7 inches to simplify. 6 ft 9 in. +1 ft 7 in. —————— 7 ft 7 in. The expression cannot be further simplified.
Add 1 hour 5 minutes and 6 hours 57 minutes.	Add like units. 1 h 5 min +6 h 57 min —————— 7 h 62 min Now simplify by changing 62 minutes into hours and minutes. 7 h 62 min +1 h 2 min —————— 8 h 2 min Simplified answer.
Subtract 5 pounds 14 ounces from 7 pounds 2 ounces.	It will be necessary to borrow. 7 lb 2 oz −5 lb 14 oz Because 1 pound = 16 ounces, borrowing 1 pound adds 16 ounces to the 2 ounces already there. 6 lb 18 oz 7 lb 2 oz −5 lb 14 oz —————— 1 lb 4 oz Simplified answer.
Multiply 5 feet 6 inches by 7.	Multiply each part by 7 and simplify. 5 ft 6 in. × 7 —————— 35 ft 42 in. 3 ft 6 in. —————— 38 ft 6 in. Simplified answer.

Examples	Solutions
Divide 13 pounds 15 ounces by 3.	Divide each part by 3, converting any remainder to the next unit.

$$
\begin{array}{r}
\quad\quad 4\text{ lb}\quad\ 10\text{ oz} \\
3\overline{)\ 13\text{ lb}\quad 14\text{ oz}} \\
\underline{12\text{ lb}}\quad\quad\quad \\
\underline{1\text{ lb} = 16\text{ oz}} \\
30\text{ oz} \\
\underline{30\text{ oz}} \\
0
\end{array}
$$

Now try these.

Add 6 pounds 12 ounces and 7 pounds 4 ounces.	14 pounds
Subtract 1 hour 57 minutes from 3 hours 5 minutes.	1 hour 8 minutes
Multiply 6 feet 5 inches by 9.	57 feet 9 inches.
Divide 15 quarts 6 pints by 6.	2 quarts 2 pints

Please read this section again or talk with an instructor if you had difficulty with these problems.

Exercises Part I

Answer to these exercises can be found on page 480.

Simplify these denominate numbers.

1. 5 feet 17 inches
2. 5 weeks 8 days
3. 1 quart 5 pints
4. 2 pounds 31 ounces
5. 3 yards 39 feet
6. 2 hours 72 minutes
7. 5 gallons 5 quarts
8. 20 minutes 63 seconds
9. 2 gallons 5 quarts 3 pints
10. 2 yards 4 feet 14 inches

Perform the indicated operations.

11.
$$
\begin{array}{r}
4\text{ ft}\ \ 6\text{ in.} \\
+5\text{ ft}\ \ 7\text{ in.} \\
\hline
\end{array}
$$

12.
$$
\begin{array}{r}
3\text{ h}\ \ 27\text{ min} \\
+2\text{ h}\ \ 50\text{ min} \\
\hline
\end{array}
$$

13.
$$
\begin{array}{r}
3\text{ gal}\ \ 1\text{ qt} \\
-1\text{ gal}\ \ 3\text{ qt} \\
\hline
\end{array}
$$

14.
$$
\begin{array}{r}
6\text{ lb}\ \ 2\text{ oz} \\
-2\text{ lb}\ \ 5\text{ oz} \\
\hline
\end{array}
$$

15.
$$
\begin{array}{r}
3\text{ qt}\ \ 1\text{ pt} \\
\times\quad\quad 5 \\
\hline
\end{array}
$$

16.
$$
\begin{array}{r}
2\text{ yd}\ \ 2\text{ ft} \\
\times\quad\quad 9 \\
\hline
\end{array}
$$

17. $5\overline{)\ 6\text{ ft}\ \ 3\text{ in.}}$

18. $4\overline{)\ 10\text{ lb}\ \ 5\text{ oz}}$

19.
$$
\begin{array}{r}
3\text{ qt}\ \ 1\text{ pt} \\
+5\text{ qt}\ \ 1\text{ pt} \\
\hline
\end{array}
$$

20.
$$
\begin{array}{r}
5\text{ ft}\ \ 3\text{ in.} \\
-2\text{ ft}\ \ 7\text{ in.} \\
\hline
\end{array}
$$

Simplify these denominate numbers.

1. 2 yards 5 feet
2. 3 pounds 39 ounces
3. 2 gallons 6 quarts 3 pints
4. 2 yards 5 feet 16 inches

Perform the indicated operation.

5. $\begin{array}{r} 4 \text{ ft} \quad 3 \text{ in.} \\ +3 \text{ ft} \quad 10 \text{ in.} \\ \hline \end{array}$

6. $\begin{array}{r} 6 \text{ lb} \quad 3 \text{ oz} \\ -2 \text{ lb} \quad 7 \text{ oz} \\ \hline \end{array}$

7. $\begin{array}{r} 2 \text{ qt} \quad 1 \text{ pt} \\ \times \qquad 3 \\ \hline \end{array}$

8. $4\overline{)5 \text{ lb} \quad 4 \text{ oz}}$

REVIEW MATERIAL

**Review Skill
Conversions
in the English
System**

(a) Change 3 feet to inches. Use the formula, 1 foot = 12 inches; therefore,

$$\frac{1 \text{ ft}}{12 \text{ in.}} = 1 \quad \text{and} \quad \frac{12 \text{ in.}}{1 \text{ ft}} = 1$$

Multiply by the unit fraction that cancels feet, leaving inches.

$$3 \text{ ft} = 3 \cancel{\text{ft}} \times \frac{12 \text{ in.}}{1 \cancel{\text{ft}}}$$

$$= 36 \text{ in.}$$

(b) Change 50 inches to feet and inches.

$$50 \text{ in.} = 50 \cancel{\text{in.}} \times \frac{1 \text{ ft}}{12 \cancel{\text{in.}}}$$

$$= \frac{50 \text{ ft}}{12}$$

$$= 4 \text{ ft with a remainder of 2 in.}$$

$$= 4 \text{ ft } 2 \text{ in.}$$

(c) Change 100 inches to yards.

$$100 \text{ in.} = 100 \cancel{\text{in.}} \times \frac{1 \cancel{\text{ft}}}{12 \cancel{\text{in.}}} \times \frac{1 \text{ yd}}{3 \cancel{\text{ft}}}$$

$$= \frac{100 \text{ yd}}{36}$$

$$= 2\frac{28}{36} \text{ yd}$$

$$= 2\frac{7}{9} \text{ yd}$$

Exercise A   Answers to this exercise can be found on page 480.

1. Change 5 minutes to seconds.
2. Change 120 minutes to hours.

3. Change 15 pints to quarts.
4. Change 7 quarts to gallons and quarts.
5. Change 5 pounds to ounces.

Exercise B No answers are given.

1. Change 6 minutes to seconds.
2. Change 240 minutes to hours.
3. Change 13 pints to quarts.
4. Change 9 quarts to gallons and quarts.
5. Change 2 pounds to ounces.

REVIEW EXERCISES

Refer to the section listed if you have difficulty. Answers to these exercises can be found on page 480.

Section 20.2
1. Change 3 yards to feet.
2. Change 5 feet to inches.
3. Change 96 inches to yards.
4. Change 19 feet to yards and feet.

Section 20.3
5.
$$\begin{array}{r} 5\text{ h }30\text{ min} \\ +2\text{ h }50\text{ min} \\ \hline \end{array}$$
6.
$$\begin{array}{r} 3\text{ gal }2\text{ qt} \\ -1\text{ gal }3\text{ qt} \\ \hline \end{array}$$

7. $5\overline{)12\text{ lb }6\text{ oz}}$
8.
$$\begin{array}{r} 2\text{ yd }3\text{ ft} \\ \times\quad\quad 8 \\ \hline \end{array}$$

CHAPTER POST-TEST

This test should determine if you have mastered the topics in Chapter 20. Answers to this test can be found on page 480.

1. Change 5 pounds to ounces.
2. Change 34 ounces to pounds.
3. Change 108 inches to yards.
4. Change 16 feet to yards and feet.

Perform the indicated operation.

5.
$$\begin{array}{r} 3\text{ gal }2\text{ qt} \\ +1\text{ gal }3\text{ qt} \\ \hline \end{array}$$
6.
$$\begin{array}{r} 5\text{ h }30\text{ min} \\ -2\text{ h }50\text{ min} \\ \hline \end{array}$$

7. $6\overline{)13\text{ lb }2\text{ oz}}$
8.
$$\begin{array}{r} 2\text{ yd }2\text{ ft} \\ \times\quad\quad 9 \\ \hline \end{array}$$

UNIT FIVE
MEASUREMENT
REVIEW AND POST-TEST

Conversion Formulas

```
1 kilo = 1,000 units

100 centi = 1 unit

1,000 milli = 1 unit

Meters (m) measure length.
Liters (ℓ) measure volume.
Grams (g) measure weight.
```

METRIC

```
1 in. = 2.54 cm

3.3 ft = 1 m

.621 mi = 1 km

1 oz = 28 g

1 lb = 454 g

1.06 qt = 1 ℓ
```

ENGLISH TO METRIC

$$F = \frac{9}{5}C + 32 \qquad C = \frac{5}{9}(F - 32)$$

ENGLISH

```
1 ft = 12 in.

1 yd = 3 ft

1 mi = 5,280 ft
```

```
1 min = 60 s

1 h = 60 min

1 d = 24 h

1 wk = 7 d

1 yr = 52 wk = 365 d
```

```
1 lb = 16 oz

1 ton = 2,000 lb
```

```
1 qt = 2 pt

1 gal = 4 qt
```

Conversion and Unit Fractions

Conversion Procedures

1. Use the conversion formula that relates the two units of the problem to form a unit fraction.
2. Multiply by the unit fraction that has the old unit in the denominator and the new unit in the numerator.

For example, change 10 centimeters to inches.

1. Because 1 inch = 2.54 centimeters there are two unit fractions.

$$\frac{1 \text{ in.}}{2.54 \text{ cm}} = 1 \quad \text{and} \quad \frac{2.54 \text{ cm}}{1 \text{ in.}} = 1$$

2. To convert centimeters and leave inches, use the unit fraction that has centimeters in the denominator.

$$10 \text{ cm} = 10 \text{ cm} \times \frac{1 \text{ in.}}{2.54 \text{ cm}}$$

$$= 3.9 \text{ in.} \quad \text{To the nearest tenth.}$$

Before working the Unit Five Post-Test work (or rework)

Chapter 19 Review Exercises, page 452
Chapter 20 Review Exercises, page 477

Study the sections for which you cannot work the sample problems.

UNIT FIVE POST-TEST

Answers to this test can be found on page 480.

Perform the indicated changes.

1. Change 120 seconds to minutes.
2. Change 3 feet to inches.
3. Change 59 ounces to pounds and ounces.
4. Change 420 grams to ounces.
5. Change 1,500 meters to kilometers.
6. Change 15 millimeters to centimeters.

Perform the indicated operations.

7.   3 gal 1 qt
    −1 gal 3 qt

8.   6 lb 5 oz
    +2 lb 15 oz

9. $3 \overline{)\, 5 \text{ ft } 2 \text{ in.}}$

10.   2 h 14 min
    ×   9

ANSWERS

Unit Five Pre-Test 1. 30 ounces 2. .162 kilometers 3. 160 millimeters 4. 6 minutes 5. 3 pounds 12 ounces 6. 9 pounds 1 ounce 7. 2 feet 5 inches

Chapter 19

Section 19.1 1. 2.54 centimeters 2. 5 inches 3. 10 kilometers 4. 16.5 feet 5. 5 liters

Section 19.2   1. 6.3 kilograms   2. 1.5 kiloliters   3. .0123 meters   4. 73 meters
5. 73 grams   6. 53.2 millimeters   7. 200,000 centigrams
8. .23 centimeter   9. .08 kilometer   10. 2,100 liters

Exercise A   1. 2 grams   2. 4,000 millimeters   3. 6.52 kiloliters
4. 200 centigrams   5. 5 centimeters

Review Exercises   1. 12.7 centimeters   2. 326.7 feet   3. 10 liters   4. .6 kiloliter
5. .0053 meter   6. .0012 meter   7. 50 millimeters

Post-Test   1. 2,724 grams   2. 6 liters   3. .07 kiloliter   4. 160 centimeters
5. .6 centigram

Chapter 20

Section 20.2   1. 21 feet   2. 28 yards   3. 24 inches   4. $1\frac{1}{3}$ feet   5. 2 yards

6. 51 feet   7. 36,960 feet   8. 29 yards 2 feet   9. 6 feet 7 inches

10. 5 yards 2 feet   11. 2 feet 5 inches   12. $\frac{1}{12}$ hour   13. 2 hours

14. 4 pints   15. $4\frac{1}{2}$ quarts   16. 4 gallons   17. 4 gallons 2 quarts

18. 48 ounces   19. 5 pounds 12 ounces   20. 64,000 ounces

Section 20.3   1. 6 feet 5 inches   2. 6 weeks 1 day   3. 3 quarts 1 pint
4. 3 pounds 15 ounces   5. 16 yards   6. 3 hours 12 minutes
7. 6 gallons 1 quart   8. 21 minutes 3 seconds   9. 3 gallons 2 quarts 1 pint
10. 3 yards 2 feet 2 inches   11. 10 feet 1 inch   12. 6 hours 17 minutes
13. 1 gallon 2 quarts   14. 3 pounds 13 ounces   15. 17 quarts 1 pint

16. 24 yards   17. 1 foot 3 inches   18. 2 pounds $9\frac{1}{4}$ ounces

19. 9 quarts or 2 gallons 1 quart
20. 2 feet 8 inches

Exercise A   1. 300 seconds   2. 2 hours   3. $7\frac{1}{2}$ quarts   4. 1 gallon 3 quarts

5. 80 ounces

Review Exercises   1. 9 feet   2. 60 inches   3. $2\frac{2}{3}$ yards   4. 6 yards 1 foot

5. 8 hours 20 minutes   6. 1 gallon 3 quarts   7. 2 pounds $7\frac{3}{5}$ ounces

8. 24 yards

Post-Test   1. 80 ounces   2. $2\frac{1}{8}$ pounds   3. 3 yards   4. 5 yards 1 foot

5. 5 gallons 1 quart   6. 2 hours 40 minutes   7. 2 pounds 3 ounces
8. 24 yards

Unit Five   1. 2 minutes   2. 36 inches   3. 3 pounds 11 ounces   4. 15 ounces
Post-Test   5. 1.5 kilometers   6. 1.5 centimeters   7. 5 gallons   8. 9 pounds 4 ounces

9. 1 foot $8\frac{2}{3}$ inches   10. 20 hours 6 minutes

UNIT 6
introduction to algebra

UNIT SIX PRETEST

This test is given *before* you read the material to see what you already know. If you cannot work each sample problem, you need to read the indicated chapters carefully. Answers to this test can be found on page 509.

Chapter 21   Perform the indicated operations.

1. $7 - 12$
2. $(-2) \times 3$
3. $14 \div (-2)$
4. $(-3) \times (-7)$
5. $-7 + 3$
6. $-7 - 3$

Chapter 22   Evaluate.

7. $-3 + 9 - 2$
8. $-7 + 5 - (-3)$
9. $(-2)^3$
10. $(5 + 2) \div 7$
11. $2 + [(5 + 9) \times 5]$
12. $3F + C$ when $F = 5$ and $C = -2$
13. $6(F + C)$ when $F = -6$ and $C = 2$

$$... -5, -4, -3, -2, -1, 0, 1, 2, 3, 4, 5, ...$$

If you are familiar with reading a thermometer, you have worked with signed numbers. Believe it or not, you have probably used them on a number of occasions.

1. −$20 represents the balance of your checking account after you write a check for $50 with only $30 in the bank. It represents $20 *in the hole* (an overdraw).
2. −6° represents the temperature on the morning you find your car radiator frozen. It represents 6° *below* zero.
3. −7 points represents your score in a card game after *losing* the bid you base on the ace of spades you think is a sure win. It represents a *loss* of 7 points.

Number Line For each positive number there is a corresponding negative number. Using a marking device similar to a thermometer, I can represent **signed numbers** (both positive and negative) on a line.

First, draw a line.

Select a point, and label it with zero.

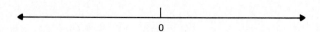

0

Moving to the right from zero (the *positive* direction) mark off equally spaced points and label them with the whole numbers.

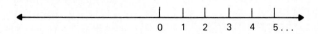

0   1   2   3   4   5 . . .

Next, mark off equally spaced points in the opposite (the *negative*) direction and label them with the corresponding negative numbers.

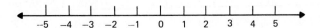

-5 -4 -3 -2 -1   0   1   2   3   4   5

Fractions and mixed numbers are located on the number line by using markings like those found on a ruler. The distances between the whole numbers are divided into equal parts.

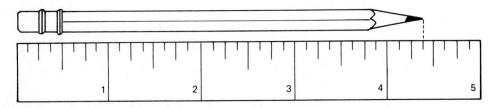

This pencil measures 4 inches plus an additional 3 of 8 equal parts of a whole inch; that is, $4\frac{3}{8}$ inches.

Loss and Gain   To understand the arithmetic of signed numbers, think of a positive number as a *gain* and a negative number as a *loss*.

The number -5 is used to represent a *loss* of 5. For example,

1. A *withdrawal* of $5 from a bank account.
2. A *drop* of 5° in the temperature.
3. A *loss* of 5 yards by a football team.

The number $+5$, or simply 5, is used to represent a *gain* of 5. For example,

1. A *deposit* of $5 in a bank account.
2. A *rise* of 5° in the temperature.
3. A *gain* of 5 yards by a football team.

Relate signed numbers to ideas you understand and you will have little difficulty with this material.

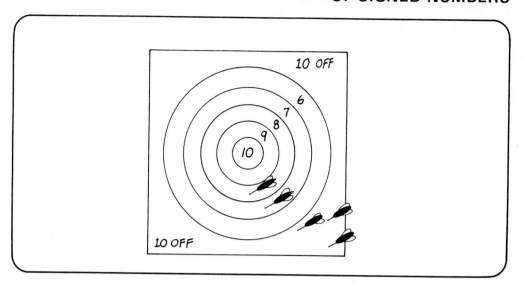

Addition If you can total the score on the dart board above, you can add signed numbers. First, I add the score for the darts in the 8-point and 7-point regions.

$$8 + 7 = 15 \quad \text{points}$$

The two *gains* add to a total *gain* of 15 points.

Next, I add a 10-OFF score, using -10 to represent the loss. Because a gain of 15 cancels all the loss points, leaving 5 gain points to spare,

$$15 + (-10) = 5 \quad \text{points}$$

To this score, I add the next 10-OFF. This time my gain points only partially cover my loss, leaving me 5 points in the hole, indicated by -5.

$$5 + (-10) = -5 \quad \text{points}$$

Finally, with 5 points in the hole, or -5, I add the last 10-OFF score. It should be clear that if I am in the hole 5 points, losing 10 more points puts my score at 15 points in the hole, or -15 points.

$$-5 + (-10) = -15 \quad \text{points}$$

Addition of Signed Numbers

Because you should think of a positive number as a gain and a negative number as a loss, addition can be thought of in the following manner.

ADDITION	THOUGHT PROCESS
$8 + 7 = 15$	Two gains add to a total *gain* of 15.
$15 + (-10) = 5$	Gain and loss, but *gain* is more by 5.
$5 + (-10) = -5$	Gain and loss, but *loss* is more by 5.
$-5 + (-10) = -15$	Two losses add to a total *loss* of 15.

Addition is as easy as keeping score in a game.

In the following examples, see if you can write the correct answer before looking at the solution; then read the solution for the correct response.

Examples	Solutions
$4 + (-6) = ?$	Think: Gain of 4, add loss of 6. The loss is more by 2. Using the negative to indicate the loss, the answer is -2.
$5 + 6 = ?$	11, adding the gains.
$-2 + (-6) = ?$	Think: Loss of 2, add loss of 6. The total of the losses is 8. Using the negative to indicate the loss, the answer is -8.
$11 + (-7) = ?$	4, gain is more by 4.
$(-7) + 4 = ?$	Think: Loss of 7, add gain of 4. Loss is more by 3; therefore, the answer is -3.
$(-2) + (-3) = ?$	Adding the losses, the answer is -5.
$(-3) + 11 = ?$	8, gain is more by 8.
$0 + (-2) = ?$	-2, beginning with no points and adding a loss of 2.
$3 + (-3) = ?$	0, losing and gaining the same amount.

Now try these problems—no hints given.

$12 + 4 = ?$	16
$12 + (-4) = ?$	8
$(-12) + 4 = ?$	-8
$(-12) + (-4) = ?$	-16

Ready to continue?

Subtraction   Subtraction of a positive number is a loss. Therefore, subtraction can be thought of in the same loss-gain way.

Subtraction of a Positive Number

Whether or not the first value is a loss or gain, in each case the subtraction of a positive number can be thought of as a loss of that quantity.

SUBTRACTION	THOUGHT PROCESS
$7 - 3 = 4$	Gain of 7, loss of 3; *gain* is more by 4.
$4 - 6 = -2$	Gain of 4, loss of 6; *loss* is more by 2.
$-2 - 5 = -7$	Two losses, with a total *loss* of 7.

Subtraction of a negative number is a gain. The illustration might help explain this concept.

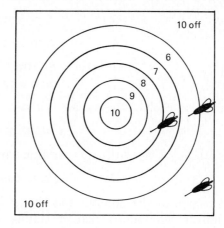

Suppose the last dart thrown was the one in the 10-OFF region. You quickly calculate the score.

$$8 + 6 = 14$$

$$14 + (-10) = 4 \quad \text{points}$$

But then the dart in the penalty region falls off the board. According to the rules, suppose this dart is not counted—it neither gains nor loses any points for you.
Your score can be adjusted by removing the penalty points.

$$-(-10)$$
removing   penalty

Removing the penalty points means that you *add* the 10 points to the score of 4, giving 14 points again.
Therefore,

$$4 - (-10) \quad \text{is the same as} \quad 4 + 10$$

removing penalty   gain

Double Negative Law

The following principle is commonly called the **double negative law**: To subtract a negative number is the same as adding the corresponding positive number.

$$5 - (-3) = 5 + 3 \qquad -7 - (-4) = -7 + 4$$
$$= 8 \qquad\qquad\qquad = -3$$

Gains and Losses

The 5 in each of these cases is a gain.

5	A positive number is a gain.
$3 + 5$	Adding a positive number is a gain.
$3 - (-5)$	Subtracting a negative number is a gain by the double negative law.

The 2 in each of these cases is a loss.

-2	A negative number is a loss.
$5 + (-2)$	Adding a negative number is a loss.
$5 - 2$	Subtracting a positive number is a loss.

The sign of the number and the arithmetic operation preceding the number indicate whether to treat the number as a loss or gain.

Your turn again!

Examples	Solutions
Would the 7 in each of these problems be considered as a loss or gain? (a) $5 - 7$ (b) $5 + 7$ (c) $5 + (-7)$ (d) $5 - (-7)$	 (a) Loss, subtracting positive. (b) Gain, adding positive. (c) Loss, adding negative. (d) Gain, double negative law.
Would the 5 in each of these problems be considered as a loss or gain? (a) $(-5) + 2$ (b) $-5 - 2$ (c) $2 - 5$ (d) $2 - (-5)$ (e) $2 + (-5)$ (f) $5 - 2$ (g) $-2 + 5$	 (a) Loss, negative number. (b) Loss, negative number. (c) Loss, subtracting positive. (d) Gain, double negative. (e) Loss, adding negative. (f) Gain, positive number. (g) Gain, adding positive.
$5 + 7 = ?$	Gain of 5, gain of 7; total gain of 12. Answer: 12
$6 + (-8) = ?$	Gain of 6, loss of 8; loss is more by 2. Answer: -2
$-5 + 4 = ?$	Loss of 5, gain of 4; loss is more by 1. Answer: -1
$-3 + 4 = ?$	Loss of 3, gain of 4; gain is more by 1. Answer: 1
$-3 + (-4) = ?$	Loss of 3, loss of 4; total loss of 7. Answer: -7

Examples	Solutions
$5 - 7 = ?$	Gain of 5, loss of 7; loss is more by 2. Answer: -2
$-3 - 5 = ?$	Loss of 3, loss of 5; total loss of 8. Answer: -8
$2 - (-5) = ?$	Gain of 2, gain of 5 (double negative). Answer: 7
$-3 - (-2) = ?$	Loss of 3, gain of 2; loss is more by 1. Answer: -1
$0 - 2 = ?$	Starting with zero, loss of 2; this gives a loss of 2. Answer: -2
$2 - 0 = ?$	Gain of 2, loss of zero; this gives a gain of 2. Answer: 2

Now try these problems—no hints given.

$5 + (-9)$	-4
$(-5) + (-9)$	-14
$5 - 9$	-4
$-5 + 9$	4
$-5 - 9$	-14
$5 - (-9)$	14
$-5 - (-9)$	4

If you had difficulty with these last problems, please talk with an instructor or read this section again.

Exercises Part I   Compute the additions and subtractions. Answers to these exercises can be found on page 509.

1. $6 - 3$ 2. $3 + 4$ 3. $-6 + 8$
4. $-6 - (-6)$ 5. $3 - 8$ 6. $-4 - 7$
7. $2 - 6$ 8. $0 - 3$ 9. $-5 + 7$
10. $8 - (-2)$ 11. $-4 - 3$ 12. $-6 + 1$
13. $8 + 4$ 14. $-5 - 6$ 15. $-7 + 2$

16. After the seventh play in a card game the score for Herman was 25 points. On the next play Herman lost 57 points. What is his new score?

17. At noon the temperature was 68°. If the temperature dropped 70° by midnight, what was the temperature at that time?

Compute the additions and subtractions.

1. $9 - 2$
2. $9 + 2$
3. $-9 - 2$
4. $-9 + 2$
5. $-9 - (-2)$
6. $-5 - 7$
7. $-4 + 2$
8. $-4 + 9$
9. $-8 - 3$
10. $8 - 12$

SECTION 21.3
MULTIPLICATION AND DIVISION
OF SIGNED NUMBERS

Sue feels she can lose 5
pounds a month. If she works
at that rate, she can reach
her goal of a loss of 20 pounds
in 4 months.

Multiplication   Multiplication represents repeated addition; that is,

8×2 means eight 2's

means $2 + 2 + 2 + 2 + 2 + 2 + 2 + 2$

means 16

Therefore, $8 \times 2 = 16$.

The product of two positive numbers is positive.

What does $8 \times (-2)$ mean?

$8 \times (-2)$ means eight -2's

means $(-2) + (-2) + (-2) + (-2) + (-2) + (-2) + (-2) + (-2)$

means -16 (adding up the losses)

Therefore, $8 \times (-2) = -16$

When the negative number is given first, change the order of the multiplication.

$(-8) \times 2 = 2 \times (-8)$ which means two -8's

which means $(-8) + (-8)$

which means -16

Therefore, $(-8) \times 2 = -16$

If one factor is positive and the other is negative, the product is negative. Consider $(-1) \times 8$.

$$(-1) \times 8 = 8 \times (-1)$$
$$= \text{eight } -1\text{'s}$$
$$= -8$$

Therefore, $-8 = (-1) \times 8$.

A negative sign in front of any number can be replaced by -1, which is then multiplied times the number.

$$-8 = (-1) \times 8$$
$$-6 = (-1) \times 6$$

and

$$-(-16) = (-1) \times (-16)$$

Finally, what does $(-8) \times (-2)$ mean? The result can be based on the preceding information.

First, write -8 as $(-1) \times 8$.

$$(-8) \times (-2) = (-1) \times 8 \times (-2)$$

Next, multiply $8 \times (-2)$, which equals -16 because the product is negative when one factor is positive and the other is negative.

$$(-8) \times (-2) = (-1) \times \underbrace{8 \times (-2)}$$
$$= (-1) \times \quad (-16)$$

But $(-1) \times (-16)$ is the same as having a negative sign in front of the number (-16); that is, $(-1) \times (-16) = -(-16)$.

$$(-8) \times (-2) = (-1) \times (-16)$$
$$= -(-16)$$
$$= 16 \quad \text{The double negative law.}$$

Therefore, $(-8) \times (-2)$ equals 16.

 When both signs are negative, the product is positive.

Multiplication of Signed Numbers

$$8 \times 2 = 16$$
$$8 \times (-2) = -16$$
$$(-8) \times 2 = -16$$
$$(-8) \times (-2) = 16$$

To multiply two signed numbers,

1. Multiply the values as if they were positive numbers.
2. Give this arithmetic product a sign based on the following rules:
   a. The sign is positive if the signs of the two factors are like (both positive or both negative).

   $$8 \times 2 = 16 \quad \text{and} \quad (-8) \times (-2) = 16$$

   b. The sign is negative if the signs of the two factors are different (one negative and one positive).

   $$8 \times (-2) = -16 \quad \text{and} \quad (-8) \times 2 = -16$$

Examples	Solutions
In $2 \times (-3)$, what is the sign of the product?	Negative, because the signs are different.
In $(-2) \times (-3)$, what is the sign of the product?	Because both signs are like, the product is positive.
$2 \times 5 = ?$	10
$(-3) \times 4 = ?$	-12, the answer is negative because the signs are different.
$(-5) \times (-3) = ?$	15, the answer is positive because the signs are like.
$5 \times (-4) = ?$	-20
$(-6) \times 0 = ?$	0, if one factor is zero, then the product is zero.
$0 \times (-3) = ?$	0

Now try these—no hints given.

$(-3) \times 2 = ?$	-6
$6 \times (-5) = ?$	-30
$(-7) \times (-8) = ?$	56
$(-5) \times 4 = ?$	-20

Ready to continue?

Division   Consider $36 \div 9$.

$$9 \overline{)\,36\,}^{\,?}$$

Because $9 \times 4 = 36$,

$$9 \overline{)\,\begin{matrix}4\\36\\\underline{36}\end{matrix}\,}$$

Therefore, $36 \div 9 = 4$.
    Consider $10 \div (-2)$.

$$-2 \overline{)\,10\,}^{\,?}$$

Because $(-2) \times (-5) = 10$,

$$-2 \overline{)\,\begin{matrix}-5\\10\\\underline{10}\end{matrix}\,}$$

Therefore, $10 \div (-2) = -5$.
    Consider $(-36) \div 4$.

$$4 \overline{)\,-36\,}^{\,?}$$

Because $4 \times (-9) = -36$,

$$4 \overline{)\,\begin{matrix}-9\\-36\\\underline{-36}\end{matrix}\,}$$

Therefore, $(-36) \div 4 = -9$.

With division, if the signed numbers have different signs, the quotient is negative.

$$10 \div (-2) = -5 \quad \text{and} \quad (-36) \div 4 = -9$$

Consider $(-36) \div (-9)$.

$$-9 \overline{)\,-36\,}^{\,?}$$

Because $(-9) \times 4 = -36$,

$$-9 \overline{)\,\begin{matrix}4\\-36\\\underline{-36}\end{matrix}\,}$$

Therefore, $(-36) \div (-9) = 4$.

> With division, if the signed numbers have like signs, the quotient is positive.
>
> $$36 \div 9 = 4 \quad \text{and} \quad (-36) \div (-9) = 4$$

Try these.

Examples	Solutions
$(-6) \div 2 = ?$	Think of this in two steps. First, the arithmetic quotient is 3. Next, the sign of the quotient is negative because the signs are different. The answer is -3.
$(-24) \div (-6) = ?$	4 (like signs)
$49 \div (-7) = ?$	-7 (different signs)
$(-38) \div 2 = ?$	-19 (different signs)
$27 \div 0 = ?$	No answer, because division by zero is undefined.
$0 \div (-4) = ?$	0, because $(-4) \times 0 = 0$.

Now try these problems—no hints given.

$(-5) \div 1$	-5
5×1	5
$(-5) \div (-1)$	5
$5 \div (-1)$	-5
$(-5) \times (-1)$	5

If you had difficulty with these last problems, please talk with an instructor or read this section again.

**Exercises
Part I**

Compute the divisions and multiplications. Answers to these exercises can be found on page 509.

1. 5×5 2. $(-3) \times (-4)$ 3. $(-7) \times 2$
4. $7 \times (-4)$ 5. $10 \div 2$ 6. $(-10) \div 2$
7. $10 \div (-2)$ 8. $(-10) \div (-2)$ 9. $5 \div 1$
10. $(-6) \div (-3)$ 11. $(-861) \div 3$ 12. $(-221) \div (-17)$

13. Sybil receives $6 for every hour she works as a tutor. If she tutored 12 hours last week, how much money did she receive?

14. A 12-member club lost $96 on their bake sale. If each member equally shared the loss, write each member's loss as a signed number.

**Exercises
Part II**

Compute the divisions and multiplications.

1. $(-4) \times (-5)$ 2. $(-4) \times 9$ 3. $5 \times (-2)$
4. $15 \div (-3)$ 5. $(-15) \div 5$ 6. $(-16) \div (-8)$

Refer to the section listed if you have difficulty. Answers to these exercises can be found on page 509.

Section 21.2 Add or subtract, as indicated.

1. $9 + 2$ 2. $9 - 2$ 3. $2 - 9$
4. $-2 - 9$ 5. $-4 + 9$ 6. $-17 + 12$
7. $-9 - (-2)$

Section 21.3 Multiply or divide, as indicated.

8. 3×4 9. $(-3) \times 4$ 10. $(-3) \times (-4)$
11. $(-6) \div 2$ 12. $6 \div (-3)$ 13. $5 \times (-9)$

CHAPTER POST-TEST

This test should determine if you have mastered the topics in Chapter 21. Answers to this test can be found on page 509.

Perform the indicated operations.

1. $2 \times (-3)$ 2. $5 - 9$ 3. $-3 + 5$
4. $(-6) \div (-3)$ 5. $(-2) \times 3$ 6. $-5 - 9$
7. $3 + (-5)$ 8. $6 \div (-3)$ 9. $(-2) \times (-3)$
10. $5 - (-9)$ 11. $(-6) \div 3$ 12. 2×3
13. $-5 - (-9)$ 14. $6 \div 3$ 15. $-2 - 3$

22
substitution and evaluation

SECTION 22.1
SIMPLE EXPRESSIONS

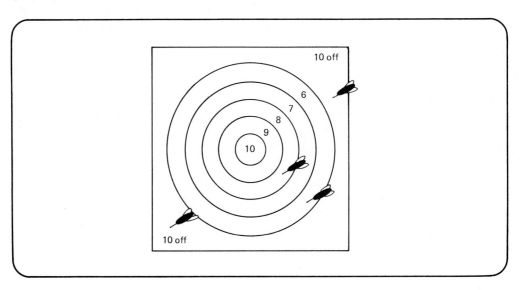

What is the score on the dart board?

The points can be listed using the expression

$$-10 + 8 + 6 - 10$$

Left to right you should think of this expression as a loss of 10, followed by a gain of 8, followed by a gain of 6, and finally a loss of 10.

Evaluating left to right,

$$\underbrace{-10 + 8}_{} \overset{\downarrow}{+\, 6} - 10 \qquad \text{First two combined.}$$

$$= \underbrace{-2 \quad + 6}_{} \qquad\qquad \text{Then 6.}$$

$$= \underbrace{4 \qquad\quad - 10}_{} \qquad \text{Then 10.}$$

$$= \qquad\qquad -6$$

Therefore, the score is 6 points in the hole, or -6.

EXAMPLE 1: Evaluate $-2 + 1 - 3 - (-6)$.
Calculating left to right,

$$\underbrace{-2 + 1}_{} \overset{\downarrow}{-\, 3} - (-6)$$

$$= \underbrace{-1 \quad - 3}_{} $$

$$= \underbrace{-4 \quad - (-6)}_{}$$

$$= \underbrace{-4 \quad + \quad 6}_{} \qquad \text{Double negative.}$$

$$= \qquad\quad 2$$

Therefore, $-2 + 1 - 3 - (-6) = 2$.

EXAMPLE 2: Evaluate $(-2) \times 3 \times (-2) \times (-4)$.
Calculating left to right,

$$\underbrace{(-2) \times 3}_{} \overset{\downarrow}{\times (-2)} \times (-4)$$

$$= \underbrace{-6 \quad \times (-2)}_{}$$

$$= \underbrace{12 \quad\quad \times (-4)}_{}$$

$$= \qquad\qquad -48$$

Therefore, $(-2) \times 3 \times (-2) \times (-4) = -48$.

EXAMPLE 3: Evaluate $(-2) \times 3 \times 2 \times (-4) \times 3$.
From left to right,

$$\underbrace{(-2) \times 3}_{} \overset{\downarrow}{\times 2} \times (-4) \times 3$$

$$= \underbrace{-6 \quad \times 2}_{}$$

$$= \underbrace{-12 \quad\quad \times (-4)}_{}$$

$$= \underbrace{48 \quad\quad \times 3}_{}$$

$$= \qquad\qquad 144$$

Therefore, $(-2) \times 3 \times 2 \times (-4) \times 3 = 144$.

Notice that no new rules are used in these examples. You are just working through an expression containing more than two numbers. Practice by working these exercises.

Answers to these exercises can be found on page 509.

Evaluate.

1. $-4 - 3 + 2$ 2. $8 + 4 + 6$
3. $-6 + 1 - 5$ 4. $-5 - 6 - 7$
5. $-6 + 3 - 8 + 14$ 6. $-2 + 1 - 3 - (-6)$
7. At noon the temperature was 80°. By 2 o'clock the temperature had risen 6°. From 2 o'clock to midnight the temperature dropped 25°. What was the temperature at midnight?
8. $3 \times (-4) \times (-1)$
9. $(-4) \times (-1) \times (-5)$
10. $(-1) \times (-5) \times 2 \times (-3) \times (-1)$
11. 5^3 (*Hint:* $5^3 = 5 \times 5 \times 5$)
12. $(-5)^3$ (*Hint:* $(-5)^3 = (-5) \times (-5) \times (-5)$)
13. 6^2
14. $(-2)^6$
15. $(-2)^5$
16. $(-3)^4$

Evaluate.

1. $-5 - 3 - 7 - (-4)$ 2. $7 + 2 - 4 - 5$
3. $-3 + 1 - 3 - (-6)$ 4. $(-1) \times (-5) \times 6 \times (-7)$
5. $(-5) \times 11 \times (-2)$ 6. 4^3
7. $(-4)^3$

SECTION 22.2
COMBINING THE OPERATIONS

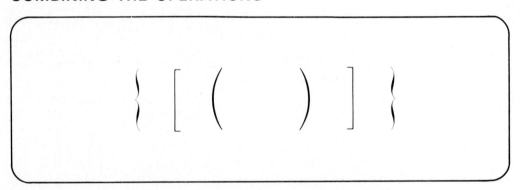

Grouping symbols (parentheses, brackets, and braces) are used to indicate clearly the meaning of an arithmetic or formula expression.

There is another possible way to calculate the expression

$$2 + 3 \times 6$$

other than the normal left-to-right method.

The groupings symbols shown below clearly indicate only one possible order of operations.

$$2 + (3 \times 6)$$

> **Order of Operations**
>
> Operations within grouping symbols should be performed first.

$$2 + \underbrace{(3 \times 6)}$$
$$= 2 + \quad 18 \qquad \text{Parentheses first.}$$
$$= 20$$

Therefore, $2 + (3 \times 6) = 20$.

> **Order of Operations**
>
> Operations within grouping symbols should be performed first, starting with the innermost grouping symbol.

There are many possible ways to evaluate

$$2 + 3 \div 9 - 6$$

The grouping symbols below indicate only one way.

$$2 + [3 \div \underbrace{(9 - 6)}]$$
$$= 2 + [3 \div \quad 3 \quad] \qquad \text{Innermost first.}$$
$$= 2 + 1 \qquad \text{Grouping next.}$$
$$= 3$$

Therefore, $2 + 3 \div (9 - 6) = 3$.

Evaluate the following examples.

Examples	Solutions
$(6 + 9) - 3 = ?$	$\underbrace{(6 + 9)} - 3$ $= \quad 15 \quad - 3$ $= 12$
$6 + (9 \div 3) = ?$	$6 + (9 \div 3)$ $= 6 + \quad 3$ $= 9$
$(6 + 5) \times (4 \div 2) = ?$	$(6 + 5) \times (4 \div 2)$ $= \quad 11 \quad \times \quad 2$ $= 22$
$[6 + (5 \times 4)] \div 2 = ?$	$[6 + \underbrace{(5 \times 4)}] \div 2$ $= [6 + \quad 20 \quad] \div 2$ $= \quad 26 \quad \div 2$ $= 13$

Examples	Solutions
Now try these problems—no hints given.	
$(5 - 2) \div 3 = ?$	1
$2 + [5 - (6 \div 2)] = ?$	4

If you had difficulty, please talk with an instructor or read this section again.

Exercises Part I Answers to these exercises can be found on page 509.

Evaluate.

1. $(4 + 6) \div 2$
2. $4 + (6 \div 2)$
3. $(3 + 6) \times (6 \div 2)$
4. $[6 + (4 \times 2)] - 2$
5. $[(6 + 4) \times 2] - 2$
6. $(6 + 4) \times (2 - 2)$
7. $6 + [(4 \times 2) - 2]$

Exercises Part II Evaluate.

1. $(4 - 6) \div 2$
2. $(2 + 5) \times (5 \div 5)$
3. $[5 + (2 \times 3)] - 4$
4. $6 + [2 - (4 \div 2)]$

SECTION 22.3
FORMULAS AND SUBSTITUTION

$$A = P(1 + RT)$$

Variables In algebra, letters are used to represent or to stand for numbers. The letters are called **variables**.

One advantage of using variables and mathematical symbols is that they provide a simple and short way to express lengthy verbal statements.

Verbal Expression	Algebraic Expression
(a) Thirteen more than some unknown number.	$N + 13$
(b) Twenty-five decreased by some number.	$25 - T$

Verbal Expression	Algebraic Expression
(c) Four multiplied by the sum of ten and a number.	$4(10 + y)$
(d) The quotient of some number and negative nine.	$\dfrac{x}{-9}$

Formulas You have probably seen variables used in formulas. Here, again, the advantage of using variables is simplicity.

Consider the formula

$$R = \frac{D}{T}$$

This is the formula for determining the average rate of speed of an object. Written verbally it states that the average *rate* of speed (R) equals the *distance* traveled (D) divided by the amount of *time* it took to travel that distance (T). Notice how simple the expression looks algebraically.

The preceding examples are intended to give you a hint as to how variables are used in real-life applications. However, at this point you are not ready to jump into work with formulas without the foundation of the remainder of this section.

Symbolizing Operations

The operations are symbolized in algebra much the same as they are in arithmetic, except for multiplication.

1. Addition: $+$
2. Subtraction: $-$
3. Multiplication: ( ), $\cdot$, or no operation sign
4. Division: $\div$ or the fraction bar

To be certain that there are no misunderstandings, the symbol X will not be used in algebra to express multiplication because it could possibly be misinterpreted as a variable.

Each of the following expresses the multiplication of the variable a times the variable b.

$\left. \begin{array}{l} a(b) \\ (a)(b) \end{array} \right\}$ Parentheses in the absence of any other sign of operation.

$a \cdot b$ Dot notation.

ab Written close in the absence of any other sign of operation.

Each of the following expresses the multiplication of the number seven and the variable a.

$$7(a)$$

$$(7)(a)$$

$$7 \cdot a$$

or more simply, $7a$, written close in the absence of any other operation sign.

Rules of Order   Sometimes to indicate a complicated order of operations, you are forced to use too many symbols. This in itself becomes confusing.

Therefore, mathematicians have agreed upon the following rules to clarify order of operation without using grouping symbols. These rules are particularly useful for formulas.

Rules of Order

Follow these steps in the order presented.

1. Do all operations in parentheses, brackets, and braces.
2. Next, do all multiplications and divisions in the order in which they appear from left to right.
3. Finally, do all additions and subtractions in the order in which they appear from left to right.

Under these rules multiplication and division are considered operations of higher priority than the operations of addition and subtraction.

EXAMPLE 1:   Evaluate $-4 + 3 \times 2$.

Following the stated rules, you cannot just work left to right. Scan the expression for multiplications and divisions and do them first.

$$-4 + \underbrace{3 \times 2}$$
$$= -4 + \quad 6$$
$$= \quad\quad 2$$

Using the rules of order, $-4 + 3 \times 2 = 2$.

EXAMPLE 2:   Evaluate $6 - 4 \div 2$.

Scan the expression for multiplications and divisions.

$$6 - \underbrace{4 \div 2}$$
$$= 6 - \quad 2$$
$$= 4$$

Evaluating Algebraic Expressions   When a number value is assigned to each variable appearing in an algebraic expression or formula, then that expression can be simplified and expressed as a single number. This process is called evaluating an algebraic expression.

EXAMPLE 3:   Evaluate $3a + b$ when $a = -2$ and $b = 8$.

Remember that $3a$ indicates a multiplication.

$$3a \quad\quad + b$$

$$= \underbrace{3(-2)} + 8 \qquad \text{Substitute values for the variables.}$$

$$= \quad -6 \quad + 8 \qquad \text{Do multiplication first,}$$

$$\qquad\qquad\qquad\quad \text{then addition.}$$

$$= 2 \qquad\qquad\qquad \text{Answer.}$$

EXAMPLE 4:   Evaluate $a - 3n$ when $a = 29$ and $n = -4$.

$$a - 3n$$

$$= 29 - 3(-4)$$ Substitute values.

$$= 29 - (-12)$$ Do multiplication first,

$$= 29 + 12$$ then do the subtraction.

$$= 41$$

EXAMPLE 5: Evaluate $\frac{1}{2}(a + b)$ when $a = 5$ and $b = 7$.

$$\frac{1}{2}(a + b)$$

Substitute values.

$$= \frac{1}{2}(5 + 7)$$

Do grouping first,

$$= \frac{1}{2}(\quad 12 \quad)$$

Then multiply.

$$= 6$$

EXAMPLE 6: If $m = 3$ and $n = 5$, find the value of $\frac{(m - n)^2}{3n}$.

$$\frac{(m - n)^2}{3n}$$

Substitute values.

$$= \frac{(3 - 5)^2}{3 \cdot 5}$$

Do grouping first.

$$= \frac{(-2)^2}{3 \cdot 5}$$

Apply exponent and multiply.

$$= \frac{4}{15}$$

Now try these exercises. See if you can work through each example before you look at the solution.

Examples	Solutions
Evaluate each for the given values of the variables.	
(a) $a + b$ when $a = -9$ and $b = 5$	(a) $\quad a + b$ $= (-9) + 5$ $= -4$
(b) $(c + d)^2$ when $c = -4$ and $d = 2$	(b) $\quad (c + d)^2$ $= (-4 + 2)^2$ $= (\quad -2 \quad)^2$ $= 4$

Examples	Solutions
(c) $3x - y$ when $\quad x = -2$ and $y = -6$	(c) $\quad 3x - y$ $\qquad = 3(-2) - (-6)$ $\qquad = -6 \ - (-6)$ $\qquad = -6 \ + \ 6$ $\qquad = 0$

Now try these—no hints given.

$5x + y$ when $x = -3$ and $y = 10$	-5
$\dfrac{k^2}{m - n}$ when $k = -4$ and $m = 5$ and $n = 3$	8

If you had difficulty, please talk with an instructor or read this section again.

Exercises Part I

Answers to these exercises can be found on page 509.

Evaluate each expression for the given values of the variables.

1. $a - b$ when $a = -3$ and $b = 5$
2. $c + 2d$ when $c = 32$ and $d = 12$
3. $\dfrac{1}{2}F$ when $F = 25$
4. $r(s + t)$ when $r = 2$, $s = 12$, and $t = 4$
5. $2L + 2W$ when $L = 14$ and $W = 12$

Exercises Part II

Evaluate each of the expressions for the given values of the variables.

1. $d + 2f$ when $d = 5$ and $f = -2$
2. $a - b$ when $a = 6$ and $b = -3$
3. $(g + H)^2$ when $g = 2$ and $H = 9$

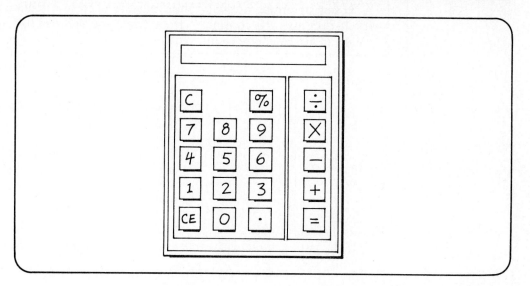

Note: If you have not already read them, please read pages 357 through 360.
How would you enter an expression such as

$$2 + 3 \times 6$$

on the calculator?

If you assume that the calculator follows the rules of order discussed in the last section, then you would enter the expression as it is written, pressing the equals key at the end. Most hand calculators (although not all) are programmed to follow the same order of operations that we have agreed to use in arithmetic and with formulas; that is, the expression will be evaluated as follows:

$$2 + \underbrace{3 \times 6}$$
$$= 2 + \quad 18$$
$$= 20$$

Multiplication first.

Addition next.

The expression would be entered in the keyboard as follows:

$$\boxed{2}, \boxed{+}, \boxed{3}, \boxed{\times}, \boxed{6}, \boxed{=}$$

and the result of 20 would appear in the display.

Always check the instructions that come with the calculator to see if these assumptions are true for your machine.

What if you wanted a different order? For example, how would you enter

$$(2 + 3) \times 6$$

on the calculator?

As we have just seen, if the expression is entered and the equals key pressed at the end, the multiplication will be evaluated first, not the addition as the parentheses indicate.

> **Intermediate Results Rule**
>
> A closing parenthesis usually indicates the need to calculate an intermediate result first, then to continue. In general, you can use the equals key to calculate an intermediate result.

Therefore, $(2 + 3) \times 6$ should be entered as follows:

$\boxed{2}, \boxed{+}, \boxed{3}$ To process the addition and get an intermediate result.

$\boxed{=}$

$\boxed{\times}, \boxed{6}$ To multiply the display value by 6.

$\boxed{=}$ To complete the processing.

The display will read 30.

If the last example was written with the multiplication in the other order,

$$6 \times (2 + 3)$$

this technique would not work.

The sequence

$$\boxed{6}, \boxed{\times}, \boxed{2}, \boxed{+}, \boxed{3}, \boxed{=}$$

would be processed according to the rules of order.

$$\underbrace{\boxed{6}, \boxed{\times}, \boxed{2}}_{12} , \underbrace{\boxed{+}, \boxed{3}}_{+ \quad 3} , \boxed{=}$$
$$\underbrace{}_{15}$$

But

$$6 \times (2 + 3) = 6 \times 5 \qquad \text{Grouping first.}$$
$$= 30$$

 To apply the intermediate result rule, evaluate the expression by making your entries according to the indicated order of operations, not in their sequence as written.

EXAMPLE 1:   How should $6 \times (2 + 3)$ be entered on a calculator?
Parentheses indicate to do the addition first.

$$\boxed{2}, \boxed{+}, \boxed{3}$$

The closing parenthesis indicates an intermediate result; use the equals key.

$$\boxed{=}$$

Now multiply this result by 6.

$$\boxed{\times}, \boxed{6}$$

Then evaluate.

$$\boxed{=}$$

Notice that in order to get the correct result, you, the user, have to be smarter than the calculator.

EXAMPLE 2:   How would you enter $5 + [6 \times (5-7)]$ on the calculator?

Do the innermost grouping first.

$$\boxed{5}, \boxed{-}, \boxed{7}$$

Get an intermediate result.

$$\boxed{=}$$

Next, multiply by 6.

$$\boxed{\times}, \boxed{6}$$

Get an intermediate result.

$$\boxed{=}$$

Finally, add 5 and evaluate.

$$\boxed{+}, \boxed{5}, \boxed{=}$$

The display should show -7.

Exercises Part I   Answers to these exercises can be found on page 509.

Use a calculator to evaluate the following.

1. $2 + 7 \times 9$ 2. $3 + 6 \div 3 + 4$
3. $2 + (7 \times 9)$ 4. $(2 + 7) \times 9$
5. $2 \times (7 + 9)$ 6. $2 \times [7 + (5 \times 4)]$
7. $2 \times [(7 + 5) \times 4]$ 8. $[7 + (8 - 4)] \times 7$

Exercises Part II
1. $2 + 8 \times 3$ 2. $4 \times (7 - 4)$
3. $2 + [7 \times (4 - 1)]$ 4. $[7 + (7 - 3)] \times 5$

REVIEW EXERCISES

Refer back to the section listed if you have difficulty. Answers to these exercises can be found on page 509.

Section 22.1   Evaluate.

1. $-2 - 5 + 2$
2. $-7 + 2 - (-3)$
3. $-2 - 3 - 4$
4. $(-3) \times (-1) \times 4$
5. $(-3) \times (-1) \times (-4)$
6. 3^2
7. $(-3)^2$

Evaluate.

    8. $(5 + 2) \div 7$

    9. $(6 + 2) \times (5 - 3)$

    10. $[(2 + 9) \times 3] - 2$

Section 22.3   Evaluate each expression for the given values of the variables.

    11. $c + 3d$ when $c = -4$ and $d = 10$

    12. $5(c - 2)$ when $c = 9$

    13. $(t - s)^2$ when $t = -4$ and $s = 2$

CHAPTER POST-TEST

This test should determine if you have mastered the topics in Chapter 22. Answers to this test can be found on page 509.

Evaluate.

    1. $-7 - 2 + 3$

    2. $-8 + 2 - (-4)$

    3. $(-5) \times 4 \times (-1)$

    4. $(-5) \times (-4) \times (-1)$

    5. $(-4)^3$

    6. $(9 + 3) \div 3$

    7. $[(5 + 3) \times 2] - 4$

    8. $H + 2R$ when $H = -2$ and $R = 5$

    9. $5(H - R)$ when $H = -2$ and $R = 5$

    10. $(H + R)^2$ when $H = 2$ and $R = 5$

UNIT SIX POST-TEST

Perform the indicated operations. Answers to this test can be found on page 509.

    1. $9 - 14$

    2. $(-3) \times 6$

    3. $15 \div (-3)$

    4. $(-7) \times (-3)$

    5. $-8 + 2$

    6. $-8 - 2$

    7. $-7 + 10 - 3$

    8. $-9 + 5 - (-2)$

    9. $(-3)^2$

    10. $(5 + 3) \div 8$

    11. $5 + [(2 + 8) \times 2]$

    12. $5H + C$ when $H = 4$ and $C = 9$

    13. $7(H - C)$ when $H = -2$ and $C = 4$

Unit Six   **1.** −5   **2.** −6   **3.** −7   **4.** 21   **5.** −4   **6.** −10   **7.** 4   **8.** 1
Pretest   **9.** −8   **10.** 1   **11.** 72   **12.** 13   **13.** −24

Chapter 21

Section 21.2   **1.** 3   **2.** 7   **3.** 2   **4.** 0   **5.** −5   **6.** −11   **7.** −4   **8.** −3   **9.** 2
   10. 10   **11.** −7   **12.** −5   **13.** 12   **14.** −11   **15.** −5   **16.** −32
   17. −2°

Section 21.3   **1.** 25   **2.** 12   **3.** −14   **4.** −28   **5.** 5   **6.** −5   **7.** −5   **8.** 5
   9. 5   **10.** 2   **11.** −287   **12.** 13   **13.** $72   **14.** −$8

Review Exercises   **1.** 11   **2.** 7   **3.** −7   **4.** −11   **5.** 5   **6.** −5   **7.** −7   **8.** 12
   9. −12   **10.** 12   **11.** −3   **12.** −2   **13.** −45

Post-Test   **1.** −6   **2.** −4   **3.** 2   **4.** 2   **5.** −6   **6.** −14   **7.** −2   **8.** −2
   9. 6   **10.** 14   **11.** −2   **12.** 6   **13.** 4   **14.** 2   **15.** −5

Chapter 22

Section 22.1   **1.** −5   **2.** 18   **3.** −10   **4.** −18   **5.** 3   **6.** 2   **7.** 61°   **8.** 12
   9. −20   **10.** 30   **11.** 125   **12.** −125   **13.** 36   **14.** 64   **15.** −32
   16. 81

Section 22.2   **1.** 5   **2.** 7   **3.** 27   **4.** 12   **5.** 18   **6.** 0   **7.** 12

Section 22.3   **1.** −8   **2.** 56   **3.** $12\frac{1}{2}$ **4.** 32   **5.** 52

Section 22.4   **1.** 65   **2.** 9   **3.** 65   **4.** 81   **5.** 32   **6.** 54   **7.** 96   **8.** 77

Review Exercises   **1.** −5   **2.** −2   **3.** −9   **4.** 12   **5.** −12   **6.** 9   **7.** 9   **8.** 1
   9. 16   **10.** 31   **11.** 26   **12.** 35   **13.** 36

Post-Test   **1.** −6   **2.** −2   **3.** 20   **4.** −20   **5.** −64   **6.** 4   **7.** 12   **8.** 8
   9. −35   **10.** 49

Unit Six   **1.** −5   **2.** −18   **3.** −5   **4.** 21   **5.** −6   **6.** −10   **7.** 0   **8.** −2
Post-Test   **9.** 9   **10.** 1   **11.** 25   **12.** 29   **13.** −42

index

Metric system, 438
　　changing units, 445
　　conversions, 440, 445
　　length, 439, 444
　　prefixes, 438
　　temperature, 447
　　volume, 440, 445
　　weight, 439, 445
Minuend, 48
Mixed arithmetic, 353
Mixed numbers, 145
　　addition, 212
　　changing to improper fraction, 151
　　division, 242
　　multiplication, 236
　　subtraction, 223
Mixed operations, 498
Multiples, 192
　　common, 193
　　least common, 193
Multiplication, 90
　　applications of, 90, 257, 346
　　decimals, 319
　　denominate numbers, 471
　　fractions, 231
　　mixed numbers, 236
　　by powers of ten, 322
　　signed numbers, 492
　　whole numbers, 76
　　by zero, 73

N

Negative numbers (*See* signed
　　numbers)
Number line, 176, 483
Numerator, 139

O

Order of operations, 268–69, 499,
　　502

P

Parentheses, 268, 498
Per, 122, 260, 348
Percent, 401
　　applications of, 416–24

decimal to, 409
　　to decimal, 408
　　fraction to, 404
　　to fraction, 405
　　parts of a percent problem, 414
　　proportion formula, 414
　　solving problems, 416–24
　　to ratio, 405
Periods, 10
Place value:
　　of decimals, 286
　　of whole numbers, 4
Positive numbers (*See* signed
　　numbers)
Powers of ten, 286
Prime factors, 185
Prime numbers, 184
Product, 71
Proper fractions, 145
Proportion, 381
　　extremes, 382
　　means, 382
　　product of means and extremes,
　　　382
　　solving problems, 382
　　writing, 387

Q

Quotient, 96

R

Ratio, 96
　　to percent, 404
　　reducing, 372
　　writing, 370
Reciprocals, 240
Reducing fractions, 159, 189
Remainder, 100, 102
Rounding:
　　decimals, 302
　　in division, 332
　　whole numbers, 19

S

Short division, 116
Signed numbers:
　　addition, 485